Recent Titles in This Series

172 **Peter E. Kloeden and Kenneth J. Palmer, Editors,** Chaotic numerics, 1994

171 **Rüdiger Göbel, Paul Hill, and Wolfgang Liebert, Editors,** Abelian group theory and related topics, 1994

170 **John K. Beem and Krishan L. Duggal, Editors,** Differential geometry and mathematical physics, 1994

169 **William Abikoff, Joan S. Birman, and Kathryn Kuiken, Editors,** The mathematical legacy of Wilhelm Magnus, 1994

168 **Gary L. Mullen and Peter Jau-Shyong Shiue, Editors,** Finite fields: Theory, applications, and algorithms, 1994

167 **Robert S. Doran, Editor,** C^*-algebras: 1943–1993, 1994

166 **George E. Andrews, David M. Bressoud, and L. Alayne Parson, Editors,** The Rademacher legacy to mathematics, 1994

165 **Barry Mazur and Glenn Stevens, Editors,** *p*-adic monodromy and the Birch and Swinnerton-Dyer conjecture, 1994

164 **Cameron Gordon, Yoav Moriah, and Bronislaw Wajnryb, Editors,** Geometric topology, 1994

163 **Zhong-Ci Shi and Chung-Chun Yang, Editors,** Computational mathematics in China, 1994

162 **Ciro Ciliberto, E. Laura Livorni, and Andrew J. Sommese, Editors,** Classification of algebraic varieties, 1994

161 **Paul A. Schweitzer, S. J., Steven Hurder, Nathan Moreira dos Santos, and José Luis Arraut, Editors,** Differential topology, foliations, and group actions, 1994

160 **Niky Kamran and Peter J. Olver, Editors,** Lie algebras, cohomology, and new applications to quantum mechanics, 1994

159 **William J. Heinzer, Craig L. Huneke, and Judith D. Sally, Editors,** Commutative algebra: Syzygies, multiplicities, and birational algebra, 1994

158 **Eric M. Friedlander and Mark E. Mahowald, Editors,** Topology and representation theory, 1994

157 **Alfio Quarteroni, Jacques Periaux, Yuri A. Kuznetsov, and Olof B. Widlund, Editors,** Domain decomposition methods in science and engineering, 1994

156 **Steven R. Givant,** The structure of relation algebras generated by relativizations, 1994

155 **William B. Jacob, Tsit-Yuen Lam, and Robert O. Robson, Editors,** Recent advances in real algebraic geometry and quadratic forms, 1994

154 **Michael Eastwood, Joseph Wolf, and Roger Zierau, Editors,** The Penrose transform and analytic cohomology in representation theory, 1993

153 **Richard S. Elman, Murray M. Schacher, and V. S. Varadarajan, Editors,** Linear algebraic groups and their representations, 1993

152 **Christopher K. McCord, Editor,** Nielsen theory and dynamical systems, 1993

151 **Matatyahu Rubin,** The reconstruction of trees from their automorphism groups, 1993

150 **Carl-Friedrich Bödigheimer and Richard M. Hain, Editors,** Mapping class groups and moduli spaces of Riemann surfaces, 1993

149 **Harry Cohn, Editor,** Doeblin and modern probability, 1993

148 **Jeffrey Fox and Peter Haskell, Editors,** Index theory and operator algebras, 1993

147 **Neil Robertson and Paul Seymour, Editors,** Graph structure theory, 1993

146 **Martin C. Tangora, Editor,** Algebraic topology, 1993

145 **Jeffrey Adams, Rebecca Herb, Stephen Kudla, Jian-Shu Li, Ron Lipsman, and Jonathan Rosenberg, Editors,** Representation theory of groups and algebras, 1993

144 **Bor-Luh Lin and William B. Johnson, Editors,** Banach spaces, 1993

(*Continued in the back of this publication*)

Chaotic Numerics

CONTEMPORARY MATHEMATICS

172

Chaotic Numerics

An International Workshop
on the Approximation and Computation
of Complicated Dynamical Behavior
July 12–16, 1993
Deakin University, Geelong, Australia

Peter E. Kloeden
Kenneth J. Palmer
Editors

American Mathematical Society
Providence, Rhode Island

Chaotic Numerics: An International Workshop on the Approximation and Computation of Complicated Dynamical Behaviour was held at Deakin University, Geelong, Australia, from July 12 to July 16, 1993, with support from the Faculty of Science and Technology of Deakin University and the Australian Mathematical Society.

1991 *Mathematics Subject Classification*. Primary 34A50, 58F13; Secondary 65L99.

Papers in this volume are final versions, which will not be published elsewhere, unless otherwise indicated by a footnote.

Library of Congress Cataloging-in-Publication Data

Chaotic numerics: an international workshop on the approximation and computation of complicated dynamical behavior, Deakin University, Geelong, Australia, July 12–16, 1993 / Peter E. Kloeden, Kenneth J. Palmer, editors.

p. cm. — (Contemporary mathematics, ISSN 0271-4132; v. 172)

Includes bibliographical references.

ISBN 0-8218-5184-5

1. Differentiable dynamical systems—Congresses. 2. Numerical analysis—Congresses. 3. Chaotic behavior in systems—Congresses. I. Kloeden, Peter E. II. Palmer, Kenneth J. (Kenneth James), 1945– . III. Deakin University. IV. Series: Contemporary mathematics (American Mathematical Society); v. 172.

QA614.8.C54 1994 94-21580

515′.352–dc20 CIP

Printed in the United States of America.

∞ The paper used in this book is acid-free and falls within the guidelines
established to ensure permanence and durability.

♻ Printed on recycled paper.

This volume was printed from copy prepared by the authors.
Some articles were typeset using AMS-TeX or AMS-LaTeX,
the American Mathematical Society's TeX macro systems.

10 9 8 7 6 5 4 3 2 1 99 98 97 96 95 94

Table of Contents

Preface ix

Numerical Dynamics 1
Jack K. Hale

Error Backward 31
Robert M. Corless

Modified Equations for ODEs 63
M.P.Calvo, A. Murua, and J. M. Sanz-Serna

The Dynamics of Some Iterative Implicit Schemes 75
H.C.Yee and P.K.Sweby

Shadowing of Lattice Maps 97
Shui-Nee Chow and Erik S. Van Vleck

Periodic Shadowing 115
Brian A.Coomes, Hüseyin Koçak and Kenneth J.Palmer

On Well-Posed Problems for Connecting Orbits in Dynamical Systems 131
W.-J.Beyn

Numerical Computation of a Branch of Invariant Circles starting at a Hopf Bifurcation Point 169
Laurent Debraux

Numerics of Invariant Manifolds and Attractors 185
Jens Lorenz

Interval Stochastic Matrices and Simulation of Chaotic Dynamics 203
P.Diamond, P.Kloeden and A.Pokrovskii

Mathematical and Numerical Analysis of a Mean-Field Equation for the Ising Model with Glauber Dynamics 217
C.M.Elliott, A.R.Gardiner, I.Kostin and Bainian Lu

Attractors for Weakly Coupled Map Lattices 243
Volker Matthias Gundlach

Effective Chaos in the Nonlinear Schrödinger Equation 253
M.J. Ablowitz and C.M. Schober

Discretisation Effect on a Dynamical System with Discontinuity 269
Xinghuo Yu

Preface

Much of what we know about specific dynamical systems is obtained from numerical experiments. Not only can we visualize the dynamics, we can also evaluate key characteristics such as bifurcation parameters, Lyapunov exponents, Hausdorff dimension and entropy. The discretization process usually has no significant effect on the results for relatively simple, well-behaved dynamics. On the other hand, acute sensitivity to changes in initial conditions is a hallmark of chaotic behaviour, so how confident can we be that the numerical dynamics reflects that of the original system? Do we always obtain a nearby attractor, homoclinic orbit or inertial manifold? Do numerically calculated trajectories always shadow a true one? In brief, what role does numerical analysis have in the study of dynamical systems? Conversely, can recent advances in dynamical systems provide new insights into our understanding of numerical algorithms?

These and related issues at the interface of dynamical systems and numerical analysis were the focus of the workshop *CHAOTIC NUMERICS: an international workshop on the approximation and computation of complicated dynamical behaviour,* which took place at Deakin University in Geelong, Australia, in July 1993. Over sixty delegates, half of whom were from outside Australia, participated. There were forty long and short lectures, including some computing and video presentations. The keynote speakers (W.-J. Beyn, S.N. Chow, J. Hale, J. Lorenz) each gave two one-hour lectures, while the invited speakers (R. Corless, C.M. Elliott, H. Koçak, J.M Sanz-Serna, T. Sauer, H. Yee) gave a one-hour lecture. The contributions in this book are based on these lectures together with several selected from those of other speakers to provide a broad overview of the different matters covered at the workshop.

While organized independently, this workshop can be seen as a natural sequel to the "*Dynamics of Numerics and Numerics of Dynamics*" conference held in Bristol in 1990. Readers may also be interested to consult its proceedings, and are strongly recommended to read Andrew Stuart's definitive exposition and survey article "*The Numerical Analysis of Dynamical Systems*" in the 1994 issue of *Acta Numerica*.

The organizers are grateful to the Faculty of Science and Technology of Deakin University and the Australian Mathematical Society for sponsorship and financial support of the "*CHAOTIC NUMERICS*" workshop. They thank the participants for their stimulating contributions, verbal and written, and the American Mathematical Society for making these contributions more widely available by publishing this book.

Peter Kloeden, Geelong
Ken Palmer, Coral Gables
February 1994

Contemporary Mathematics
Volume **172**, 1994

Numerical Dynamics

JACK K. HALE

ABSTRACT. This paper consists of two distinct parts. The first one is concerned with illustrations of how numerics can be used to test dynamics (mechanism for the onset of chaos, existence of chaos) and how dynamics can be used to explain numerics (slow motion manifolds and numerically stable equilibrium points). The second part deals with the problem of comparing the dynamics of an evolutionary equation to the dynamics of a discretization of the equation.

1. Introduction

In recent years, there have been many interesting developments concerned with dynamics and numerics. These are generally concerned with using numerics to discover dynamical properties of continuous systems, using dynamics to explain unusual phenomena that occur in numerics, and the determination of numerical schemes which will accurately reflect the dynamics of continuous systems. Of course, in the last class of problems, it is important to understand the underlying principles which sometimes lead to numerical approximations which do not accurately reflect the dynamics of the continuous system. The articles, and references therein, in the recent book of Broomhead and Iserles (1992) illustrate some of the work in these directions.

One of the primary issues in classical Numerical Analysis is the convergence on a finite time interval of the solutions of a discretization scheme to the solutions of a differential equation as the discretization parameter approaches zero. Convergence results and error bounds often are obtained by classical methods from analysis. It is unlikely that the theory of Dynamical Systems can be of any assistance in such problems. On the other hand, this latter theory coupled with the error bounds on a finite time interval can play an important role when the integration is to take place over a very long time interval.

1991 *Mathematics Subject Classification.* 34D45, 34C15, 35K57, 65L20.
Key words and phrases. Attractors, chaos, numerical methods, dynamical systems.
Partially supported by DARPA 70NANB8H0860.

In classical Numerical Analysis, attempts to understand the behavior of difference schemes over long time intervals rely primarily on stability considerations for test equations which are usually linear and often time dependent. Although major advances in understanding have been achieved through these test problems, it is becoming clear that they are inadequate for understanding the behavior of nonlinear problems.

On the other hand, the basic appeal of understanding schemes based on simple models remains. For this reason, one should try to determine a set of nonlinear test problems rich enough to capture behavior of difference schemes when applied to nonlinear problems. This is the approach that we take in this paper. We give a few illustrations of how the ideas of dynamical systems are playing a role in numerics with the hope that it will encourage more of the researchers in classical numerical analysis to explore wider applications of their work.

We should point out that another major concern of numerical analysts is efficiency. While this is extremely important, it is not our present concern.

We will consider primarily *dissipative* systems; that is, systems for which there is a global attractor. We now define these concepts.

We suppose that we have an abstract continuous evolutionary equation E_0(for example, it may represent a partial differential equation, differential delay equation, etc.) on a Banach space X and that it generates a C^0-semigroup $T_0(t)$, $t \geq 0$, on X. For any subsets A, B of a Banach space X, we define the *distance from* A *to* B as $\delta_X(A,B) = \sup_{x\in A} \inf_{y\in B} \|x-y\|_X$. For a semigroup $T(t)$ on X and subsets A, B of X, we say that A *attracts* B if $\delta_X(T(t)B, A) \to 0$ as $t \to \infty$. The semigroup $T(t)$ is *point dissipative* if there is a bounded set $B \subset X$ such that B attracts each point of X.

For some types of dissipative semigroups, the asymptotic properties of the orbits is determined by a global attractor, which we now define. A subset $\mathcal{A}_0$ of X is said to be *invariant* if $T_0(t)\mathcal{A}_0 = \mathcal{A}_0$ for all $t \geq 0$. We remark that, if $\mathcal{A}_0$ is invariant, then every point x in $\mathcal{A}_0$ has the property that $T_0(t)x$ can be defined for all $t \in \mathbb{R}$. A subset $\mathcal{A}_0$ of X is said to be a *local attractor* for the semigroup $T_0(t)$, $t \geq 0$, if $\mathcal{A}_0$ is compact, invariant and there is a neighborhood U of $\mathcal{A}_0$ such that $\delta_X(T_0(t)U, \mathcal{A}_0) \to 0$ as $t \to \infty$. The set $\mathcal{A}_0$ is a *global attractor* if it is an attractor and, in addition, $\delta_X(T_0(t)B, \mathcal{A}_0) \to 0$ as $t \to \infty$ for every bounded set $B \subset X$. For any set $B \subset X$, we define the *positive orbit* $\gamma^+(B)$ of B as $\gamma^+(B) = \cup_{t\geq 0} T(t)B$.

For the global attractor to exist, we must impose some additional properties for the semigroup. The semigroup $T(t)$ is *asymptotically smooth* if, for any bounded set $B \subset X$, there is a compact set $J \subset X$ such that J attracts B.

An important result on the existence of global attractors is the following result in Hale (1988) (see also, Babin and Vishik (1989), Ladyzhenskaya (1991), Temam (1988)).

THEOREM 1.1. *If a semigroup $T(t)$ on X is asymptotically smooth, point dissipative and positive orbits of bounded sets are bounded, then there is a global attractor $\mathcal{A}$ for $T(t)$.*

If the conditions of Theorem 1.1 are satisfied and it is known that the ω-limit set of each orbit belongs to the set E of equilibrium points (that is, the semigroup is gradient-like), then the attractor is the unstable manifold of E. More precise statements are in Section 5.

We remark that dissipative in the above sense is not the same concept as appears in the literature on numerical analysis. For an ordinary differential equation, dissipative in the sense of numerical analysis is that the norm of the difference of two solutions does not increase. A more appropriate term would probably be non-expansive, which does occur in the literature on partial differential equations.

For dissipative systems, recent work indicates that, under reasonable assumptions, sensible numerical schemes perform well. The main difficulty lies in understanding the influence of the error terms.

Even if the numerical schemes are chosen properly in the sense that they theoretically should reflect the dynamics of the differential equation, finding the solutions of the difference equations requires approximation techniques since they may be nonlinear. The method of solving the difference equations has its own dynamics which may interfere with, or even superimpose unwanted behavior to, those of the discretization schemes per se. We do not attempt to discuss this important topic here.

Let us now outline the content of these notes. The paper consists of two distinct parts. The first part, Sections 2-4, consists of examples of how numerics can be used to test dynamics and how dynamics can be used to explain numerics. More specifically, in Section 2, we give an illustration of the manner in which numerical methods can be used to test a property that is believed to occur in a continuous system. The property that is tested is the manner in which chaos arises in a delay differential equation and the presentation is based upon the work of Hale and Sternberg (1988). In Section 3, we discuss some recent results of Mischaikow and Mrozek (1993) in which they give a computer assisted proof of the existence of chaos for the Lorenz equations in an open set of parameter values containing a specific value of the parameters. In Section 4, based on the work of Carr and Pego (1989), (1990), Fusco and Hale (1989), Fusco (1990), we give an illustration of how dynamics can explain numerics by discussing a reaction diffusion equation in which there is exponentially slow motion in the movement of transition layers (which numerically appear to be stable equilibrium points).

The second part of the paper, Sections 5-10, deals with the problem of comparing the dynamics of an evolutionary equation to the dynamics of a discretization of the equation. We elaborate in more detail on the types of problems that will be considered.

Let $E_h, 0 < h \leq h_0$, be a finite dimensional approximation to a continuous evolutionary equation E_0 and let $T_h(t)$, $t \geq 0$, be the corresponding semigroup of transformations acting on $X_h \equiv P_h X \subset X$, where P_h is a continuous projection. For example, for a partial differential equation, it could represent a discretization in space, an approximation obtained by using spectral methods, etc., resulting in a finite dimensional system of ordinary differential equations. It also is possible to consider discretization in time and space which results in the consideration of lattice maps and then the parameter t in the approximation ranges over a discrete set. Let us suppose that the abstract evolutionary equation has an attractor (global attractor) $\mathcal{A}_0$. The first important question that needs to be considered is the following:

(i) Does there exist an attractor (global attractor) $\mathcal{A}_h$ for E_h and are the sets $\mathcal{A}_h$, $0 \leq h \leq h_0$, upper semicontinuous at $h = 0$; that is, does $\delta(\mathcal{A}_h, \mathcal{A}_0) \to 0$ as $h \to 0$?

In Section 5, for semigroups which depend upon parameters, we present some general conditions for the upper semicontinuity of attractors due to Hale, Lin and Raugel (1988). As we will see, upper semicontinuity is to be expected in the general situation since it is a simple consequence of attractivity of $\mathcal{A}_0$, the classical finite time error bounds on the discretized problems and the hypothesis that $\cup_{0 \leq h \leq h_0} \mathcal{A}_h$ belongs to a bounded set.

If the flow on the attractor $\mathcal{A}_0$ satisfies no other properties, then upper semicontinuity is the most that can be expected. It is natural to try to obtain some other method of comparison of $T_h(t)$ to $T_0(t)$ which will yield the set $\mathcal{A}_0$. If no hypotheses are made on the dynamics on $\mathcal{A}_0$, then the semigroups $T_h(t)$ cannot be restricted to the attractors $\mathcal{A}_h$, but must use some of the knowledge of the dynamics in a neighborhood of the attractors. This leads to the following question:

(ii) For a given bounded set B in X, is it possible to define a limit of $\{T_h(t)P_h B,\ t \geq 0, 0 < h \leq h_0\}$ in such a way as to obtain $\mathcal{A}_0$ if the set B is a sufficiently large ball?

In Section 6, we introduce such a concept due to Hale and Raugel (1993) and indicate some of the applications to dissipative systems as well as to situations where $T_0(t)$ is conservative. We have not tested this concept for numerical schemes, but believe that it is appropriate and similar results should be true.

If we impose more conditions on the flow $T_0(t)$ on the attractor $\mathcal{A}_0$, then the following questions are relevant:

(iii) Are the sets $\mathcal{A}_h$, $0 \leq h \leq h_0$, lower semicontinuous; that is, does $\delta(\mathcal{A}_0, \mathcal{A}_h) \to 0$ as $h \to 0$?

(iv) Is the flow $T_0(t)|\mathcal{A}_0$ topologically equivalent to the flow $T_h(t)|\mathcal{A}_h$; that is, is there a homeomorphism taking $\mathcal{A}_0$ onto $\mathcal{A}_h$ which preserves orbits and the sense of direction in time?

In Section 7, we present some results on the lower semicontinuity of attractors for gradient systems due to Hale and Raugel (1989). The result is essentially that hyperbolicity of the equilibrium points of $T_0(t)$ imply lower semicontinuity of the attractors for the discretized problem at $h = 0$. An application of the result requires comparison of the equilibrium points, the eigenvalues of the linearization about equilibria and the unstable manifolds. We give some illustrations which deal with explicit types of discretizations for a parabolic equation.

A gradient system is *Morse-Smale* if the equilibrium points are hyperbolic and the stable and unstable manifolds of equilibria intersect transversally. In Section 8, we consider the Morse-Smale property for various types of discretizations of one-dimensional parabolic equations. We also discuss the topological equivalence of the flows $T_h(t)$ and $T_0(t)$.

In applications, the original semigroup depends upon parameters which are subjected to variations. If we suppose that $T_\epsilon(t)$, $0 < \epsilon \leq \epsilon_0$, is a family of C^0-semigroups on X, then we must discuss the above questions relative to the parameters ϵ, h for the approximate semigroups $T_{\epsilon,h}(t)$.

Affirmative answers to any of the questions (i)-(iv) show that the long time behavior of the flows defined by the approximate evolutionary equation are giving information which is a realistic reflection of the behavior of the abstract evolutionary equation when the semigroup $T(t)$ (resp. $T_\epsilon(t)$) is fixed. If the method of approximation is fixed (that is, h is fixed), then the dynamics of the approximation and the continuous system may differ dramatically as the parameter ϵ is varied. Of course, when we are numerically integrating an equation, we do not know the qualitative properties of the flow defined by the abstract equation and we must rely upon the information that is obtained from the computer. As a consequence, it is necessary to discuss in detail simple examples to attempt to build intuition concerning the new types of phenomena that can arise as a result of approximation. Also, we must attempt to determine numerical schemes which reflect as many of properties of the flow as possible; for example, dissipativeness, gradient structure, etc. The implicit schemes or semi-implicit schemes mentioned above satisfy these properties.

In Section 9, we point out some of the spurious solutions that can occur for fixed space discretizations of the one-dimensional parabolic equation when the constant diffusion coefficient is allowed to vary. We give an intuitive explanation of how this process is equivalent to another one-dimensional problem with space dependent diffusion.

For a dissipative gradient system, the limit set of any orbit belongs to the set of equilibrium points. In Section 10, we make a few remarks about the results of Hale and Raugel (1992) on the convergence of orbits of a gradient system to a single equilibrium point and how the theory applies to the discretized system.

We remark that the types of problems discussed as well as the depth of the discussion is restricted by our personal interest, knowledge and space. The literature is so vast that no attempt has been made to be historically accurate. Our hope is that the papers cited contain sufficient references to permit the reader to obtain this as well as other important information.

Finally, the author appreciates the extensive critical comments of Luca Dieci.

2. Numerics to test for dynamics

In the analysis of models of physical systems, numerical results often exhibit very complicated and chaotic structures for certain values of the physical parameters. It is therefore natural to attempt to understand numerically the reason for the onset of this type of behavior as parameters are varied.

On the theoretical side, there are several known phenomena in dynamical systems whose existence is known to imply the existence of complicated dynamics. Therefore, the following question is natural. Given one of the above phenomena and a given physical model, is it possible to devise a numerical scheme which will test for the occurence of this phenomenon? If so, this gives a numerical way to test for the dynamics in the true system.

Using ordinary differential equations for simplicity, let us briefly describe one way that complicated dynamics can occur. Suppose that the equation has a periodic orbit γ and consider the Poincaré mapping π that takes a transversal section S at a point of γ into itself. The periodic orbit corresponds to a fixed point p_0 of π. The orbit is said to be *hyperbolic* if the linear map $D\pi(p_0)$ has no eigenvalue with modulus one. In this case, there are unstable manifolds $W^u(p_0)$ and stable manifolds $W^s(p_0)$ of p_0. These manifolds are said to *intersect transversally* at a point q if the tangent spaces of these manifolds at q span the space. It has been known for a long time that the existence of transversal intersection of the stable and unstable manifolds implies that the system exhibits chaotic dynamics (see, for example, Šil'nikov (1967), Smale (1967) for the first complete discussions and Wiggins (1988) for more details, applications and references). If we suspect this to be the phenomenon that occurs in our physical model, then we ask: Is it possible to detect the existence of a transverse homoclinic orbit from numerical experiments?

Of course, the first difficulty in trying to detect such orbits is to have a very efficient scheme for accurately determining periodic orbits which are are unstable. If we suppose that this has been done, then it is natural to accurately determine the local unstable manifold of either the periodic orbit or the fixed point of a corresponding Poincaré map. Efficient schemes have been devised to do this for two dimensional maps. Of course, these schemes apply also to periodically forced second order differential equations by using the Poincaré map defined by taking an initial point to its value along the solution at time equal to the period of the forcing. The typical representative map is the Hénon planar map with an unstable fixed point for which the stable and unstable manifolds having dimension one.

In higher dimensions, the accurate determination of the local unstable manifold involves much more effort. On the other hand, the local unstable manifold is stable in the sense that orbits with initial data very close to the data for the periodic orbit converge quickly to the unstable manifold and therfore should give some global information about the flow. A good candidate for the initial data on the unstable manifold is the approximate value previously obtained for the unstable periodic orbit. The scenario for the orbit for this initial value problem should be as follows: the orbit will stay for a long time near the true periodic orbit and then it should follow the unstable manifold after it moves away from the periodic orbit. The amount of time that this orbit stays close to the periodic orbit depends of course upon the accuracy of the approximation.

How can we then detect a transversal homoclinic orbit? If there is a transversal homoclinic orbit, then our orbit which is almost on the unstable manifold must return very close to the periodic orbit on several different random time intervals and the amount of time that it stays close also should not follow any specific pattern. If we remember the form of our original approximation to the unstable periodic orbit, then we should be able to recognize this phenomenon.

At the same time that these computations are being performed, we also should estimate the size and multiplicity of the dominant stable and unstable characteristic multipliers because these determine some of the complexity of the flow in a neighborhood of the transversal intersection of the stable and unstable manifolds.

Of course, it would be very desirable to have a friendly and easy to use numerical program which will incorporate all of the above concepts.

Hale and Sternberg (1988) have carried out the above procedure (not the friendly program) to clarify numerical results that have been obtained for a special delay equation. We consider the equation

$$\dot{x}(t) = -ax(t) + b\frac{x(t-\tau)}{1+x^n(t-\tau)}, \tag{2.1}$$

where n is an even integer and a, b, τ are positive constants. This equation often is used as a model in the study of diseases of the blood (Glass and Mackey, 1977). For a, b, n fixed with $n \geq 10$, it was observed numerically that, by increasing the parameter b, there was a sequence of bifurcations - a Hopf bifurcation followed by a sequence of period doubling bifurcations and then chaotic behavior. Farmer (1982) did extensive computations in the chaotic regime and concluded that the embedding dimension of the attractor (that is, the smallest finite dimensional space in which it is possible to embed the attractor) should be at least four.

There has been no theory and even the computations did not indicate the underlying reason for chaos, why the embedding dimension is at least four, or the behavior of the flow for different values of the parameter. The computations of Hale and Sternberg (1988) are an attempt to come closer to an answer to all of these questions.

Due to the fact that there was a Hopf bifurcation, an original approximation of the periodic orbit was easy to obtain. We construct the bifurcation branch that corresponds to the original Hopf bifurcation by a homotopy method used by many persons, including Allgower and Chien, Keller, Georg, and Doedel.

We then used the approximate periodic orbit to follow the unstable manifold as indicated above. If we are close to the value of the parameter where period doubling occured, then the limit of the unstable manifold will be the period doubling orbit. In this way, we also detect period doubling as well as connections between periodic orbits. We continue to vary the parameters and observe the successive period doublings and connections from the unstable manifold.

For specific values of the parameters $a = 1$, $n = 10$, $\tau = 2$, the above type of computations were performed on equation (2.1) with the following conclusions. There is a value $b_0 \in (1.74, 1.77)$ such that, for $b = b_0$, there is a homoclinic tangency between the stable and unstable manifolds of the periodic orbit on the Hopf bifurcation branch. For $b = 1.77$, there is a transverse homoclinic orbit.

As the computations were being performed, we calculated also the number of characteristic multipliers of the Poincaré map of the periodic orbit that were outside the unit circle and the number that had maximum modulus and were inside the unit circle. In the parameter range above, there always was a unique multiplier outside the unit circle and a unique one inside with the above property. This suggest that the embedding dimension of the attractor should be three and not four as Farmer suggested. On the other hand, if the parameter b is increased slightly, we observe that, for $b = 1.8$, there is a saddle-node bifurcation from the Hopf branch and the unstable manifold becomes two dimensional. We did not develop a numerical scheme to follow the complete two dimensional unstable manifold. However, since the unstable manifold is locally stable, we could observe the transverse homoclinic orbit. The conclusion about the embedding dimension of the attractor is still valid as long as there is a unique multiplier inside the unit circle of maximum modulus. Increasing the parameter b to 1.93, we observed that there were two dominant multipliers inside the unit circle, both complex. This suggests that the embedding dimension should be at least four in agreement with the numerical observations of Farmer (1982).

If we wanted to model the behavior of this system in the parameter region for which the embedding dimension is four, an appropriate model for maps would be an extension of the Hénon map to a four dimensional map for which the eigenvalues about fixed points are allowed to be complex.

3. Computer assisted proofs in dynamics

Computer experiments of dissipative dynamical systems have resulted in the discovery of several new types of qualitative behavior; the chaotic behavior in the Lorenz equations being one of the most famous. These examples present a challenging problem in the theory of dynamical systems - classify the new phenomena and replace the numerical evidence with rigorous proofs. In a recent paper, Mischaikow and Mrozek (1993) have managed to show rigorously that chaotic dynamics does occur in the Lorenz equation for certain values of the parameters. The technique combines abstract existence results on topological invariants (Conley index) with finite, computer assisted computations necessary to verify the assumptions of the theorems in concrete cases.

More specifically, consider the Lorenz equations

$$\begin{aligned} \dot{x} &= s(y-x) \\ \dot{y} &= Rx - y - xz \\ \dot{z} &= xy - qz. \end{aligned} \tag{3.1}$$

To describe the result, we need some additional notation. If $f : \mathbb{R}^n \to \mathbb{R}^n$ is a homeomorphism and $N \subset \mathbb{R}^n$, then *the maximal invariant set of* N is defined by $\mathrm{Inv}(N, f) = \{x \in N : f^n(x) \in N \forall \text{ integers } n\}$. Let Σ_2 be the set of doubly infinite sequences on two symbols. The specific result is contained in the following theorem.

THEOREM 3.1. *Let*

$$P = \{\,(x, y, z) : z = 53\}.$$

For all parameter values in a sufficiently small neighborhood of $(s, R, q) = (45, 54, 10)$, *there exists a Poincaré section* $N \subset P$ *such that the Poincaré map* g *induced by (3.1) on* N *is Lipschitz and well defined. Furthermore, there exists an integer* d *and a continuous surjection* $\rho : Inv(N, g) \to \Sigma_2$ *such that*

$$\rho \circ g^d = \sigma \circ \rho,$$

where $\sigma : \Sigma_2 \to \Sigma_2$ *is the shift map; that is,* g^d *is semi-conjugate to the shift map* Σ_2.

The proof of this theorem has five distinct components:

(1) Algebraic invariants based on the Conley index theory which guarantee the semi-conjugacy to the shift map.
(2) An extension of these maps to multivalued maps.
(3) A theory of finite representable multivalued maps which, when combined with the mentioned invariants, serves to bridge the gap between the continuous dynamics of Lorenz equations and the finite dynamics of the computer. Of course, the Conley index must be preserved.
(4) The numerical computations of the finite multivalued map of interest.
(5) The combinatorial computations of the Conley index for the multivalued map.

The algebraic invariants in item 1. involve cohomological information generated by a continuous function g. In principle, these conditions should be verifiable in a finite number of steps (on a computer) since g is Lipschitzian and we can use the simplicial approximation theorem. However, the computations must be performed in such a way that they apply not only to the simplicial approximation, but also to all nearby maps including the original map g. Hence, it is more natural and technically simpler to employ multivalued maps. This is the reason for considering item 2.

Let us describe the ideas behind the multivalued maps. On a computer, we must replace infinite sets of mathematical objects by a finite subset of objects and code its elements with natural numbers. The elements of the selected finite subset of objects are referred to as representable objects relative to the selected coding. Let us be specific for the above example of the Lorenz equation. For any $x \in \mathbb{R}^3$, $s \in \mathbb{R}$, we let $B(x,s) = \{y \in \mathbb{R}^3 : \|y - x\| < s\}$. The set of *representable numbers* $\mathcal{R}$ is given by the standard floating point representable coordinates. The set of *representable vectors* $\mathcal{V}$ in $\mathbb{R}^3$ is the collection of triples $v = (v_1, v_2, v_3)$, where $v_i \in \mathcal{R}$, $i = 1, 2, 3$. For a given $\eta \in \mathcal{R}$, the set of *representable cubes* $\mathcal{C}_\eta$ is the collection of cubes $\mathcal{B}(v, \eta) = \{w \in \mathcal{V} : \|w - v\| < \eta\} \equiv B(v, \eta) \cap \mathcal{V}$. The set of *representable sets* $\mathcal{S}$ is the collection of $S = \cup_{j=1}^{p} \mathcal{B}(v_j, \eta)$ for some p and set $\{v_j\} \subset \mathcal{V}$.

Now suppose that $\delta > 0$ is given $\delta \in \mathcal{R}$ and $M_0 \subset M$ are subsets of $\mathbb{R}^3$. Let $f : M_0 \to M$ be a Lipschitz continuous map with Lipschitz constant L. Also, suppose that, for any $v \in \mathcal{V}$, there is an $f_0(v) \in \mathcal{V}$ such that $\|f(v) - f_0(v)\| < \delta$. Define the *representable multivalued map* $\mathcal{F} : M_0 \cap \mathcal{R} \to \mathcal{S}$ by the relation

$$\mathcal{F}(v) = \cap\{\mathrm{convex} S \in \mathcal{S} : \mathcal{B}(f_0(v), \delta + L\eta) \subset S\}.$$

We remark that, if $x \in M_0$, then $f(x) \in \mathcal{F}(v)$ if $x \in B(v, \eta)$. Therefore, for δ and η small, the multivalued map $\mathcal{F}$ is a good approximation to the original function f.

We recall that the Conley index is defined for isolating neighborhoods; that is, compact sets N such that the maximum invariant set in N does not intersect the boundary of N. After an appropriate definition of an isolating neighborhood for multivalued maps was given (see Kaczyński and Mrozek (1993)), it was shown by Mrozek (1993) that any isolating neighborhood N for the multivalued map $\mathcal{F}$ as above also is an isolating neighborhood for the function f and the Conley indices are the same.

As a consequence of these remarks, a computer assisted proof of Theorem 3.1 is feasible, but highly nontrivial. To obtain the result with the computing power available, the actual implementation of the numerics involved writing the Poincaré map as a composition of 22 finite representable multivalued maps. The total computation time was 80 hours on a Sun Sparc2 workstation.

4. Dynamics to explain numerics

If the coexistence of two phases at the transition temperature is kept under observation for a long time, then one observes that the system is not exactly in equilibrium and a very slow evolution driven by surface tension is taking place. Theoretically, one should see eventually a spatially homogeneous state, but the time for settling down is so long that what one actually observes is "motion towards a stable state" - metastability - slow motion manifolds. The complexity of the spatial distribution of the two phases keeps decreasing but appears to be stable for very long periods of time with intermittent periods of fast motion when there are small inclusions of one of the two regions embedded in the other phase.

Let us assume that the flow is determined from an evolutionary equation with a gradient structure and there is a global attractor for the flow. If the equilibrium points are hyperbolic, then we have remarked above that the flow on the attractor is the union of the unstable manifolds of the equilibria. From this theory, we see clearly what must be done in order to understand the dynamics of a particular problem. It also gives a scenario for the behavior of a specific orbit corresponding to given initial data. We know that the ω-limit set of any orbit must be in the attractor. Therefore, for any neighborhood U of the attractor, a solution must arrive in U in a finite time. If U is a very small neighborhood, then the solution will be close to an unstable manifold of an equilibrium point. It will remain close to some solution on this unstable manifold until it gets close to another unstable manifold, etc., until it eventually converges to some equilibrium point. For typical initial data, we expect that the solution will converge to a stable equilibrium point and not to a saddle. Therefore, if an orbit belongs to the basin of attraction of a stable equilibrium, then it first comes close to the attractor and then cascades along the unstable manifolds as it approaches the stable equilibrium.

From the preceding remarks, it follows that the existence of slow motion manifolds for gradient systems is closely related to the speed of the motion on the unstable manifolds of equilibria. For a model of a given physical system, if we could determine the speed of this motion and observe that it is very slow, then we would have an explanation of the above phenomenon.

For a simple reaction diffusion model, this program has been completed by Fusco and Hale (1989), Fusco (1989), Carr and Pego (1989), (1990), using ideas from singular perturbations and the theory of integral manifolds. We describe this in more detail following Fusco and Hale.

Consider the equation

$$u_t = \epsilon^2 u_{xx} + f(u) \quad \text{in } (0,\, 1), \tag{4.1}$$

with the boundary condition

$$u_x = 0 \quad \text{at } x = 0,\ x = 1\,, \tag{4.2}$$

where $\epsilon > 0$ is a small parameter.

System (4.1), (4.2) defines a gradient system for which every bounded positive orbit converges to an equilibrium point. If f satisfies a dissipative condition

$$\limsup_{|u|\to\infty} \frac{f(u)}{u} \le -\delta < 0, \text{ for } u \in R, \tag{4.3}$$

then there is a global attractor. The above theory applies and it becomes necessary to determine the structure of the flow on the unstable manifolds of equilibria.

Let us assume further that $f(u) = u - u^4$. In this simple situation, we are led to conjectures about the structure of solutions on the unstable manifolds from theoretical considerations. If we consider any initial function which does not have steep gradients, then for some time the solution will follow the flow given by the ODE

$$\dot{u} = f(u).$$

Therefore, at any point where the initial function is positive (resp. negative), the solution will go rather quickly to $+1$ (resp. -1) and very sharp transition layers will be created. At this time, the diffusion terms begin to play a role and the transition layers will move slowly if ϵ is small.

At this point, we recall some theoretical facts that we know about equilibria of (4.1), (4.2) and the dimensions of their unstable manifolds. The only stable equilibria are $+1$ and -1. If an equilibrium point has k zeros in $(0, 1)$, then the dimension of the unstable manifold has dimension k and the eigenvectors of the corresponding linearized problem about this equilibrium corresponding to the positive eigenvalues have at most k zeros. Finally, for ϵ small, the equilibria have sharp transition layers.

From these remarks, it is reasonable to suppose that the points on the unstable manifold of the equilibria should look approximately like the equilibria except that the transition layers are translated to different points. This reasoning should be valid provided that the transition layers are not too close together or too close to the boundary. When this is the case, we expect to be close to the unstable manifold of another equilibrium point and the transition time to this new manifold may occur at a different rate. The problem now is to make these remarks mathematically precise.

Again, the idea is simple. We must first determine the approximate shape of the transition layers. It is intuitively clear that these layers should be very close to a translate of the solution of the following problem:

$$\epsilon^2 v_{xx} + f(v) = 0, \qquad x \in (-\infty, +\infty)$$

$$\lim_{x \to \pm\infty} v(x) = \pm 1, \qquad v(0) = 0.$$

Let us now define what we expect to be the approximate unstable manifold for an equilibrium point of index n (the dimension of the unstable manifold is n).

Let $0 < \psi_1 < \psi_2 < \cdots < \psi_n < 1$ be n variables; $\psi = (\psi_1, \dots, \psi_n)$; $\Gamma = \{\psi : 0 < \psi_1 < \psi_2 < \cdots < \psi_n < 1\}$ and $\psi_0 = -\psi_1$, $\psi_{n+1} = 2 - \psi_n$. Let $\eta_i = (\psi_{i+1} + \psi_i)/2$, $\zeta_i = (\psi_{i+1} - \psi_i)/2$ for $0 \le i \le n$, and let $U(\cdot, \cdot) : \Gamma \times [0, 1] \to \mathbb{R}$ be defined by

$$U(\psi, x) = (-1)^{i+1} v(x - \psi_i),$$

for $\eta_{i-1} \le x \le \eta_i$, $1 \le i \le n$. The function v is a a solution of the equation

$$\epsilon^2 v_{xx} + f(v) = 0, \quad x \in (-\infty, +\infty)$$

which satisfies

$$\lim_{x \to \pm\infty} v(x) = \pm 1, \quad v(0) = 0.$$

This definition implies that $U(\psi, \cdot)$ is a continuous function with a piecewise continuous first derivative that jumps at the η_i. The map $\psi \to U(\psi, \cdot)$ defines an n-dimensional manifold $\bar{W} \in H^1(0, 1)$. The idea now is to use integral manifold theory to show that there is an exact integral manifold near $\bar{W}$ which is the unstable manifold of the equilibrium of index n.

There are several difficulties involved. First the functions on the manifold do not satisfy the boundary conditions and are not smooth. One must make some modification of $\bar{W}$ before it is a good candidate for an invariant manifold of the equation. To find this modification, Fusco and Hale (1989) introduce coordinates around the manifold: $u = U(\psi, \cdot) + V$, where V is normal to $\bar{W}$ at the point ψ. If $\bar{W}$ is to be an integral manifold, then there must be a function $\psi \to c(\psi) \in \mathbb{R}^n$ and a function $V(\psi)$ so that the following partial differential equation is satisfied:

$$\Sigma_1^n (U_i + V_i) c_i = \epsilon^2 (U_{xx} + V_{xx}) + f(U + V) \tag{4.4}$$

as well as boundary conditions and regularity conditions. The subscripts i on U and V denote partial derivatives with respect to ψ_i.

Fusco and Hale (1989) now linearize these equations (that is, throw away the terms $V_i c_i$ and linearize f about U) and solve for functions $(c(\psi, \epsilon), \tilde{V}(\psi, \epsilon))$. The function $\tilde{V}(\psi)$ defines a new manifold $\tilde{W} = \{ u = U(\psi, \epsilon, \cdot) + \tilde{V}(\psi, \epsilon)\}$ on which the functions are smooth and satisfy the boundary conditions. The approximate velocity of the transition layers are given by the equation

$$\dot{\psi} = c(\psi, \epsilon), \tag{4.5}$$

where $\psi = (\psi, \ldots, \psi)$, $c(\psi, \epsilon) = (c_1(\psi, \epsilon), \ldots, c_n(\psi, \epsilon))$, satisfy

$$c_i(\psi, \epsilon) = \epsilon \mu^2 K (e^{-\frac{\mu}{\epsilon}(\psi_{i+1} - \psi_i)} - e^{-\frac{\mu}{\epsilon}(\psi_i - \psi_{i-1})})$$

for $i = 1, \ldots, n$ and μ and K are positive constants and we have put $\psi_0 = -\psi_1, \psi_{n+1} = 2 - \psi_n$. This differential equation exhibits the exponentially slow motion of the transition layers. In fact linearization about the equilibrium solution of (4.5) yields n eigenvalues which are exponentially small. The motion of the transition layers move with the same small exponential rate proportional to the distance between transition layers.

There is a major additional problem - to prove that there is an exact integral manifold near $\tilde{W}$. Using the classical approach of the theory of integral manifolds, Fusco (1989) accomplished this but the analysis was nontrivial and required several ideas and estimates.

Carr and Pego (1989) also have proved the existence of the unstable manifolds above with the mentioned properties. The basic approach was the same except the manner in which the approximate manifold was obtained is different. These differences need to be assessed because we know that the way in which good initial approximations are obtained in singular perturbation problems plays a very important role.

For further developments of these ideas for the Cahn-Hilliard equation, see Alikakos, Bates and Fusco (1991), Alikakos and Fusco (1992) and, for a reaction diffusion equation coupled with an ODE, see Kuwamura, Ei and Mimura (1992).

5. Upper semicontinuity of attractors

In this section, we summarize some general results of Hale, Lin and Raugel (1988) on the upper semicontinuity of attractors of semigroups which depend upon parameters. Specific applications will be given in later sections. For related results, see Kloeden and Lorenz (1986), (1990), Babin and Vishik (1989).

Let $T_0(t), t \geq 0$, be a C^s-semigroup on a Banach space X with $s \geq 0$. Let $h > 0$ be a parameter and let $\{X_h, h > 0\}$ be a family of subspaces of X such that

$$\lim_{h \to 0} \delta_X(x, X_h) = 0 \quad \text{for any } x \in X. \tag{5.1}$$

We assume also that there is a continuous projection $P_h : X \to X_h$ for each $h > 0$. Let $T_h(t), t \geq 0$, be a C^s-semigroup on X_h with $s \geq 0$.

Let us suppose that, for $h \geq 0$, $\mathcal{A}_h$ is a compact global attractor for $T_h(t)$. Recall that $\{\mathcal{A}_h, h \geq 0\}$ is uppercontinuous at $h = 0$ if and only if $\delta_X(\mathcal{A}_h, \mathcal{A}_0) \to 0$ as $h \to 0$. Following Hale and Raugel (1993), we introduce the limit of the attractors,

$$\tilde{\omega}(\mathcal{A}_.) = \cap_{h_0>0} \operatorname{Cl}_X \cup_{0<h\leq h_0} \mathcal{A}_h.$$

Hale and Raugel (1993) make the following observation.

LEMMA 5.1. *If $\{\mathcal{A}_h, h \geq 0\}$ is upper semicontinuous at $h = 0$, then*

$$\tilde{\omega}(\mathcal{A}_.) \subset \mathcal{A}_0. \tag{5.2}$$

If (5.2) is satisfied and $Cl_X \cup_{h>0} \mathcal{A}_h$ is compact, then $\{\mathcal{A}_h, h \geq 0\}$ is upper semicontinuous at $h = 0$.

We say that the semigroups $T_h(t)$ *approximate* $T(t)$ *uniformly on bounded sets of* $\mathbb{R}^+ \times X$, if for any bounded set $U \subset X$ and any compact interval $I \equiv [t_0, t_1] \subset \mathbb{R}^+ \equiv [0, \infty)$, there are a constant $h(I, U) > 0$ and a function $\eta(h, I, U)$ defined for $0 < h \leq h(I, U)$ such that

$$\lim_{h \to 0} \eta(h, I, U) = 0 \tag{5.3}$$

and, for any $0 < h \leq h(I, U)$ and any $u \in U$, the functions $T(t)u$, $T_h(t)P_h u$ are defined for $t \in I$, and

$$\|T(t)u - T_h(t)P_h u\|_X \leq \eta(h, I, U) \quad \text{for } t \in I. \tag{5.4}$$

The following elementary result is in Hale, Lin and Raugel (1988).

THEOREM 5.2. *For $0 \leq h \leq h_0$, assume that $T_h(t)$ has a global compact attractor $\mathcal{A}_h$ and there is a constant M, independent of h, such that $\|x_h\|_X \leq M$ for all $x_h \in \mathcal{A}_h$. If, for any bounded set $U \in X$, $T_h(t)$ approximates $T(t)$ on U uniformly on compact sets of $[t_0, \infty)$, then the family $\{\mathcal{A}_h, h \geq 0\}$ is upper semicontinuous at $h = 0$; that is, $\delta(\mathcal{A}_h, \mathcal{A}_0) \to 0$ as $h \to 0$.*

The idea for the proof is very simple. Since $\mathcal{A}_0$ is the global attractor for $T_0(t)$, for any bounded set U in X and for any $\eta > 0$, there is a $t_\eta > 0$, such that $T_0(t)U \in \mathcal{N}(\mathcal{A}_0, \eta)$ for $t \geq t_\eta$, where $\mathcal{N}(\mathcal{A}_0, \eta)$ is the η neighborhood of $\mathcal{A}_0$. We now choose U large enough to contain all of the sets $\mathcal{A}_h, h \geq 0$. Since $T_h(t)$ approximates $T(t)$ on U uniformly on $[0, t_\eta]$, we can choose h sufficiently small so that $\|T(t_\eta)u - T_h(t_\eta)P_h u\|_X \leq \eta/2$ for all $u \in U$. We can now use the semigroup property to get the upper semicontinuity.

It is possible to obtain a similar result for local attractors, but it is more difficult to state.

THEOREM 5.3. *Assume that $T_0(t)$ has a compact local attractor $\mathcal{A}_0$ and that N_1 is a neighborhood of $\mathcal{A}_0$ such that $\delta(T(t)N_1, \mathcal{A}_0) \to 0$ as $t \to \infty$. Also, suppose that there are constants $h_0 > 0$, $\delta_0 > 0$, $t_0 \geq 0$ and two open neighborhoods N_2, N_3 of $\mathcal{A}_0$ with $N_1 \subset N_2 \subset \mathcal{N}(N_2, \delta_0) \subset N_3$, such that, for $0 < h \leq h_0$,*

(i) $T_0(t)N_1 \subset N_2$ for $t \geq 0$,
(ii) $T_h(t)(N_1 \cap X_h) \subset N_2$ for $0 \leq t \leq t_0$,
(iii) for any $x_h \in \mathcal{N}(N_2, \delta_0) \cap X_h$, there exists $t(x_h) > 0$ such that $T_h(t)x_h \in N_3$ for $0 \leq t \leq t(x_h)$.

If $T_h(t)$ approximates $T(t)$ uniformly on $N_3 \times I$, where I is an arbitrary compact set of $[t_0, \infty)$ and each $T_h(t)$ is asymptotically smooth, then there is an $h_0 > 0$ such that, for $0 < h \leq h_0$, $T_h(t)$ admits a local compact attractor $\mathcal{A}_h$ which attracts $N_1 \cap X_h$ and the family $\{\mathcal{A}_h, h \geq 0\}$ is upper semicontinuous at $h = 0$; that is, $\delta(\mathcal{A}_h, \mathcal{A}_0) \to 0$ as $h \to 0$.

Remark 5.1. If $\mathcal{A}$ is a compact local attractor under the semigroup $T(t)$, $t \geq 0$, then $\mathcal{A}$ is stable and there always exist neighborhoods N_1, N_2 satisfying (i) in Theorem 5.3.

Hale, Lin and Raugel (1988), Hale and Raugel (1989) have given applications to the Galerkin approximation of sectorial evolutionary equations, semidiscretization in time for some parabolic problems, spectral methods applied to the two dimensional Navier-Stokes equation and discretization of a linearly damped hyperbolic equation. Dettori (1990) has used the above setting to show upper semicontinuity of attractors for spectral decompositions.

Example 5.1 Semidiscretization in space. As in Hale, Lin and Raugel (1988), Hale and Raugel (1989), we consider the parabolic equation

$$(5.5)\qquad \begin{gathered} u_t - \Delta u = -f(u) - g \quad \text{in } \Omega, \\ u = 0 \quad \text{in } \partial\Omega, \end{gathered}$$

where Ω is a bounded smooth domain in $\mathbb{R}^n$, $n = 1, 2, 3$, $g \in L^2(\Omega)$, and $f \in C^2(\mathbb{R}; \mathbb{R})$ and satisfies the following conditions:

$$(5.6)\qquad \limsup_{|s|\to\infty} \frac{-f(s)}{s} \leq 0,$$

and

$$(5.7)\qquad |\frac{\partial^2 f}{\partial s^2}(s)| \leq C(|s|^\gamma + 1),$$

where $0 \leq \gamma < \infty$ if $n = 2$, and $0 \leq \gamma \leq 1$ if $n = 3$.

If we let $u(t; u_0)$ be the solution of (5.5) with $u_0 \in H_0^1(\Omega)$ and define $T_0(t)u_0 = u(t; u_0)$, then $T_0(t) : H_0^1(\Omega) \to H_0^1(\Omega)$ is a C^1-semigroup. In fact, $T_0(t)$ is a gradient system with Lyapunov function

$$V_0(\varphi) = \int_\Omega [\frac{1}{2}|\nabla\varphi|^2 + F(\varphi) + g\varphi],$$

where $F(s) = \int_0^s f$ since $V_0(\varphi)$ is equivalent to the norm on $H_0^1(\Omega)$ and

$$\dot{V}_0(u(t)) = -\int_\Omega u_t^2 \leq 0.$$

Choose subspaces X_h as above, $X_h \subset H_0^1(\Omega)$ and define $A_h : X_h \to X_h$ by the relation

$$(A_h w_h, v_h)_{L^2(\Omega)} = (\nabla w_h, \nabla v_h)_{L^2(\Omega)} \ \forall\ v_h \in X_h.$$

Define $Q_h : L^2(\Omega) \to X_h$ and $P_h : H^1(\Omega) \to X_h$ by the relations

$$(v - Q_h v, v_h) = 0 \ \forall\ v \in L^2(\Omega), v_h \in X_h,$$

$$(\nabla(v - P_h v), \nabla v_h) = 0 \ \forall\ v \in H_0^1(\Omega), v_h \in X_h.$$

For the equation

$$(5.8)\qquad u_{ht} + A_h u_h = -Q_h f(u_h) - Q_h g$$

with $u_h(0) = u_{0h} \in X_h$, we obtain a C^1-semigroup which is a gradient system with Lyapunov function

$$V_h(\varphi_h) = \int_\Omega [\frac{1}{2}|\nabla\varphi_h|^2 + F_h(\varphi_h) + Q_h g\varphi_h],$$

where $F_h(s) = \int_0^s Q_h f$ and

$$\dot{V}_h(u_h(t)) = -\int_\Omega u_{ht}^2 \leq 0.$$

THEOREM 5.4. *The attractors* $\mathcal{A}_h, h \geq 0$ *are upper semicontinuous at* $\epsilon = 0$.

This result is contained in Hale, Lin and Raugel (1988) and the basic estimate for the proof is contained in the following statement: For any $r_0 > 0$, there exists a positive constant $c(r_0)$ such that, for any $u_0 \in B_X(0, r_0)$ ($B_X(0, r_0)$ is the ball in X of radius r_0 and center 0), and, for any $t \geq 0$, we have

$$\|T_h(t)P_h u_0 - T_0(t)u_0\|_{H_0^1(\Omega)} \leq \frac{h}{t} c(r_0) e^{c(r_0)t}.$$

Example 5.2 Semidiscretization in time. Hale, Lin and Raugel (1988) consider the following discretization of (5.5) in time. Let $k > 0$, $t_n = nk$, $n \geq 0$, $u_n = u(t_n)$, $A = -\Delta$ with Dirichlet boundary conditions and consider the equations

$$(5.9) \qquad u_{n+1} = (1 - (1-\theta)kA)(1 + \theta kA)^{-1} u_n + k(1 + \theta kA)^{-1}(-f(u_n) - g),$$

where

$$\frac{1}{2} < \theta \leq 1.$$

If we let $T_k u_0 = u_1$, where u_1 is given in (5.9), then

$$T_k : H_0^1(\Omega) \to H_0^1(\Omega)$$

is a continuous mapping and therefore a discrete semigroup.

THEOREM 5.5. *There exists a* $k_0 > 0$ *such that, for* $0 < k \leq k_0$, *each of the maps* T_k *has a compact local attractor* $\mathcal{A}_k$ *and the sets* $\{\mathcal{A}_k, 0 < k \leq k_0\} \cup \{\mathcal{A}_0\}$ *is upper semicontinuous at* $k = 0$.

Remark 5.2. In (5.9), we can replace $f(u_n)$ by $f(\theta u_{n+1} + (1-\theta)u_n)$.

Remark 5.3. In Example 5.1, the discretized system was shown to be gradient. In Example 5.2, we believe this is true but do not have a proof at this time. If it is gradient, then the local attractor becomes a global attractor since it is possible to prove that the equilibrium set is upper semicontinuous (see Hale and Raugel (1989)).

Preservation of the gradient structure by the approximating system is very important. For ODE, Humphries and Stuart (1992), Stuart and Humphries and Stuart (1992a), (1992b) have given a large class of multi-step Runga-Kutta schemes for which the gradient structure is preserved. We mention also the following interesting one-step θ-scheme. Consider the ordinary differential equation

$$(5.10) \qquad \dot{u} = f(u),$$

where $u \in \mathbb{R}^p$ for some p. Consider the following approximate scheme

$$(5.11) \qquad u_{n+1} = u_n + h[(1-\theta)f(u_n) + \theta f(u_{n+1}),$$

where $u_n = u(nh)$. The following result is due to Stuart and Humphries (1992a).

THEOREM 5.6. *Suppose that* $f = -\nabla F$. *The scheme (5.11) is gradient if* $\theta \in [\frac{1}{2}, 1]$.

6. More general limits

The results in this section are taken from Hale and Raugel (1993). In Section 5, we introduced the ω-limit set of the attractors $\mathcal{A}_h$ and related it to upper semicontinuity. Further properties of $\tilde{\omega}_{(0,h_0]}(\mathcal{A}.)$ are contained in Hale and Raugel (1993). In particular, conditions are given for this set to be invariant under $T_0(t)$ and to be connected. Although this concept is weaker that Hausdorff continuity (see Hale and Raugel (1993)), it is still too strong to have $\tilde{\omega}_{(0,h_0]}(\mathcal{A}.) = \mathcal{A}_0$ for all semigroups. The simple example $\dot{x} = -x(h + (x-1)^2)$ shows this fact.

We now introduce another concept from Hale and Raugel (1993) which allows us to obtain more informaton about $\mathcal{A}_0$ from the semigroups $T_h(t)$ without making any hypotheses about the flow on $\mathcal{A}_0$.

Definition 6.1. For a given set $B \subset X$, the *ω-limit set of B with respect to the family of semigroups $T_h(t)$, $t \geq 0$, and projections* P_h, $h \in (0, h_0]$, is denoted by $\hat{\omega}_{(0,h_0]}(B)$ and is defined in the following way: a point $x \in \hat{\omega}_{(0,h_0]}(B)$ if and only if there are sequences $\{h_n\} \subset (0, h_0]$, $\{t_n\} \subset [0, \infty)$, $\{x_n\} \subset B$ such that $h_n \to 0$, $t_n \to \infty$, $T_{h_n}(t_n)P_{h_n}x_n \to x$ as $n \to \infty$.

An equivalent definition of $\hat{\omega}_{(0,h_0]}(B)$ is as follows.

Definition 6.1bis. For a given set $B \subset X$, the *ω-limit set of B with respect to the family of semigroups $T_h(t)$, $t \geq 0$, and projections P_h*, $h \in (0, h_0]$, is

$$\hat{\omega}_{(0,h_0]}(B) = \cap_{\delta>0} \mathrm{Cl} \cup_{\bar{h}\in((0,h_0]\times\mathbb{R}^+)\cap\mathcal{N}_{\mathbb{R}^+\times\mathbb{R}^+}(0,\delta)} T_h(t)P_hB,$$

where

$$\bar{h} = (h, \frac{1}{t}), \quad \mathcal{N}_{\mathbb{R}^+\times\mathbb{R}^+}(0, \delta) = \{\bar{h} = (h, \frac{1}{t}) \in \mathbb{R}^+ \times \mathbb{R}^+ : h \leq \delta, \frac{1}{t} \leq \delta\}.$$

The set $\hat{\omega}_{(0,h_0]}(B)$ in the general situation can be larger than $\tilde{\omega}_{(0,h_0]}(\mathcal{A}.)$ and, in some situations, coincides with $\mathcal{A}_0$. If $B \supset \cup_{h\in(0,h_0]}\mathcal{A}_h$ for some $h_0 > 0$, then $\hat{\omega}_{(0,h_0]}(B) \supset \tilde{\omega}_{(0,h_0]}(\mathcal{A}.)$.

The set $\hat{\omega}_{(0,h_0]}(B)$ does not depend upon $T_0(t)$. Therefore, it is possible to consider the situation where each $T_h(t)$ is dissipative and the limit $T_0(t)$ is not. As a consequence, we should be comparing $\hat{\omega}_{(0,h_0]}(B)$ with the ω-limit set $\omega_0(B) \equiv \mathcal{A}_0(B)$ of B with respect to $T_0(t)$ defined by

$$\mathcal{A}_0(B) = \cap_{\tau>0}\mathrm{Cl}_X \cup_{t\geq\tau} T_0(t)B.$$

Although many properties of $\hat{\omega}_{(0,h_0]}(B)$ are discussed in Hale and Raugel (1993), we mention only a few. We need the following hypothesis:

(6.1) For any $t_0 > 0$ and any bounded set $U \subset [t_0, \infty) \times X$ and any $\eta > 0$, there is a $\delta_0 = \delta_0(\eta, U) > 0$ such that, for $h \in (0, h_0]$ and $(t, x) \in U$, we have

$$\|T_h(t)P_hx - T_0(t)x\|_X \le \eta\,.$$

THEOREM 6.1. *Suppose that (6.1) is satisfied.*

(i) Let B be any set in X. If $T_0(t)$ is a C^0-semigroup on X, then $T_0(t)\hat{\omega}_{(0,h_0]}(B) \subset \hat{\omega}_{(0,h_0]}(B)$.

(ii) If $t_0 \ge 0$ is a fixed constant and we suppose in addition that

$$B_1 \equiv \cup_{h\in(0,h_0]} \cup_{t\ge t_0} T_h(t)P_hB$$

is bounded, and either that

(H_1) $\quad$ *$T_0(t)$ is asymptotically smooth and $\cup_{t\ge0} T_0(t)B_1$ is bounded*

or that

(H_2) $\quad$ *$T_0(t)$ is a C^0-group on $\mathbb{R}$,*

then $\hat{\omega}_{(0,h_0]}(B)$ is invariant under $T_0(t)$.

COROLLARY 6.2. *Let B be a bounded set in X. Suppose that $T_0(t)$ is a C^0-semigroup on X and that the ω-limit set $\omega_0(B) \equiv \mathcal{A}_0(B)$ of B with respect to $T_0(t)$ exists, is nonempty and attracts B. If the hypotheses of Theorem 6.1 are satisfied and $\hat{\omega}_{(0,h_0]}(B) \subset B$, then*

$$\hat{\omega}_{(0,h_0]}(B) = \mathcal{A}_0(B)\,.$$

THEOREM 6.3. *Let B be a bounded set in X and suppose that $T_0(t)$ is a C^0-semigroup on X and that the ω-limit set $\omega_0(B) \equiv \mathcal{A}_0(B)$ of B with respect to $T_0(t)$ exists, is nonempty and attracts B. If, moreover, (6.1) holds and either $\mathcal{A}_0(B) \subset \mathring{B}$ (the interior of B) or $\mathcal{A}_0(B)$ attracts also a neighborhood of B, then*

$$\hat{\omega}_{(0,h_0]}(B) = \mathcal{A}_0(B)\,.$$

In particular, if $T_0(t)$ has a global attractor $\mathcal{A}_0$, if $\mathcal{A}_0 \subset B$ and (6.1) holds, then

$$\hat{\omega}_{(0,h_0]}(B) = \mathcal{A}_0\,.$$

In the case where each of the semigroups, including the one for $h = 0$, has a global attractor and the rate at which it attracts a bounded set is exponential and uniform in h, we do not gain much additional information by considering the set $\hat{\omega}_{(0,h_0]}(B)$; that is, we may as well consider the limits of the attractors. This concept becomes more important when there is no global attractor for the semigroup at $h = 0$. In particular, this concept becomes relevant when the semigroup at $h = 0$ has a first integral.

Applications of the above results to the case where $T_0(t)$ has a first integral are given in Hale and Raugel (1993); in particular, for the delay differential equation

$$\dot{x}(t) = -(1+\epsilon)f(x(t)) + f(x(t-1))\,,$$

where $\epsilon \geq 0$ is a parameter, $f \in C^1(\mathbb{R}, \mathbb{R})$, there is positive constant δ such that $f'(x) \geq \delta$ for all x, and $f(0) = 0$; for the ordinary differential equation

$$\begin{aligned} \dot{\xi} &= f(\xi) - \eta \\ \dot{\eta} &= \epsilon(\gamma\xi - \eta), \end{aligned} \tag{6.2}$$

where $\epsilon > 0, \gamma > 0, 0 < a < 1/2$, are constants and $f(u) = u(1-u)(u-a)$. The result for (6.2) is used to show that $\hat{\omega}_{(0,h_0]}(B) = \mathcal{A}_0$ for the Fitzhugh-Nagumo equation with large diffusion. In this same paper, it is shown how to adapt the above limit to obtain $\hat{\omega}_{(0,h_0]}(B) = \mathcal{A}_0$ for the problems of thin domains for a parabolic equation considered by Hale and Raugel (1992) and the linearly damped hyperbolic equation discussed by Hale and Raugel (1992).

7. Lower semicontinuity for gradient systems

In this section, we summarize some general results of Hale and Raugel (1989) on the lower semicontinuity of attractors of gradient semigroups which depend upon parameters. For related results, see Babin and Vishik (1989), Kapitanskii and Kostin (1991).

Let us recall the definition of a gradient system (see, for example, Hale (1988).

Definition 7.1. A C^r-semigroup $T(t), t \geq 0$, on a Banach space X is said to be a *gradient system* if there exists a Liapunov function for $T(t)$; that is, there is a continuous function $V : X \to \mathbb{R}$ with the property that

(1) $V(x)$ is bounded below
(2) $V(x) \to +\infty$ as $\|x\|_X \to +\infty$,
(3) $V(T(t)x)$ is nonincreasing in t for each $x \in X$,
(4) if x is such that $V(T(t)x) = V(x)$ for all $t \in \mathbb{R}$, then x is an equilibrium point; that is, $T(t)x = x$ for all $t \in \mathbb{R}$.

We let E denote the set of equilibrium points of $T(t)$. If $T(t)$ is a C^1-semigroup, then an equilibrium point φ is said to be *hyperbolic* if $\sigma(DT(t)\varphi) \cap \{z \in \mathbb{C} : |z| = 1\} = \emptyset$ for any t, where $\sigma(DT(t)\varphi)$ is the spectrum of the operator $DT(t)\varphi$. If φ is a hyperbolic equilibrium point, then there exists a neighborhood U of φ such that

$$W^u_{\text{loc}}(\varphi) \equiv W^u(\varphi; U) \equiv \{y \in U : T(-t)y \in U, t \geq 0, T(-t)y \to \varphi \text{ as } t \to +\infty\}.$$

is a submanifold of X and coincides with the set $\{y \in U : T(-t)y \in U, t \geq 0\}$. This set is referred to as the *local unstable manifold* of φ. The *unstable set* $W^u(\varphi)$ of φ is defined to be

$$W^u(\varphi) = \{y \in X : T(-t)y \text{ is defined for } t \geq 0 \text{ and } T(-t)y \to \varphi \text{ as } t \to +\infty\}.$$

If we assume that $T(t)$ has a global compact attractor $\mathcal{A}$ and each of the equilibrium points is hyperbolic, then

$$\mathcal{A} = \cup_{\varphi \in E} W^u(\varphi).$$

To state the result of Hale and Raugel (1989) on the lower semicontinuity of attractors, we need several hypotheses.

$(\mathrm{H}.1)_0$ $T_0(t)$ is a C^1-gradient system which is asymptotically smooth;

$(\mathrm{H}.2)_0$ the set E_0 of equilibrium points is bounded;

$(\mathrm{H}.3)_0$ each element of E_0 is hyperbolic.

Hypotheses $(\mathrm{H}.1)_0$, $(\mathrm{H}.2)_0$, imply that $T_0(t)$ has a global attractor $\mathcal{A}_0$ and the hypothesis $(\mathrm{H}.3)_0$ implies that E_0 contains a finite number of elements $\varphi_{j,0}, j = 1, 2, \ldots, N_0$, and

$$\mathcal{A}_0 = \cup_{\varphi_{j,0} \in E_0} W^u(\varphi_{j,0}).$$

We now make some continuity hypotheses on the semigroups $T_h(t)$, $h \neq 0$.

$(\mathrm{H}.4)_\epsilon$ for $h \neq 0$, $T_h(t)$ is a C^1-semigroup on X_h and admits a local attractor $\mathcal{A}_h$ attracting $U_0 \cap X_h$, where U_0 is a fixed open neighborhood of $\mathcal{A}_0$ in X;

$(\mathrm{H}.5)_\epsilon$ if E_h denotes the set of equilibrium points of $T_h(t)$, then there exists an open neighborhood W_0 of E_0 in X, $W_0 \subset U_0$, such that $W_0 \cap E_h = \{\varphi_{1,h}, \ldots, \varphi_{N_0,h}\}$ where each $\varphi_{j,h}$ is hyperbolic and $\varphi_{j,h} \to \varphi_{j,0}$ as $h \to 0$.

We define the local unstable sets

$$W^u_{\mathrm{loc},h}(\varphi_{j,h}) \equiv \{y \in U_{j,h} \cap X_h : T_h(-t)y \in U_{j,h} \cap X_h, t \geq 0, \lim_{t \to \infty} T_h(-t)y = \varphi_{j,h}\},$$

where $U_{j.h}$ is a neighborhood of $\varphi_{j.h}$ in X with $U_{j,h} \subset U_0$. Since $W^u_{\mathrm{loc},h}(\varphi_{j,h}) \subset U_0$, it follows that $W^u_{\mathrm{loc},h}(\varphi_{j,h}) \subset \mathcal{A}_h$. As above, there is a neighborhood $U_{j,h}$ of $\varphi_{j,h}$ such that $W^u_{\mathrm{loc},h}(\varphi_{j,h})$ is a submanifold of X_h and it is finite dimensional since $\mathcal{A}_h$ is compact. We further assume

$(\mathrm{H}.6)_h$ $\delta_X(W^u_{\mathrm{loc},0}(\varphi_{j,0}), W^u_{\mathrm{loc},h}(\varphi_{j,h})) \to 0$ as $h \to 0$;

$(\mathrm{H}.7)_h$ for any $\eta > 0$, $\tau > 0$, $t_0^* > 0$, there exist real numbers $\delta^* \equiv \delta(\eta, \tau, t_0^*) > 0$, $h^* \equiv h(\eta, \tau, t_0^*) > 0$, such that, for $0 < h \leq h^*$,

$$\|T_h(t)y - T_0(t)x\|_X < \eta \quad \text{for } t_0^* \leq t \leq \tau,$$

provided that $x \in \mathcal{A}_0$, $y \in \mathcal{A}_h$ and $\|y - x\|_X \leq \delta^*$.

The following result is due to Hale and Raugel (1989).

THEOREM 7.1. *Under the above hypotheses $(H.1)_0$ to $(H.7)_h$, the sets $\mathcal{A}_h$ are lower semicontinuous at $h = 0$; that is, $\delta_X(\mathcal{A}_0, \mathcal{A}_h) \to 0$ as $h \to 0$.*

If we combine Theorem 5.2 with Theorem 7.1, it is possible to prove the following interesting result.

THEOREM 7.2. *For $0 \leq h \leq h_0$, assume that each $T_h(t)$ is a C^1-gradient system, asymptotically smooth with the set E_h of equilibrium points bounded (thus, $T_h(t)$ has a global compact attractor $\mathcal{A}_h$), and, for any bounded set $U \in X$,* *suppose that $T_h(t)$ approximates $T(t)$ on U uniformly on compact sets of $[t_0, \infty)$. If, in addition, all of the equilibrium points of $T_0(t)$ are hyperbolic, then the attractors $\mathcal{A}_h$ are continuous at $h = 0$; that is,*

$$\sup(\delta_X(\mathcal{A}_h, \mathcal{A}_0), \delta_X(\mathcal{A}_0, \mathcal{A}_h)) \to 0 \quad \text{as } h \to 0.$$

In Hale and Raugel (1989), for the discretizations in Examples 5.1 and 5.2 above, it is shown that hyperbolicity of the equilibriums for the semigroup $T_0(t)$ implies that the attractors are lower semicontinuous and, thus, continuous.

8. Morse-Smale discretizations for 1-D parabolic equations

In this section, we consider the time and space discretization of the parabolic equation (4.1), (4.2) and discuss the problem of transversality of the stable and unstable manifolds as well as the topological equivalence of the flows.

Let us recall that two submanifolds N, M of a Banach space X are *transversal* if either $N \cap M = \emptyset$ or, for every point $q \in N \cap M$, the tangent spaces T_qN and T_qM of respectively N and M at q span the space X.

A gradient system is said to be Morse-Smale if all of the equilibrium points are hyperbolic and the stable and unstable manifolds of the equilibrium points intersect transversally.

The following fundamental result is due to Henry (1985), Angenent (1986) and uses in a significant way the integer valued Lyapunov function given by the number of zeros of a function in $[0, 1]$.

THEOREM 8.1. *For (4.1), (4.2), if $\varphi, \psi \in E_\epsilon$ are hyperbolic, then $W^u_\epsilon(\varphi)$ is transversal to $W^u_\epsilon(\psi)$. Thus, (4.1), (4.2) is Morse-Smale if and only if all of the equilibrium points are hyperbolic.*

We now turn to the problem of determining when certain discretizations of (4.1), (4.2) are Morse-Smale.

8.1. Discretizations in time.

We consider first the *semi-implicit Euler scheme*

$$u^{n+1} - u^n = \epsilon^2 h u^{n+1}_{xx} + h f(u^n) \quad \text{in } (0, 1) \tag{8.1}$$

with the boundary conditions (4.2).

We consider (8.1) in the space $H^1((0, 1))$. For any $\varphi \in H^1((0, 1))$, the equation

$$\psi - \epsilon^2 h \psi_{xx} = \varphi + h f(\varphi) \tag{8.2}$$

defines a C^1-mapping $T_{\epsilon,h} : H^1((0, 1)) \to H^1((0, 1))$; that is, $T_{\epsilon,h}$ is a C^1 discrete dynamical system on $H^1((0, 1))$. Furthermore, the regularity theory of elliptic equations implies that $T_{\epsilon,h}$ is a compact map. Also, for every $\epsilon > 0$ and for every $h > 0$, system (8.1) is a gradient system. In fact, it is possible to construct a Liapunov function to see that the ω-limit set of $\{T^n_{\epsilon,h}\varphi, n \geq 0\}$ exists and is a fixed point of $T_{\epsilon,h}$. Finally, the set F_ϵ of fixed points of $T_{\epsilon,h}$ is independent of h and coincides with the set of equilibrium points of (4.1), (4.2). Therefore, it is bounded and we deduce that the global attractor $\mathcal{A}_{\epsilon,h}$ exists. If each fixed point is hyperbolic, then

$$\mathcal{A}_{\epsilon,h} = \cup_{\varphi \in F_\epsilon} W^u_{\epsilon,h}(\varphi),$$

where $W^u_{\epsilon,h}(\varphi)$ is the unstable manifold of φ.

Oliva, de Oliveira and Solà-Morales (1993) have shown that the stable and unstable manifolds of hyperbolic fixed points of $T_{\epsilon,h}$ must intersect transversally. As a consequence, the dynamical system defined by $T_{\epsilon,h}$ is Morse-Smale if each fixed point is hyperbolic. The method of proof exploits an integer valued Lyapunov function.

It is possible to show that the unstable manifold $W^u_{\epsilon,h}(\varphi)$ converges, as $h \to 0$, in the C^1-sense to the unstable manifold W^u_ϵ of the equilibrium point φ of (4.1), (4.2). With this remark and results of Hale, Magalhães and Oliva (1984), Raugel (1990), we conclude that there is an $h_0(\epsilon)$ such that, for $0 < h \leq h_0(\epsilon)$, the flow defined by (8.1), (4.2) is topologically equivalent (on the attractors) to the flow defined by (4.1), (4.2). Therefore, for each fixed ϵ, we can assert that the dynamics that is seen in the discretization in t is the same as the dynamics that is seen in the continuous system if the spacing h is sufficiently small.

We remark that similar results hold for the implicit scheme

$$u^{n+1} - u^n = \epsilon^2 h u^{n+1}_{xx} + hf(u^{n+1}) \quad \text{in } (0,1) \tag{8.3}$$

although the proofs are somewhat more complicated due to the way that the nonlinear function f enters into defining the map $T_{\epsilon,h}$.

We do not discuss the explicit scheme,

$$u^{n+1} - u^n = \epsilon^2 h u^n_{xx} + hf(u^n) \quad \text{in } (0,1)$$

since it is a well known fact that they are very bad for numerical integration. However, it is interesting to understand the reasons that spurious solutions may occur in such schemes (see, for example, Elliott and Stuart (1992), Griffiths and Mitchell (1992)).

8.2. Space discretization. Suppose that J is a positive integer and that the interval $[0,1]$ is partitioned by the points j/J, $j = 0,1,2,\ldots,J$. If $u(x,t)$ is a solution of (4.1), (4.2), we let $u_j(t) = u(j/J,t)$, $j = 1,2,\ldots,J$ and, for convenience, define $u_{-1}(t) = u_0(t)$, $u_{J+1}(t) = u_J(t)$. If we discretize the Laplacian according to this partition, then (4.1), (4.2) becomes the system of ordinary differential equations

$$\dot{u}_j = r^2(u_{j+1} - u_j) + r^2(u_{j-1} - u_j) + f(u_j), \quad j = 0,1,2,\ldots,J, \tag{8.4}$$

where $r = \epsilon/J$.

If we suppose that the function f is dissipative (see (4.3)), then, for each ϵ, J, the system (8.4) defines a semigroup $T_{\epsilon J}(t)$ on $H^1((0,1))$ which is a gradient system and has a global attractor $\mathcal{A}_{\epsilon J}$.

Fusco and Oliva (1988) have shown that, if φ, ψ are equilibrium points which are hyperbolic, then $W^u_{\epsilon J}(\varphi)$ is transversal to $W^u_{\epsilon J}(\psi)$. As a consequence, the system (8.4) is Morse-Smale if and only if all of the equilibrium points are hyperbolic. Using the ideas from Hale, Magalhães and Oliva (1984), Raugel (1990), it is possible also to show that there is a $J_0(\epsilon)$ such that the flow defined by $T_{\epsilon J}(t)$ is topologically equivalent on the attractor to the flow defined by $T_\epsilon(t)$ on the attractor if J is sufficiently large if all of the equilibrium points are hyperbolic.

8.3. Discretization in time and space. In this section, we briefly discuss the *semi-implicit scheme* of discretization in time and space

$$u_j^{n+1} - u_j^n = \frac{\epsilon^2 h}{J^2}(u_{j+1}^{n+1} - 2u_j^{n+1} + u_{j-1}^{n+1}) + hf(u_j^n) \quad \text{in } (0,1) \tag{8.5}$$

and the *implicit scheme* of discretization in time and space

$$u_j^{n+1} - u_j^n = \frac{\epsilon^2 h}{J^2}(u_{j+1}^{n+1} - 2u_j^{n+1} + u_{j-1}^{n+1}) + hf(u_j^{n+1}) \quad \text{in } (0,1). \tag{8.6}$$

For a discussion of the explicit scheme, see Elliott and Stuart (1992), Griffiths and Mitchell (1992).

Magalhães and Oliva (1992) have shown that there exists an $h_0 > 0$ such that, for $0 < h \leq h_0$, the equation (8.5) defines a C^1 discrete dynamical system $\tilde{T}_{\epsilon,h}$ for which there exists a global attractor $\tilde{\mathcal{A}}_{\epsilon,h}$. Also, the ω-limit set of any orbit is a fixed point of $\tilde{T}_{\epsilon,h}$. Furthermore, if the fixed points are hyperbolic, then the stable and unstable manifolds intersect transversally. As a consequence, the system is Morse-Smale if the fixed points are hyperbolic.

Using the same type of argument as in the previous section, we may use the results of Hale, Magalhães and Oliva (1984), Raugel (1990) to conclude that there are positive constants $J_0(\epsilon), h_0(\epsilon)$ such that, for $0 < J \leq J_o(\epsilon)$, $0 < h \leq h_0(\epsilon)$, the flow defined by (8.5) is topologically equivalent to the flow defined by (4.1), (4.2). Therefore, for each fixed ϵ, we can assert that the dynamics that is seen in the discretization in time and space is the same as the dynamics that is seen in the continuous system if the spacings J and h are sufficiently small.

For a discussion of the existence of the attractor and the convergence of orbits to the set of fixed points for this type of numerical procedure as well as others for problems in several space variables, see Elliott and Stuart (1992). We reemphasize that these 'good' procedures preserve the gradient structure.

9. Space discretization and spurious solutions

As in Section 7.3, we discretize the Laplacian to obtain the system (7.4), which we rewrite here:

$$\dot{u}_j = r^2(u_{j+1} - u_j) + r^2(u_{j-1} - u_j) + f(u_j), \quad j = 0, 1, 2, \ldots, J, \tag{9.1}$$

where $r = \epsilon/J$.

For definiteness, let us assume that $f(u) = u - u^3$. Let us now fix the number of points in the partition; that is, the integer J, the size of the partition, and let $\epsilon \to 0$. We have the following result.

THEOREM 9.1. *For definiteness, let us suppose that $f(u) = u - u^3$. For fixed J, there exist positive constants $\epsilon_0(J)$, $\epsilon_1(J)$ such that, for $\epsilon > \epsilon_1(J)$, the global attractor $\mathcal{A}_r$ of (9.1) is one dimensional consisting of exactly three hyperbolic equilibrium points P_{1r}, P_{2r}, P_{3r}, with P_{1r}, P_{3r} being stable and P_{2r} being a saddle point with a one dimensional unstable manifold. As $\epsilon \to \infty$, the attractor $\mathcal{A}_r$ approaches the straight line segment joining the point $u_j = -1$ for all j to the point $u_j = 1$ for all j.*

For $0 < \epsilon \leq \epsilon_0(J)$, the equation (9.1) has exactly 3^J equilibrium points and 2^J of these are asymptotically stable.

The proof of the first statement is easily supplied by observing that for r large, the equilibrium points must be close to the diagonal on which all of the coordinates are equal. For the second part, for $r = 0$, the equilibrium points are hyperbolic and one can apply the Implicit Function Theorem and observe the stability properties by checking the Jacobian for $r = 0$.

The last part of Theorem 9.1 demonstrates a behavior for the solutions of (9.1) which is completely different from the behavior of the solutions of (4.1), (4.2). There are 2^J stable solutions of (9.1) and only two for (4.1), (4.2). The additional stable solutions must be considered as *spurious solutions*. This last observation with the proof outlined above is in Elliott and Stuart (1992) (see also, Bence, Merriman and Osher (1991) for an estimate of $\epsilon_0(J)$) (see Stephens and Shubin (1981) for the viscous Burgers' equation with small dissipation).

Elliott and Stuart (1992) give a possible explanation for the appearance of these spurious stable solutions. It is based on the exponentially slow movement of the transition layers (meta-stable states) on the unstable manifolds as described in the Section 4. To quote from Elliott and Stuart (1992): 'The resolution of the numerical method is insufficient to capture the tiny propagation speeds for the transition layers and hence these meta-stable states are stabilized by the discretization and become steady solutions of the numerical method.' Also, 'The computed solution is smooth and there is nothing obvious which tells us that it is spurious - it is only the fact that we know a priori that genuine steady solutions have equidistributed zeros that enables to rule out the computation as spurious.'

We now point out the close connection that equations (9.1) have to another reaction diffusion problem of the form (4.1), (4.2), but with variable diffusion coefficient. This observation seems to be another way to explain the appearance of spurious solutions. We consider the equation

$$u_t + (a_\nu(x)u_x)_x = f(u) \tag{9.2}$$

with the homogeneous Neumann boundary condition (4.2). If the diffusion coefficient a_ν is assumed to be very large when the parameter ν is small, then the solutions of (9.2), (4.2) on the attractor behave essentially as the solutions of the ordinary differential equation $\dot{\xi} = f(\xi)$ where ξ is approximately the average of u (see, for example, Hale (1986), Conway, Hoff and Smoller (1978)). Therefore, it is reasonable to suppose that, if, for ν small, the diffusion coefficient a_ν is assumed to be very large on a finite number of intervals I_j whose complement in $[0,1]$ is very small, and in most of this complement the diffusion coefficient is very small, then the solutions of (9.2), (4.2) on the attractor should behave essentially as the solutions of a system of ordinary differential equations for a vector ξ whose j^{jh} component ξ_j should correspond approximately to the average of the solution on I_j. The vector field for the jth component should be $f(\xi_j)$ plus some diffusive coupling; that is, the same type of equation as (9.1). These intuitive remarks have recently been made precise by Fusco (1987), Carvalho and Pereira (1992). We now describe their results.

For $\nu \in (0,\nu_0)$, we are going to define a positive diffusion coefficient $a_\nu \in C^2([0,1],\mathbb{R})$. Let $s=(s_0,s_1,\dots,s_n)$, $0=s_0<s_1<\dots<s_n=1$ be a partition of $[0,1]$ and let $\ell=(\ell_1,\dots,\ell_{n-1})$, $\beta=(\beta_0,\dots,\beta_n)$ be two sequences of positive constants and let $\ell_1',\dots,\ell_{n-1}'$, $\beta_0',\dots,\beta_n'$ be functions of ν that approach $\ell_1,\dots,\ell_{n-1},\beta_0,\dots,\beta_n$ from above as $\nu\to 0$ and Also, let $\ell_0=\ell_0'=0=\ell_n=\ell_n'$ and let $e=(e_1,\dots,e_n)$ be another sequence of positive constants. We define $a_\nu=a_\nu(s,\ell,\beta)$ in the following manner:

$$
\begin{aligned}
&a_\nu(x)\geq \frac{e_i}{\nu} && s_{i-1}+\nu\ell_{i-1}'\leq x\leq s_i-\nu\ell_i'\\
&a_\nu(x)\geq \nu\beta_i && s_i-\nu\ell_i'\leq x\leq s_i+\nu\ell_i'\\
&a_\nu(x)\leq \nu\beta_i' && s_i-\nu\ell_i\leq x\leq s_i+\nu\ell_i
\end{aligned}
\tag{9.3}
$$

for $i=0,1,\dots,n$.

If we let

$$\xi_j(t)=\frac{1}{s_j-s_{j-1}}\int_{s_{j-1}}^{s_j}u(x,t)dx,\quad j=1,2,\dots,n,$$

where $u(x,t)$ is a solution of (9.2), (4.2) and if we assume that the equilibrium points of (9.2), (4.2) are hyperbolic, then it is shown in Carvalho and Pereira (1992) (a similar result is in Fusco (1987)) that, for ν sufficiently small, the flow on the attractor of (9.2), (4.2) is topologically equivalent to the flow defined by the system of ordinary differential equations

$$
\begin{aligned}
\dot\xi_1&=r_j^2(\xi_2-\xi_1)+f(\xi_1),\\
\dot\xi_j&=r_j^2(\xi_{j+1}-\xi_j)+r_{j-1}^2(\xi_{j-1}-\xi_j)+f(\xi_j),\quad 2\leq j\leq n-1,\\
\dot\xi_n&=r_{n-1}^2(\xi_{n-1}-\xi_n)+f(\xi_n),
\end{aligned}
\tag{9.4}
$$

where

$$r_j^2=\frac{\beta_j}{2\ell_j(s_j-s_{j-1})},\quad j=1,2,\dots,n.\tag{9.5}$$

If we assume that the constants β_j and the points s_j of the partition of $[0,1]$ are chosen in such a way that all of the r_j are equal, then we obtain the equation (9.1). As a consequence, we see that the space discretization of (4.1), (4.2) leads to the same flow as the one obtained from the continuous system (9.1), (4.2) for ν small. This is not unexpected since space discretization is similar to replacing the function on a spatial interval by its average on that interval. If we want a contionuous system to have the property that it is described by the averages over some intervals, then the diffusion coefficient should be large on that interval.

Theorem 9.1 is valid even in the case of several space variables. To illustrate, we consider the simple domain $\Omega = (0,1) \times (0,1)$ and the equation

$$u_t - \Delta u = f(u) \quad \text{in } \Omega \tag{9.6}$$

with the boundary conditions

$$\frac{\partial u}{\partial n} = 0 \quad \text{in } \partial\Omega, \tag{9.7}$$

where n is the unit outward normal.

If we discretize the Laplacian with a uniform square grid of size $1/J$, let $u_{j,k}(t) = u(j/J, k/J, t)$ and define $u_{-1,k} = u_{0,k}, u_{J+1,k} = u_{J,k}$, $u_{j,-1} = u_{j,0}, u_{j,J+1} = u_{j,J}$, then we obtain the system of ODE

$$\begin{aligned} \dot{u}_{j,k} =& r^2(u_{j+1,k} - u_{j,k}) + r^2(u_{j-1,k} - u_{j,k}) \\ &+ r^2(u_{j,k+1} - u_{j,k}) + r^2(u_{j,k-1} - u_{j,k}) + f(u_{j,k}), \end{aligned} \tag{9.8}$$

where $r^2 = \epsilon^2/J^2$. For (9.8), Theorem 9.1 is valid.

It also is possible to relate the dynamics of (9.8) to the dynamics of a continuous problem with appropriate diffusion coefficients. Let us consider the equation

$$u_t - (a_\nu u_x)_x - (b_\nu u_x)_x = f(u) \quad \text{in } \Omega \tag{9.9}$$

with the boundary condition (9.7), where $a_\nu = a_\nu(s, \ell, \beta)$, $b_\nu = b_\nu(\bar{s}, \bar{\ell}, \bar{\beta})$ are functions of the form (9.3). It has been shown by Carvalho (1992) that, for ν small, the flow on the attractor for (9.9), (9.7) is equivalent to the flow on the attractor for the ODE

$$\begin{aligned} \dot{u}_{j,k} =& r_{j,k}^2(u_{j+1,k} - u_{j,k}) + r_{j-1,k}^2(u_{j-1,k} - u_{j,k}) \\ &+ r_{j,k}^2(u_{j,k+1} - u_{j,k}) + r_{j,k-1}^2(u_{j,k-1} - u_{j,k}) + f(u_{j,k}), \end{aligned} \tag{9.10}$$

for some positive constants $r_{j,k}^2$ given explicitly in terms of the parameters defining a_ν, b_ν. This is under the assumption that the system (9.9), (9.7) is Morse-Smale; that is, the stable and unstable manifolds intersect transversally. If all of the $r_{j,k}$ are equal, then we obtain (9.8).

Similar remarks must apply also to general domains. A precise proof is not easy because delicate eigenvalue analysis is required to obtain the equations (9.10).

10. Remarks on convergence

In this section, we only want to point out the following result.

THEOREM 10.1. *For each of the systems (8.1) for h small, (8.4) for J large, and (8.5) for h small and J large, the ω-limit set of any orbit is a single equilibrium point.*

The proof of this fact follows from general results of Hale and Raugel (1992) on convergence of gradient-like systems. It is only necessary to show that, if φ is an equilibrium point (resp. fixed point) of any of these equations (resp. maps), then the linear operator defined by the derivative of the semigroup evaluated at φ has 1 as an eigenvalue of multiplicity at most one. Since this occurs for $T_\epsilon(t)$, it means that we need only to verify that the equilibrium points (resp. fixed points) of $T_{\epsilon h}(t)$ are close to those of $T_\epsilon(t)$ and that the eigenvalues in a compact set K with $0 \notin K$ of the corresponding linearization of the perturbed and unperturbed operators at $t = 1$ are close if h is small.

REFERENCES

Alikakos, N. D., P. W. Bates and G. Fusco, *Slow motion for the Cahn-Hilliard equation in one space dimension*, J. Differential Equations **90** (1991), 81-135.

Alikakos, N. D. and G. Fusco, *Equilibrium and dynamics of bubbles for the Cahn-Hilliard equation*, Preprint (1992).

Angenent, S. B., *The Morse-Smale property for a semilinear parabolic equation*, J. Differential Equations **62** (1986), 427-442.

Babin, A. V. and M. I. Vishik, *Attractors of Evolution Equations*, Russian edition, Nauka, Moscow (1992).

Bence, J., B. Merriman and S. Osher, *On spurious numberical solutions of fast reaction, slow diffusion equations*, In preparation.

Broomhead, D. S. and A. Iserles, *Dynamics of Numerics and Numerics of Dynamics*, Clarendon Press, Oxford (1992).

Carr, J. and R. L. Pego, *Metastable patterns in solutions of $u_t = \epsilon^2 u_{xx} - f(u)$*, Comm. Pure Appl Math. (1989).

Carr, J. and R. L. Pego, *Invariant manifolds and metastable patterns in solutions of $u_t = \epsilon^2 u_{xx} - f(u)$*, Proc. Royal Soc. Edinburgh (1990).

Carvalho, A., *Infinite Dimensional Dynamics Described by Ordinary Differential Equations*, Ph.D. Thesis, Ga. Tech. (1992).

Carvalho, A. and A. L. Pereira, *A scalar parabolic equation whose asymptotic behavior is dictated by a system of ordinary differential equations*, J. Differential Equations (1992), to appear.

Dettori, L., *Spectral approximations of attractors of a class of semi-linear parabolic equations*, Preprint, Univ. Parma, Italy.

Elliott, C. M. and A. M. Stuart, *The global dynamics of discrete semilinear parabolic equations*, IMA Preprint Series **973** (1992).

Farmer, J. D., *Chaotic attractors of an infinite dimensional system*, Physica **4D** (1982), 366-393.

Fusco, G., *A system of ODE which has the same attractor as a scalar PDE*, J. Differential Equations **69** (1987), 85-110.

Fusco, G., *A geometric approach to the dynamics of $u_t = \epsilon^2 u_{xx} + f(u)$ for small ϵ*, Lect. Notes Physics **359** (1990), 53-73.

Fusco, G. and J. K. Hale, *Slow-motion manifolds, dormant instability, and singular perturbations*, Dyn. Diff. Equations **1** (1989), 75-94.

Fusco, G. and W.M. Oliva, *Jacobi matrices and transversality*, Proc. Roy. Soc. Edinburgh **109A** (1988), 231-243.

Glass, L. and M. Mackey, *Oscillation and chaos in physiological control systems*, Science **197** (1977), 287-289.

Griffiths, D. F. and A. R. Mitchell, *Spurious behavior and nonlinear instability in discretized partial differential equations*, The Dynamics of Numerics and the Numerics of Dynamics (Eds. D. S. Broomhead and A. Iserles, Oxford (1992).

Hale, J. K., *Dynamics and numerics*, The Dynamics of Numerics and the Numerics of Dynamics (Eds. D. S. Broomhead and A. Iserles, Oxford (1992).

Hale, J. K., X.-B. Lin and G. Raugel, *Upper semicontinuity of attractors for approximations of semigroups and partial differential equations*, Math. Comp. **50** (1988), 89-123.

Hale, J. K., L. Magalhães and W.M. Oliva, *An Introduction to Infinite Dimensional Dynamical Systems - Geometric Theory*, Appl. Math. Sci. **47** (1984), Springer-Verlag.

Hale, J. K. and G. Raugel, *Lower semicontinuity of attractors of gradient systems and applications*, Ann. Math. Pura Appl. **154** (1989), 281-326.

Hale, J. K. and G. Raugel, *Reaction-diffusion equation on thin domains*, J. Math. Pures Appl. **71** (1992), 33-95.

Hale, J. K. and G. Raugel, *A damped hyperbolic equation on thin domains*, Trans. Am. Math. Soc. **329** (1992), 185-219.

Hale, J. K. and G. Raugel, *Convergence in gradient-like systems with applications to PDE*, ZAMP **43** (1992), 63-124.

Hale, J. K. and G. Raugel, *Limits of semigroups depending on parameters*, Resenhas **1** (1993), Univ. São Paulo, Brazil.

Hale, J. K. and N. Sternberg, *Onset of chaos in differential delay equations*, Comp. Physics **77** (1988), 221-239.

Henry, D., *Some infinite dimensional Morse-Smale systems defined by parabolic differential equations*, J. Differential Equations **59** (1985), 165-205.

Humphries, A. R. and A. M. Stuart, *Runge-Kutta methods for dissipative and gradient dynamical systems*, Num. Anal. Project Manuscript NA-92-18, Stanford, (1992a).

Kaczyński, T. and M. Mrozek, *Conley index for discrete multivalued dynamical systems*, in preparation (1993).

Kloeden, P. E. and J. Lorenz, *Stable attracting sets in dynamical systems and in their one-step descretizations*, SIAM J. Numer. Anal. **23** (1986), 986-995.

Kloeden, P. E. and J. Lorenz, *A note on multistep methods and attracting sets of dynamical systems*, Numer. Math. **56** (1990), 667-673.

Kostin, I. N., *Approximation of attractors for non-gradient systems*, preprint (1992).

Kuwamura, M., S.-I. Ei and M. Mimura, *Very slow dynamics for some reaction-diffusion systems of the activator-inhibitor type*, Japan J. Indust. Appl. Math. **9** (1992), 35-77.

Ladyzhenskaya, O., *Attractors for Semigroups and Evolution Equations*, Lezioni Lincee, Cambridge Univ. Press (1991).

Magalhães, L. and W.M. Oliva, preprint (1992).

Mischaikow, K. and M. Mrozek, *Isolating neighborhoods and chaos*, CDSNS 116 (1993).

M. Mrozek, *Topological invariants, multivalued maps and computer assisted proofs in dynamics*, in preparation (1993).

Oliva, W.M., J.C.F. de Oliveira and J. Solà-Morales, *An infinite dimensional Morse-Smale map*, to appear (1993).

Raugel, G., *Stabilité d'une équation parabolique de Morse-Smale perturbée de manière singulière en une équation hyperbolique*, C. R. Acad. Sci. Paris **310** (1990), 85-88.

Stuart, A. M. and A. R. Humphries, *Model problems in numerical stability theory for initial value problems*, Num. Anal. Project Manuscript NA-92-16, Stanford (1992a).

Stuart, A. M. and A. R. Humphries, *An analysis of local error control for dissipative, contractive and gradient dynamical systems*, Num. Anal. Project Manuscript NA-92-18, Stanford (1992b).

Šil'nikov, L. P., *On the Poincaré-Birkoff problem. Math. USSR-Sbornik* **3** (1967), 353-371.

Smale, S., *Differentiable dynamical systems*, Bull. Amer. Math. Soc. **53** (1967), 747-817.

Temam, R., *Infinite Dimensional Dynamical Systems in Mechanics and Physics*, Springer-Verlag (1988).

Wiggins. S, *Global Bifurcations and Chaos*, Springer Verlag (1988).

CENTER FOR DYNAMICAL SYSTEMS AND NONLINEAR STUDIES, GEORGIA INSTITUTE OF TECHNOLOGY, ATLANTA, GA 30332

E-mail address: hale@math.gatech.edu

Contemporary Mathematics
Volume **172**, 1994

Error Backward

ROBERT M. CORLESS

ABSTRACT. Numerical simulations of mathematical models can suggest that the models are chaotic. For example, one can compute an orbit and its associated finite-time Lyapunov exponents, and these computed exponents can be positive. It is not clear how far these suggestions can be trusted because, as is well known, numerical methods can introduce spurious chaos or even suppress actual chaos. However, comparison with physical experiment often shows that the numerics are right after all. In an effort to explain this success, we give examples here of the use of three types of *backward error analysis* (omitting exposition of a fourth, more familiar, type of backward error analysis known as shadowing). The types are, first, *the method of modified equations*, from which we learn about the differential equation the numerical method 'really' solved; second, *defect control*, which isn't so much analysis as construction of a useful solution in the first place; and, finally, a new analysis which combines the first two.

As an example of existing theoretical results which become more useful in view of the existence of defect-controlled numerical methods, we give a small theorem, similar to one due to Kloeden and Lorenz and to a lemma of Yoshizawa, which uses Lyapunov's direct method to establish that attracting sets are faithfully reported by defect-controlled numerical methods.

1. Introduction

The technique of *backward error analysis* is very popular in certain fields of numerical analysis. Ideally, one shows that the numerical method has produced the *exact* solution of a *nearby problem*, one that is just as good a model of the underlying physical phenomena as was originally specified. Any conclusions that could have been drawn from the exact solution to the specified problem can be drawn from the exact solution to this nearby problem.

This approach is not very popular with most pure mathematicians and computer scientists. One suspects that this is because they are not comfortable with

1991 *Mathematics Subject Classification.* Primary 58F30, 65L99; Secondary 65G99, 34D30.
This work was supported by NSERC.
This paper is in final form and no version of it will be submitted for publication elsewhere

changing the problem; mathematicians because there is usually no context for the problem to change in, and computer scientists because they wish to write general-purpose codes which can solve any problem in a given class, and this task would be more complicated if the problem context had to be taken into account as well.

However, if the problem context is *not* taken into consideration, one can be led astray by rigorous mathematics and previous numerical experience with stable problems. For example, the work in [**1**] does not clearly distinguish between unstable numerical methods and ill-conditioned problems, ignores the problem context, and hence falsely suggests that most numerical studies of chaotic dynamical systems are fatally flawed, being completely contaminated by rounding errors. That this is false can be seen from the exceptional agreement between numerical simulations and physical experiment for a great many chaotic systems [**35**]. The work in [**1**] is mathematically correct, and indeed local errors such as roundoff errors and truncation errors can and do cause trajectories to wander unpredictably; but then, so do physical disturbances. One can draw an analogy with the work of [**1**] and early work in numerical linear algebra which showed that forward error bounds for the solution of linear systems by Gaussian elimination (with pivoting) were very large. These error bounds were so pessimistic as to imply that Gaussian elimination could not possibly be useful in practice; the contrary fact that Gaussian elimination actually *does* work well in practice cannot be explained by looking at these bounds. However, the success of Gaussian elimination can be explained by considering backward error [**20,49**]. As we will see here, the success of numerical methods for the solution of chaotic ordinary differential equations can also be explained by a similar backward error analysis.

Mathematical modelling of real phenomena *always* requires approximation and neglect of small effects. One neglects, for example, the effect of the gravitational attraction of Jupiter on one's earthbound experiment. As a deeper example, in fluid mechanics the Navier-Stokes equations are themselves only an approximate description of fluid phenomena—for example, one might consider the Burnett equations, or perhaps the Ladyzhenskaya equations, as being a more accurate description of real fluids. Similarly, one ignores 'small' stochastic terms in ordinary differential equation models of many phenomena, or 'small' non-autonomous perturbations of physics experiments (such as the effect of passing trucks).

So a numerical analysis of methods of solving ordinary differential equations which put truncation and roundoff errors on the same basis as modelling, measurement, and data errors would be a completely successful analysis. If we could say of any particular numerical simulation of a continuous dynamical system that it was the exact solution for a physically equivalent dynamical system model, we would have exactly as much information as if we had the exact solution of the specified model. We then have to study the effects of perturbations, of course, but we have to do this even if we know the exact solution of the specified problem.

This idea is related to the idea of 'shadowing', but is less deep. Nevertheless there are some interesting mathematical and computational questions associated with it, as we will see. In 'shadowing', one is allowed to change the initial conditions but not the equations—which makes sense if the physical laws giving the equations are well-tested but the initial conditions are not known very well—whereas the more general 'backward error' approach allows both to change. For a more detailed discussion of shadowing and defect control, see [**8,9**].

These ideas are *not* related to the concept of structural stability [**36**]. A structurally stable system is one whose trajectories are topologically conjugate to those of perturbed systems. Verification that a given system is structurally stable is very difficult, and we usually want more quantitative information anyway—that is, we usually need a quantitative characterization of how much effect a perturbation of a given (small) size has. This is not to say that the idea of structural stability is not useful or interesting, as indeed it is, but here we concentrate on computable quantities.

In this paper we discuss three approaches to backward error analysis for ordinary differential equations: the *method of modified equations*, which gives good results for compact time intervals asymptotically as the step-size or tolerance goes to zero; the idea of *defect control* which controls the numerical stepsize to give an exact solution to an equation that is within a user-specified tolerance of the original equation; and finally a combination of the two approaches which helps explain the influence of the underlying discretization on a defect-controlled method.

2. The Method of Modified Equations

What follows is a very brief example of the method of modified equations. For a fuller explanation and applications, see [**21,22**]. The basic idea of this method is, given a numerical solution of a specified problem, to find a *modified problem* whose exact solution is closer to the numerical solution of the specified (original) problem. Such problems are not unique.

We look for such modified problems because they are often easier to understand qualitatively than the difference equations that define the numerical method being used. Thus, the modified equation helps us to analyze the behaviour of the numerical method. The purpose of a modified equation, then, is *explanation* of the numerical results.

To find such a modified problem, we start with an expression for the 'local truncation error' of the numerical method applied to the specified problem. We then manipulate this expression algebraically as described in [**21**] to find the modified problem. For low-order, fixed time-step methods this is very easy; the method can be extended to higher-order methods but the calculations can get tedious. Work is in progress on the automation of this process using computer algebra. Some variable-step methods can also be analyzed with the method of modified equations [**22**]. For example, consider the problem

$$x' = \frac{dx}{dt} = f(t,x) = x^2 - t \tag{2.1}$$

which is used in the excellent pedagogical text [**28, p. 242**]. This problem is solved numerically in [**28**] by several fixed time-step methods. A second-order Runge-Kutta method (RK2) and the classical fourth-order Runge-Kutta method (RK4) in particular give incorrect asymptotic behaviour—each has an apparently smooth spurious solution ("designed to fool people who trust computers") which the authors of [**28**] are unable to explain in detail but use to good advantage to encourage healthy skepticism on the part of the student. The authors of [**28**] do give a partial explanation, showing when the numerical solution becomes unstable and drifts away from the true solution. Using the method of modified equations, we will see a *qualitative* explanation of the resulting smooth spurious solution. However, we cannot give a complete, quantitative explanation.

Note that (2.1) is a solvable Riccati equation, whose solution is expressible in terms of Airy functions (substitution of $x = -w'/w$ gives the Airy equation $w'' = tw$). Almost all solutions tend asymptotically to $x = -\sqrt{t}$. We will compare this later with the solution by RK2 and with the solution of certain modified equations.

RK2 is the two-stage, second-order Runge-Kutta scheme given below.

$$\begin{aligned} k_1 &= f(t_n, x_n) \\ k_2 &= f(t_n + \tfrac{1}{2}h, x_n + \tfrac{1}{2}hk_1) \\ x_{n+1} &= x_n + hk_2 \,. \end{aligned} \tag{2.2}$$

We use the local truncation error to construct a modified equation, as in [**21**]. The local truncation error is defined as the difference between the numerical solution x_{n+1} and the *local exact solution* $u(t)$ which satisfies the differential equation exactly but starts at the initial condition $u(t_n) = x_n$. By convention this quantity is divided by the stepsize h. Thus the local truncation error is

$$\mathrm{LTE} = \frac{u(t_n + h) - x_{n+1}}{h} \,.$$

Expanding this in a Taylor series about t_n we get

$$\begin{aligned} \mathrm{LTE} &= \frac{[x_n + hx'_n + \frac{h^2}{2}x''_n + \cdots] - [x_n + hk_2]}{h} \\ &= x'_n - k_2(h, x_n, t_n) + \tfrac{h}{2}x''_n + \tfrac{h^2}{6}x'''_n + \cdots \end{aligned}$$

The *order* of the method is the degree of the first nonzero term in this Taylor series once expanded; in this case, RK2 has order 2.

The method of modified equations starts with the equation

$$0 = v' - k_2(h, v, t) + \tfrac{h}{2}v'' + \tfrac{h^2}{6}v''' + \cdots \tag{2.3}$$

whose solution, if it existed, would exactly interpolate the numerical solution x_{n+1} (because the local truncation error would be zero). This equation is a singular perturbation of the original equation (note that $k_2(h,v,t) = v^2 - t + O(h)$). This modified equation is more difficult to deal with than we would like, so we simplify it by two means: first, we truncate it to some finite order $O(h^p)$ (higher than $O(h^2)$ here, because otherwise we would get no new information: we already know that the solution of the the original equation is $O(h^2)$-close to the numerical solution). Second, we manipulate the series to find equivalent $O(h^p)$ modified equations which do not contain higher derivatives v'', v''', etc.

In our example, suppose we choose $p = 4$, and rename v as x. Then our singularly-perturbed modified equation is

$$\begin{aligned} &x' + \tfrac{1}{2}hx'' + \tfrac{1}{6}h^2x''' + \tfrac{1}{24}h^3x'''' \\ &\quad - (x^2 - t) - h(x^3 - xt - 1/2) - h^2(x^2-t)^2/4 = O(h^4)\,. \end{aligned} \tag{2.4}$$

To eliminate the higher derivatives, we first differentiate equation (2.4) to get

$$\begin{aligned} &x'' + \tfrac{1}{2}hx''' + \tfrac{1}{6}h^2x'''' \\ &\quad - (2xx' - 1) - h(3x^2x' - x't - x) + h^2(x^2-t)(2xx'-1)/2 = O(h^3), \end{aligned} \tag{2.5}$$

$$\begin{aligned} &x''' + \tfrac{1}{2}hx'''' \\ &\quad - (2(x')^2 + 2xx'') - h(6x(x')^2 + 3x^2x'' - x''t - 2x') = O(h^2), \end{aligned} \tag{2.6}$$

and

$$x'''' - (4x'x'' + 2x'x'' + 2xx''') = O(h)\,. \tag{2.7}$$

One can use $O(h)$ approximations to x', x'', and x'''—obtained by truncating equations (2.4–2.6)—in equation (2.7) to obtain an expression for x'''' that is accurate to $O(h)$ though containing t and x only. Similarly one can find $O(h^2)$ expressions for x''', $O(h^3)$ approximations for x'', and finally the following $O(h^4)$ expression for x':

$$\begin{aligned} x' = (x^2 - t) &+ h^2(-3/4x^4 + 5/6x^2t - 1/12t^2 + 1/3x) \\ &+ h^3x(5x^4 - 6x^2t + t^2 - 2x)/4\,. \end{aligned} \tag{2.8}$$

The exact solution of this equation is $O(h^4)$-close to the numerical solution of equation (2.1) obtained by RK2 on finite time intervals. See Table 2.1.

Notice the modified equation is different from the original equation by a term of $O(h^2)$. This is because RK2 is a second-order method. In fact, it is the real reason we call RK2 a second-order method. The fact that the 'residual' (*i.e.* the difference between the equation we wanted to solve and the one we *did* solve) is $O(h^2)$ means, by the Gröbner-Alexeev nonlinear variation of constants formula [**47**], that the *global error* is also $O(h^2)$, on finite time intervals.

If we apply RK2 to equation (2.1) with a small fixed step-size h, then the solution we get will be more accurately followed by solutions of (2.8), for some finite time. We can compare the true solution of (2.1) with the numerical solution generated by RK2. We can also compare the true solution of (2.8), found by

Table 2.1: Order of Global Error for $0 \le t \le 1$

$\log_{10} h$	equation (2.1)	equation (2.8)
−1.00	2.77	5.04
−1.22	2.64	4.83
−1.44	2.55	4.69
−1.67	2.49	4.60
−1.89	2.43	4.52
−2.11	2.39	4.45
−2.33	2.35	4.34
−2.56	2.32	4.11
−2.78	2.29	3.81
−3.00	2.27	3.53

using Matlab's ODE45 routine using a tolerance of 1.E−10, with the numerical solution of (2.1) by RK2 (note that we are comparing the numerical solution of one equation with the true solution of another). The results are summarized in Table 2.1. We see that the true solution of (2.8) is much closer to the numerical solution of (2.1) than the true solution of (2.1) is; that is, equation (2.8) explains the behaviour of the numerical solution of (2.1) by RK2. Note that the numbers from the RK2 solution of equation (2.1) are slightly better than $O(h^2)$-close to the exact solution of (2.1), while those *same* numbers are roughly $O(h^4)$-close to the exact solution of the modified equation (2.8).

REMARK. If we wished instead to analyze RK4, then an $O(h^5)$ modified equation, obtained by the same general method, is

$$x' = (x^2 - t) + h^4\left(\tfrac{1}{80} - \tfrac{1}{24}x^4t + \tfrac{3}{40}x^2t^2 + \tfrac{1}{15}xt - \tfrac{1}{24}x^6 + \tfrac{1}{120}t^3\right). \qquad (2.9)$$

The modified equation differs from (2.1) by a term of $O(h^4)$ because RK4 is a fourth-order method.

In Figure 2.1 we see the asymptotic value of the exact solution of (2.1) compared with the solution by RK2 with $h = 0.3$. Also shown is the solution of the modified equation (2.8), computed numerically by ODE45 with a tolerance of 1.E–10. We see that after a while, the RK2 solution drifts away from the true solution and finds a spurious, smooth solution; likewise the solution of the modified equation is asymptotically different from the true solution and qualitatively (but not quantitatively) similar to that of RK2. In fact this spurious solution behaviour occurs for large times, for $t = O(1/h^2)$ in RK2 solution and $t = O(1/h)$ in the RK4 solution [**28**], and thus the method of modified equations

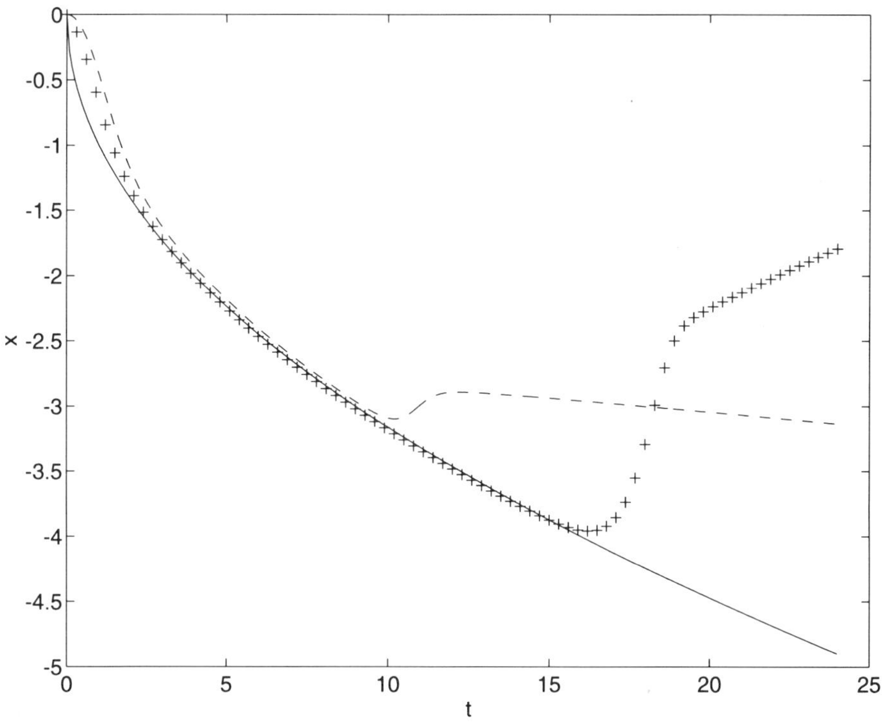

FIGURE 2.1. Modified Equations for the Hubbard and West example. The solid line is an asymptotic estimate of the exact solution to equation (2.1). The dashed line is a good numerical solution of the modified equation (2.8). The plus signs are the results of applying RK2 with fixed time-step $h = 0.3$ to equation (2.1).

as presented here can not, in fact, explain the apparently smooth spurious solution. This conclusion is general: finite-order modified equations may not help in examining long-time asymptotic behaviour of numerical methods. In the next section, we examine the case of *infinite*-order modified equations.

It is important to note at this point that the h^2 and h^3 terms are both *large* near the spurious solution—the modified equation is *not* a small perturbation of the original differential equation. We can see clearly here that the numerical method is giving us the solution of a problem nowhere near to the one we want to solve—in this case, we say the numerical method is at fault.

3. An Example due to W.-J. Beyn

In [**4**, *p.* **221**] W.-J. Beyn gives the following didactic example to show that it is impossible in general to embed an arbitrary discrete dynamical system into a continuous one. Consider Euler's method, with fixed stepsize h, applied to the simple nonlinear problem $y' = y^2$. Then the resulting discrete dynamical system is $u \to u + hu^2$, which is not a diffeomorphism (the derivative is zero at $hu = -1/2$, and the inverse map is not unique), whereas the h-flow of any continuous dynamical system must be a diffeomorphism. Hence in some sense it is impossible to embed this discrete system in a continuous one. Beyn [**4**] then remarks that this proof relies on the global behaviour of the discrete flow, and conjectures that even locally this would be impossible (*i.e.* in some u-neighbourhood of 0).

This example is indeed didactic, and we pursue it further here. What happens if we proceed with the method of modified equations on this problem, knowing that we will fail in the end? We start by computing a few terms in the h-series for the modified equation. As usual, we expand the local error in a Taylor series and set it to zero:

$$\frac{u(t+h) - u(t)}{h} - u^2(t) = u'(t) - u^2(t) + \tfrac{1}{2!}hu''(t) + \tfrac{1}{3!}h^2u'''(t) + \cdots \tag{3.1}$$

Differentiating once to eliminate u'', and again to eliminate u''', and so on, we find that a fourth-order modified equation is

$$u' = (1 - hu + \tfrac{3}{2}(hu)^2 - \tfrac{8}{3}(hu)^3)u^2 \,. \tag{3.2}$$

This leads us to suspect a very simple form for the *infinite*-order modified equation that we wish to find, *viz*

$$u' = B(hu)u^2 \,. \tag{3.3}$$

We now simplify by nondimensionalizing. Put $v = hu$ and $\tau = t/h$, and then

$$\frac{dv}{d\tau} = B(v)v^2 \tag{3.4}$$

and

$$v(\tau + 1) = v(\tau) + v^2(\tau) \,. \tag{3.5}$$

A simple Maple [**6**] program was written to compute more terms in the series for $B(v)$. Once a few more terms in the series were computed, the series was *recognized* [**29, 30**]. It turns out that this problem has already been solved, in [**34**], in which Labelle studies the general problem of interpolating discrete dynamical systems in the domain of formal power series.

The series for $B(v)$ can be constructed recursively as follows. If

$$B(v) = c_1 + c_2 v + c_3 v^2 + \cdots \tag{3.6}$$

(in a purely formal sense), then $c_1 = 1$ and

$$c_n = -\frac{1}{n-1}\sum_{i=1}^{n-1}\binom{n-i+1}{i+1}c_{n-i} \,, \quad \text{for } n > 1 \,, \tag{3.7}$$

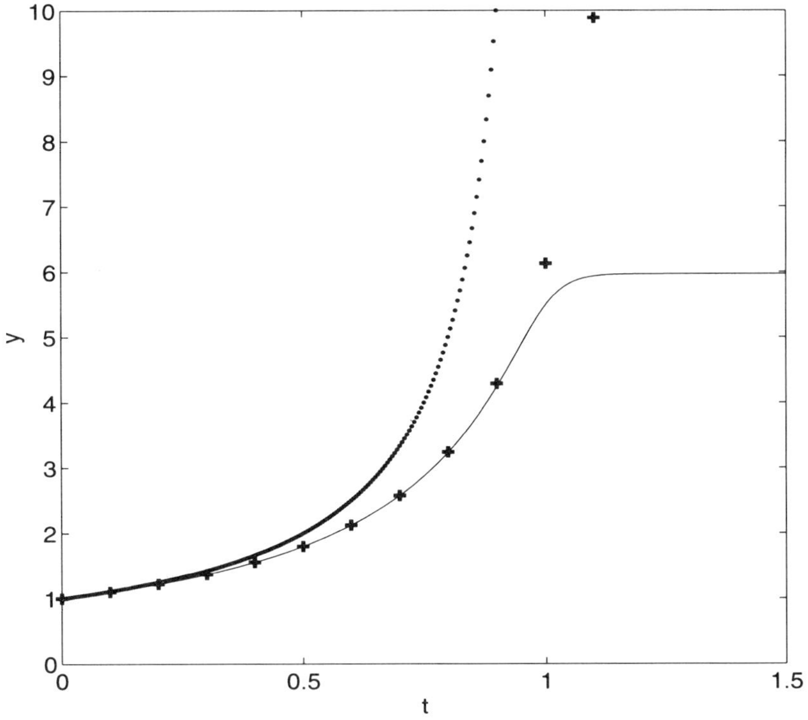

FIGURE 3.1. Seventh-degree modified equation for $y' = y^2$. The dotted line is the exact solution $y = 1/(1-t)$, while the plus signs are the results of applying Euler's method with $h = 0.1$. The solid line is a good numerical solution of $y' = (1 - hy + 3/2(hy)^2 - 8/3(hy)^3 + \cdots - 9427/210(hy)^7)y^2$. This solution nearly interpolates the Euler method solution until it reaches a spurious fixed point.

where $\binom{n}{j}$ is the binomial coefficient [**34**]. This enables efficient calculation of any desired number of terms in the series for $B(v)$.

In Figure 3.1, we see a graph of the exact solution to $y' = y^2$, together with an Euler's method solution with $h = 0.1$ and a seventh-order modified equation solution (computed numerically with Matlab's ODE45). We see that for small t, the modified solution apparently interpolates the Euler solution, but for large values of t the modified solution suddenly 'turns a corner' and settles on a fixed point. This is because the seventh-degree polynomial approximation for $B(v)$ has a zero at $v = hu \approx 0.6$, which introduces a spurious fixed point. Taking a higher-degree approximation may remove this spurious fixed point, but many other high-degree modified equations do have such spurious fixed points. Thus the finite-order modified equation may not be a good model of what the Euler solution is doing for large times.

It is unclear just how much use the series (3.6) is at this point, so we use an alternate approach to get more information about $B(v)$. Differentiation of equation (3.5) and using (3.4) gives us the following functional equation for $B(v)$:

$$B(v) = \frac{(1+v)^2}{(1+2v)} B(v+v^2)\,. \tag{3.8}$$

This equation allows us to describe $B(v)$ completely, and, together with the series, to compute it efficiently and accurately.

We take $B(0) = 1$, as we expect from the series and from consistency of Euler's method as $h \to 0$. Now consider $B(-1/4)$, for example. If $B(-1/4)$ exists, then

$$\begin{aligned} B(-1/4) &= \frac{9}{8} B(-3/16) \\ &= \left(\frac{9}{8}\right) \cdot \left(\frac{169}{160}\right) B(-39/256) \\ &= \cdots \\ &= \prod_{k=0}^{\infty} \frac{(1+v_k)^2}{(1+2v_k)} \\ &= 1.4266762676859975\ldots \end{aligned} \tag{3.9}$$

where $v_{k+1} = v_k + v_k^2$ with $v_0 = -1/4$ and the penultimate equality is tentative at this moment—pending proof that the product converges to a function which satisfies (3.8)—and where we have used the fact that $v_k \to 0^-$ as $k \to \infty$ if $v_0 = -1/4$ and hence assumed that $B(v_k) \to 1$.

Convergence of the infinite product in (3.9) is established by discovering that the asymptotic behaviour of v_k is $v_k \approx -1/(k - 1/v_0)$, for initial values v_0 in $(-1, 0)$ [**3**], and hence the terms in the product are asymptotic to $1 + O(1/k^2)$ and so the product converges absolutely (Theorem 8.1(c) in [**26**], p. 4). Note that $1 + 2v_k \neq 0$ unless $v_0 = -1/2$.

Now consider v_0 near $-1/2$. Equation (3.8) gives

$$B(v_0) = \frac{(1+v_0)^2}{1+2v_0} B(v_0 + v_0^2)\,, \tag{3.10}$$

and since $v_1 = v_0 + v_0^2$ will be near $-1/4$ if v_0 is near $-1/2$ we see that $B(v)$ has a pole at $v = -1/2$. Note that this is exactly the place where the map $u \to u + hu^2$ fails to be diffeomorphic: $hu = v = -1/2$. This is not a coincidence.

We now consider pre-images of $-1/2$ under $v \to v + v^2$; these, too, will be poles (we cannot cancel out a pole with a zero unless $(1 + v_k) = 0$ which only happens if $v_k = -1$; but all the forward images of -1 are 0). We graph the first 2000 pre-images of the pole at $v = -1/2$ in Figure 3.2. This set of preimages approaches (and densely fills out) the Julia set of the quadratic map $v \to v + v^2$ since it approaches the α-limit set of the unstable fixed point $v = 0$ [**14, p. 287**]. One sees that there is an infinite number of poles of B, and as a consequence of the arbitrary approach of the Julia set of this map to the origin [**14**] we see that

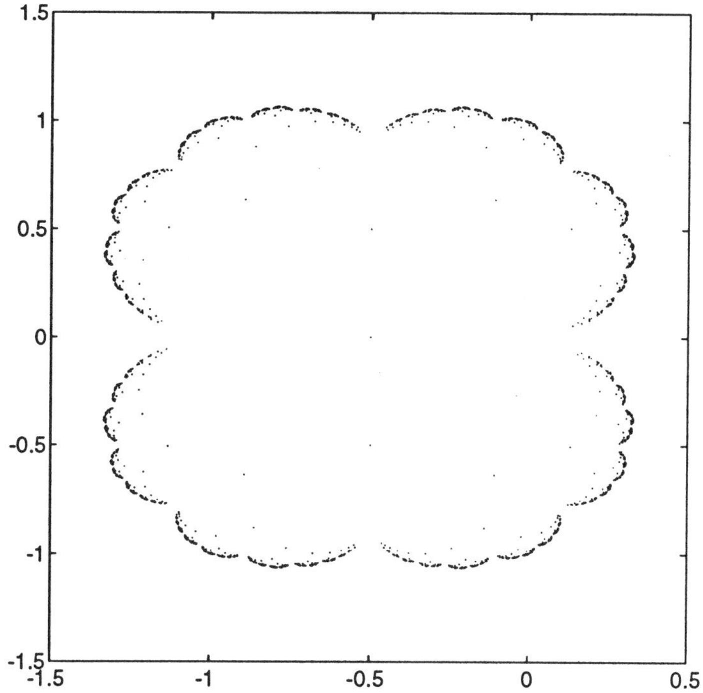

FIGURE 3.2. The first 2000 pre-images of $v = -1/2$ under the map $v \to v + v^2$. These are locations of poles of $B(v)$, and as the number of pre-images increases, we see the poles approach the Julia set of the quadratic map $v \to v + v^2$. In fact the poles are dense on that Julia set and form a natural boundary to analytic continuation of the function $B(v)$. The poles of $B(v)$ thus also come arbitrarily close to the origin.

there are poles arbitrarily close to the point of expansion for the series (3.6). Thus the radius of convergence of the series (3.6) is zero. Further, there is a *natural boundary* preventing analytic continuation of the function defined by the infinite product (3.9) to the region outside the Julia set.

There are also an infinite number of zeros of $B(v)$ close to this Julia set, as we see by looking at the pre-images of $v = -1$. The periodic points, dense in the Julia set, of the map $v \to v + v^2$ can be either poles or zeros of B, it doesn't matter which, so long as there are only poles or only zeros in a given periodic orbit. Hence $B(v)$ is unique only up to the assignment on these periodic points. This suggests that the dynamics for complex initial points or complex h will be very complicated, though the relation between the modified equation and the discrete dynamical system may reasonably be doubted in that case.

However, we are mainly interested in $B(v)$ for positive v, which is definitely outside the Julia set. This means that $v_k \to \infty$ as $k \to \infty$ (which we knew before). But if we run the iteration backwards, then we ought to be able to recover a convergent product. Define

$$u_{k+1} = \frac{2u_k}{1+\sqrt{1+4u_k}} \tag{3.11}$$

and

$$B(v) = \prod_{k=1}^{\infty} \frac{(1+2u_k)}{(1+u_k)^2} \,. \tag{3.12}$$

Note that the product starts at $k = 1$ this time, and that we have chosen a particular pre-image for each u_k by choosing u_{k+1} to be the root of $u + u^2 = u_k$ closest to zero. Note also that a numerically stable formula for this root has been used in (3.11). A similar analysis to that for the product (3.9) shows that this product converges for $v = u_0$ outside the Julia set, and by construction satisfies $B(0) = 1$ and the functional equation (3.8).

We can solve (3.4) up to quadrature by separation of variables (this provides a check for the product formulas for $B(v)$):

$$\int_{v_0}^{v_k} \frac{dv}{B(v)v^2} = \int_0^k dt = k \,. \tag{3.13}$$

It is an easy matter to verify by the change of variables $v = u + u^2$ (which can be done for real u and v so long as $v \geq -1/4$ and hence $u \geq -1/2$) that as a consequence of the functional identity (3.8),

$$\int_{v_{k-1}}^{v_k} \frac{dv}{B(v)v^2} = \int_{v_k}^{v_{k+1}} \frac{du}{B(u)u^2} = \text{constant} \,. \tag{3.14}$$

Using the asymptotics of v_k and of u_k as $k \to \infty$ we can show this constant is 1. This is easily confirmed by numerical quadrature. Hence (3.13) is satisfied. Furthermore, we can use (3.14) to investigate the asymptotics for $B(v)$ for large v, and one sees that $B(v) \sim \ln 2 \ln v/v$. This in turn tells us that $v_k \sim \exp(C2^k)$ for positive v_0. [To identify the constant requires knowledge of one particular v_n for large enough n.]

To evaluate $B(v)$ in practice it turns out that the divergent series (3.6) is useful. We can use the Maple `evalf/Sum` program, which uses Levin's u-transform [**48**] for convergence acceleration; as well, this method will sum certain divergent series. For this example, Levin's u-transform is successful for real v in $[-0.1, 0.1]$ (for settings of `Digits` ≤ 30) and this is precisely the region where convergence of the infinite products (3.9) and (3.12) is slowest. The Maple code to evaluate $B(v)$ using this algorithm is short and included in Appendix A.

An alternate approach is to use the knowledge of the asymptotic behaviour of v_k and u_k to accelerate the products themselves, using the Euler-Maclaurin formula and Aitken's Δ^2 acceleration. Also, one could use the asymptotic behaviour of $B(v)$ for large v, by using (3.8) *forward* until $v + v^2$ was large enough.

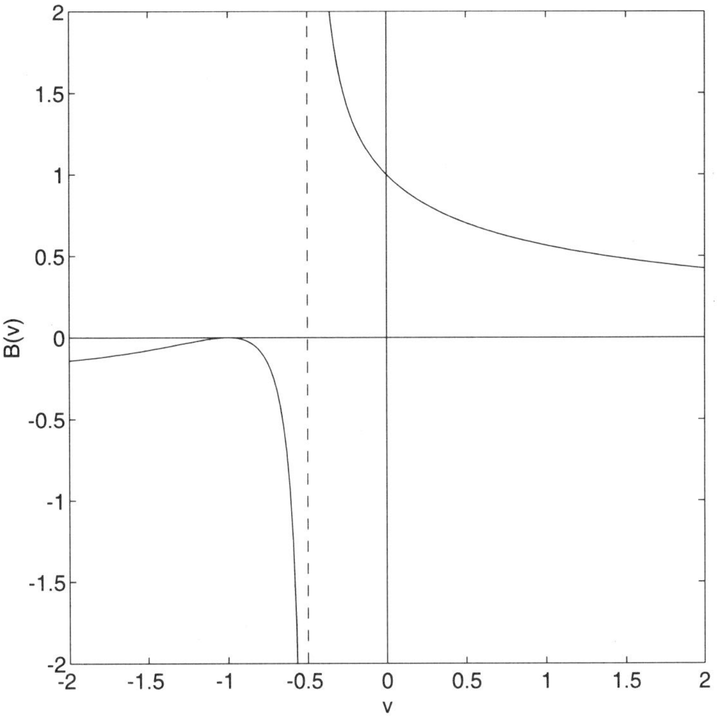

FIGURE 3.3. The graph of $B(v)$ for real v. The infinite-order modified equation is $y' = B(hy)y^2$, whose solution interpolates the Euler solution to $y' = y^2$ for initial conditions $y_0 \geq -1/(2h)$. $B(v)$ can be evaluated by use of two convergent infinite products, away from the only real pole at $v = -1/2$.

But the current technique is satisfactory, and we graph $B(v)$ in Figure 3.3. We see that there is a double zero at $v = -1$ and a pole at $v = -1/2$, and, as noted previously, that $B(v)$ decays as $v \to +\infty$.

We have thus found a differential equation (3.4) whose solution interpolates the Euler's method solution of the original problem, for initial conditions $v_0 \geq -1/2$. The singularity at $v = -1/2$ shows that this is not a dynamical system in the ordinary sense, and thus supports Beyn's original observation [**4**].

The differential equation (3.3) is a *large* perturbation of $y' = y^2$, *no matter how small* h *is*, once u gets large enough. However, on compact u-sets, we see that by taking h small enough, we can get the exact solution of a differential equation arbitrarily close to the original problem.

One can consider generalizations of this problem: (a) use a higher-order fixed-stepsize method to solve the same problem, and see what happens; (b) use the same analysis for (fixed stepsize) one-step methods on different scalar problems;

(c) use the same style of analysis for arbitrary fixed-stepsize one-step methods for vector problems; and (d) use this approach to study systems with special structure, such as Hamiltonian systems.

It turns out that (a) is straightforward. One can reproduce this analysis for *e.g.* some other low-order Runge-Kutta methods, but there are no surprises. More interestingly, Taylor series methods of arbitrary order can be easily constructed for this problem: they lead to iterations of the form

$$v_{k+1} = v_k + v_k^2 + v_k^3 + \cdots + v_k^n \tag{3.15}$$

and the functional equation for $B(v)$ is then

$$B(v) = \frac{(1 + v + v^2 + \cdots + v^{n-1})^2}{(1 + 2v + 3v^2 + \cdots + nv^{n-1})} B(v + v^2 + \cdots + v^n) \tag{3.16}$$

which can be analysed by the same methods as before. One sees that for large n the Julia sets are of similar structure, with poles and zeros dense on the Julia sets, arbitrarily close to the origin; the only real effect of the increase in n is to move the poles in the interior of the Julia set outward slightly. Thus we can see that fixed time-step methods of arbitrary order can be constructed which all fail to reproduce the correct asymptotic behaviour of the solutions, for arbitrarily small time-steps. It was observed [**23**] that this is essentially due to the lack of compactness of the problem, and hence is perhaps not surprising to the expert. Nevertheless, we will see later that a certain class of *variable*-stepsize codes can faithfully capture the correct asymptotic behaviour of the true solution, so we see that the 'fixed timestep' nature of the methods we are discussing at the moment also contributes to their poor performance on this problem.

Further, note that as v gets large, the difference between the specified problem and the modified problem also gets large. This is an important observation: one can decide that the numerical method is 'bad' if it gives the solution of a problem very different than the one we had intended, whereas the *problem* is 'bad' if its solution is very sensitive to small changes.

The generalization (b), of using this analysis on general scalar problems, is easy in theory. Given the problem

$$y' = f(y) \tag{3.17}$$

and a one-step fixed-stepsize method

$$u(t + h) = u(t) + h\varphi(h, u(t)) \tag{3.18}$$

one can expand (3.18) in a formal power series and use the methods of [**34**] to generate a formal power series for F_h where $y' = F_h(y)$ is the modified equation. Alternatively, one can attempt to use the functional relationship

$$F_h(u) = (1 + h\varphi_u(h, u))^{-1} F_h(u + h\varphi(h, u)) \tag{3.19}$$

and, if the analysis of the discrete dynamical system (3.18) is as easy as it was for the case $u \to u + hu^2$ above, one can find F_h as a collection of appropriate infinite products. This has been done only for a few other systems, since if the system (3.18) has complicated behaviour then this approach is not actually helpful.

For the case (c), extension to vector systems, equation (3.19) has a straightforward re-interpretation as a matrix functional equation and the analytical difficulties multiply. It seems possible that some simple schemes for gradient systems might profitably be analyzed in this manner, but not much more.

The question in part (d) deserves some comment. The following analysis rests heavily on an adaption of the method of modified equations for Hamiltonian systems [**38**]. Instead of searching for a modified differential equation, one searches for a modified Hamiltonian. Consider the Hamiltonian $H = \frac{1}{2}(p^2 + q^2)$ (later we will examine one not quite so trivial) for the simple harmonic oscillator. If we solve this numerically by the leapfrog scheme (which is known to be symplectic) we first write

$$\begin{aligned} p_{n+1} &= p_n - hq_n \\ q_{n+1} &= q_n + hp_{n+1}\,, \end{aligned} \tag{3.20}$$

and then expand $H = \frac{1}{2}(p^2+q^2) + hH_1(p,q) + h^2H_2(p,q) + \cdots$ and use $dp/dt = -\partial H/\partial q$ and $dq/dt = \partial H/\partial p$ to expand p and q in Taylor series about $t = 0$. When we compare with (3.20) to identify the unknown H_1, H_2, and so on, we find that (3.20) gives, in exact arithmetic, the exact solution to the problem with the Hamiltonian

$$\begin{aligned} H = {} & \tfrac{1}{2}\left(p^2+q^2\right) - \tfrac{1}{2}hpq \\ & + \tfrac{1}{12}h^2\Big(\left(p^2+q^2\right) - hpq\Big) \\ & + \tfrac{1}{60}h^4\Big(\left(p^2+q^2\right) - hpq\Big) \\ & + \tfrac{1}{280}h^6\Big(\left(p^2+q^2\right) - hpq\Big) \\ & + \tfrac{1}{1260}h^8\Big(\left(p^2+q^2\right) - hpq\Big) \\ & + \cdots \\ = {} & \left(\tfrac{1}{2}(p^2+q^2) - \tfrac{1}{2}hpq\right) f(h) \end{aligned} \tag{3.21}$$

where $f(h) = 4\tan^{-1}(h/\sqrt{4-h^2})/(h\sqrt{4-h^2})$ if $|h| < 2$. [In the presence of roundoff error, of course, the numerical method does not give the exact solution of the problem with this perturbed Hamiltonian.] From the summation of the series in (3.21) we can suspect that taking $h > 2$ will cause disaster.

If we instead use the implicit midpoint rule on this problem, we find $H = \frac{1}{2}(p^2+q^2)f(h)$, where $f(h) = \sin^{-1}(4h/(4+h^2))/h$ again if $|h| < 2$, which has a purely even power series expansion. Note that, as before, the order of the 'residual' is the same as the order of the method: here the implicit midpoint method is second order, while the leapfrog method used above is first order.

A less trivial example is the harmonic oscillator, with $H_0 = \frac{1}{2}p^2 + (1 - \cos(q))$.

Using the leapfrog method gives

$$\begin{aligned} H = \quad & H_0 - \tfrac{1}{2}hp\sin q + \tfrac{1}{12}h^2(p^2\cos q + \sin^2 q) \\ & - \tfrac{1}{12}h^3 p\sin q\cos q \\ & + \tfrac{1}{720}h^4\left[(\cos 3q - \cos q)/240 \right. \\ & \qquad\qquad \left. + (1 + 3\cos 2q)p^2/240 + p^4\cos q\right] \\ & + \cdots \end{aligned} \tag{3.22}$$

These first few terms are enough to give insight into the behaviour of the method, and in particular to see that the numerical method has changed the Hamiltonian in a nontrivial way.

To summarize the results of this section, we have seen that finite order modified equations do not always help us with the long-time behaviour of the numerical solutions as compared to those of the differential equation. In contrast, infinite-order modified equations, which are possible if the dynamics of the numerical method are simple enough, can help. In particular, we showed that the numerical solution by Euler's method can indeed be locally embedded in a dynamical system, but not a smooth one, and of course not globally. We have also seen that fixed time-step codes of *arbitrary order* can give incorrect asymptotic behaviour.

4. The Defect

We can look at a simpler type of modified equation, as follows. Modern differential equation codes provide interpolants, for various practical reasons such as efficient graphical output, 'g-stops', delays, *etc.*, and since this is the case it might pay us to think about the interpolated solution (as opposed to merely using it). One can substitute a continuously differentiable interpolant back into the differential equation, for example, and look at the *residual* or *defect* as it is called in the initial-value problem community [**18**]:

$$\delta(x) = \frac{dx}{dt} - f(x) = \varepsilon v(t), \text{ say,} \tag{4.1}$$

where $x = x(t)$ is our computed, interpolated solution and ε is our input tolerance. A trivial rearrangement of that equation gives us

$$\frac{dx}{dt} = f(x) + \varepsilon v(t) \tag{4.2}$$

and we see that if our error control mechanism can guarantee that $||v(t)|| \le 1$ then *we have the exact solution to a nearby problem* [**18**]. For a fuller discussion of this approach for chaotic problems, see [**9**]. In view of physical perturbations (such as the gravitational effects of 'the rapidly moving masses of Jupiter'– E E. (Doc) Smith, *Gray Lensman*), we can see that we have the *exact* solution of just as good a mathematical model of the physical problem as the one that was originally written down. Further, $v(t)$ is computable, and we can examine it at our leisure.

If there is some relevant *statistic* (that is to say, numerical quantity) that is insensitive to such perturbations, then we say the problem is 'well-enough conditioned'. If there is no such statistic, then the model is useless for any practical purpose, in view of physical perturbations. In **[9]** this concept is contrasted with the more usual ideas of 'well-posedness' and 'stability', and an example of a problem that is not 'well-enough conditioned' is given.

Interval arithmetic is also necessary here **[10]** if we wish a guarantee that $||v(t)|| \leq 1$, as opposed to an estimate $||v(t)|| = O(1)$. Estimating codes such as PAMETH **[19]** are roughly as efficient as the best existing codes using classical error control strategies, but are not intended for computer-aided proofs.

The technical idea of 'defect control' can be replaced by a simpler, albeit less efficient, point of view: that of merely examining the defect in a computed solution, however attained. We postpone discussion of variable-step size control using the defect until this simpler examination is made.

5. Fixed-Step Euler's method for the Lorenz Equations

The analysis of the following example is a continuation of that in **[8]**. We examine the defect that arises from solving the Lorenz equations by Euler's method.

For Euler's method, the natural interpolant is piecewise linear. This is not C^1. To get a C^1 interpolant we can use cubic Hermite interpolation but this only adds complexity to the example. It turns out that the size (*e.g.* infinity norm) of the defect for any C^1 interpolant is not much different than the size of the (discontinuous) defect that arises from the piecewise linear interpolant. Therefore, we set

$$x(t) = x_n + (t - t_n)f(x_n) \qquad \text{for } t_n \leq t < t_{n+1} \,. \tag{5.1}$$

We can sample the defect at t_{n+1}^- to get

$$\delta(t_{n+1}^-) = f(x_n) - f(x_{n+1}) \,, \tag{5.2}$$

which gives us, asymptotically as $h \to 0$, the maximum value of the defect on $t_n \leq t < t_{n+1}$. This sampling also has the advantage that we require no extra computations, for this simple method. For more complicated methods, say a higher-order Runge-Kutta method, extra stages are typically required to estimate the defect **[18]**. In the case of Adams PECE methods, a 'free' estimate is available from quantities already computed by modern codes **[27]**. For Taylor series methods we can sample at t_{n+1}^- as we did above for Euler's method, and this is simple and effective. If we wish to avoid the use of asymptotics ($h \to 0$) as above, we may sample the defect at more points, but of course this increases the cost.

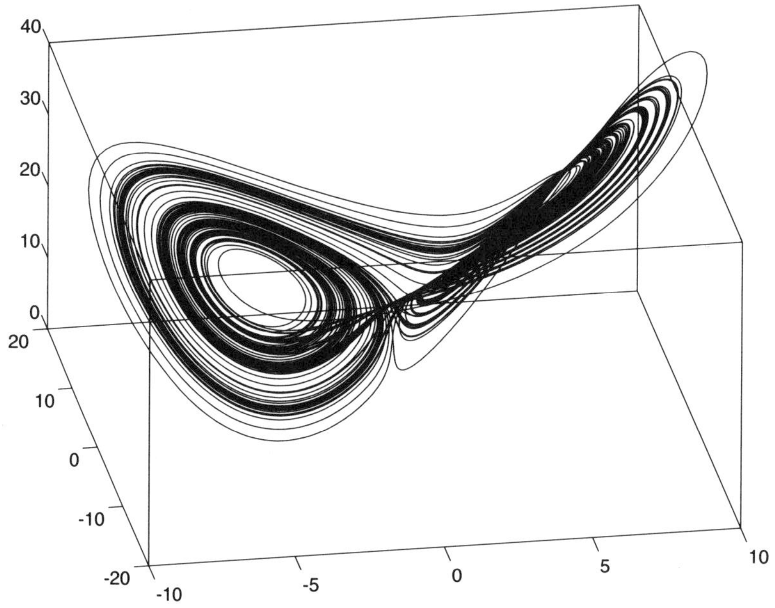

FIGURE 5.1. The solution to the Lorenz equations by Euler's method with fixed time-step $h = 0.001$. The values of the parameters in the Lorenz system used here are $\sigma = 3$, $\beta = 1$, and $\rho = 26.5$.

Figure 5.1 shows an orbit of the Lorenz system, and its defect appears in Figure 5.2. We notice two things immediately:

1) The defect is large—bigger than 1 occasionally—so Euler's method is adding a *large* perturbation to the Lorenz system. Decreasing h decreases the defect, asymptotically linearly. (This is because Euler's method is a first order method: the defect is asymptotically $O(h)$, and for a well-conditioned problem—which the Lorenz system is not, because it is chaotic—this would imply the global error was $O(h)$ as well.)
2) The defect has structure. The defect is clearly correlated with the solution—we have a miniature but distorted version of the Lorenz mask produced, plus some other structure. This has some serious implications, as correlations can be *e.g.* 'resonant'.

We wish to explain this correlation here, using a combination of the method of modified equations as earlier outlined, and the idea of the defect. If we apply the method of modified equations to $y' = f(y)$ solved by Euler's method, we find

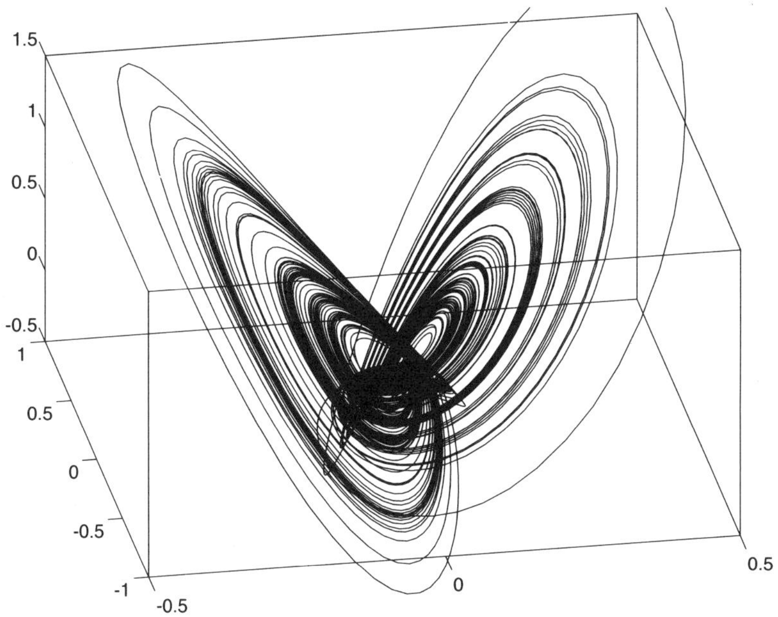

FIGURE 5.2. The defect in the Euler's method solution to the Lorenz equations with fixed time-step $h = 0.001$. Parameter values are as in Figure 5.1. We see a clear correlation between the defect $\delta(t) = y'(t) - f(y(t))$ and the solution $y(t)$. This defect is better modelled as $\delta(t, y) = -h/2 J_f(y) f(y) + \delta_2(t)$, where J_f is the Jacobian of the Lorenz system. Separate investigation shows that $\delta_2(t)$ is three orders of magnitude smaller than $\delta(t, y)$ for this value of h.

that we have actually found a second-order solution to (for example)

$$y' = (I - \tfrac{h}{2} J_f(y)) f(y) \,. \tag{5.3}$$

$J_f(y)$ is the Jacobian matrix of f evaluated at y. We can now interpolate the Euler's method solution with cubic Hermite polynomials, *matching the derivatives from* (5.3) *and not from* $y' = f(y)$. We can then substitute this C^1 function back into (5.3) and show that we have computed the *exact* solution to

$$y' = (I - \tfrac{h}{2} J_f(y)) f(y) + h^2 b(t) \,. \tag{5.4}$$

Now we know that $b(t) = O(1)$ as $h \to 0$ but we do not know if $||b(t)|| \leq 1$. In fact we find that for $h = 0.001$, as in Figures 5.1 and 5.2, the maximum value of $h^2 b(t)$ is approximately 0.004, whereas the maximum value of the defect from

(5.2), over the same time interval, is roughly 0.6. This means that more than 99.5% of the defect is accounted for by the term $-\frac{1}{2}hJ_f(y)f(y)$. This explains the correlations seen in Figure 5.2, and makes the point that a better model for the defect in equation (4.2) is $\varepsilon v(t,x)$, not just $\varepsilon v(t)$.

6. Variable-stepsize by Defect Control, and Consequences

All the above analysis has been for fixed time-step methods. But the most efficient use of the defect is to allow the time-step to vary, and keep control of some norm of the defect to make it less than the user's input tolerance. This replaces the local error controls used classically, and is more easily interpreted by the ordinary user.

Analysis of the dynamics of variable time-step numerical methods using local error control is very difficult. Some recent results include the very interesting work of Stuart and Humphries [**31, 41, 42**], where they introduce the idea of an 'algebraically stable embedded pair' and show that certain such methods have good asymptotic properties for contractive, dissipative, gradient, and Hamiltonian problems. It can be shown that a defect-controlled implementation of Euler's method satisfies the hypotheses of the theorems in these works, and thus also has these good asymptotic properties [**43**]. However, analysis of higher-order defect-controlled methods along these lines seems difficult.

Instead, we may try to discover some of the same asymptotic properties by using existing theoretical perturbation results. We sketch some relevant results here, as they are not likely to be known to many non-specialists. We do not examine here an exhaustive list of useful perturbation results—that would require a good-sized review article of its own. Instead we examine some of the simpler ones; others can be found, for example, in the exercises in [**24**].

We note first that defect control can be chosen so as to provide the exact solutions to one of a variety of equations:

$$\dot{x} = f(x) + \varepsilon v(t)\,, \tag{6.1}$$

$$\dot{x} = f(x)(1 + \varepsilon v(t))\,, \tag{6.2}$$

$$\dot{x} = f(x) + \varepsilon v(t)x\,, \tag{6.3}$$

or some other generalizations. Note that we can use $v(t,x)$ instead of $v(t)$ in the above, if we like. In equation (6.1) we say that the absolute defect is small; in equation (6.2) we say that the relative defect is small; and in equation (6.3) we say the defect is small relative to x. The user would normally select the criterion relevant to the problem at hand, and use only one of (6.1)–(6.3). In each case, ε is the user's input tolerance, and for estimating codes $||v(t)|| \approx 1$ and for interval codes $||v(t)|| \leq 1$. The latter would be used in computer-aided proofs, for example.

Perturbation Theorems. There are a great many theorems concerning the perturbation of linear systems. We quote only a theorem of Aulbach [**2**] here, but there is also work of Sell [**39**] and a more general theorem of Taniguchi [**45**] which entails both of these as corollaries; and the work in [**46**] uses the log-norm in the bounds and so may be of more interest to numerical analysts.

THEOREM (AULBACH, [**2**]). *If*

$$\dot{x} = (A(t) + \varepsilon B(t))x \tag{6.4}$$

where $\varepsilon > 0$, $||B(t)|| \leq 1$ *for all relevant* t *and* $A(t)$ *is continuous, and if there exist constants* $\alpha \in \Re$ *and* $\gamma > 0$ *such that for all solutions of* (6.4) *with* $\varepsilon = 0$ *(the unperturbed system) we have*

$$||x(t)|| \leq \gamma ||x(s)|| e^{\alpha(t-s)}, \tag{6.5}$$

then any solution of (6.4) *with* $\varepsilon > 0$ *satisfies*

$$||x(t)|| \leq \gamma ||x(s)|| e^{(\alpha \pm \gamma\varepsilon)(t-s)}, \tag{6.6}$$

where the growth/decay rate is $\alpha + \gamma\varepsilon$ *if* $t \geq s$ *and* $\alpha - \gamma\varepsilon$ *if* $t \leq s$.

This theorem shows that, for example, if the origin is an asymptotically stable fixed point of the unperturbed system, then the numerical solution (solved by defect control where the defect is kept smaller than ε relative to x) will also approach the origin, and moreover do so at approximately the right rate. Thus one could compute the Lyapunov exponent of this fixed point correct to $O(\varepsilon)$, for example. Note that the growth constant γ also plays a role.

The computation of Lyapunov exponents in general, on the other hand, is nontrivial: the main difficulty is that numerical solution of the *nonlinear* problem may *a priori* give a spurious orbit, and hence any possible computation of Lyapunov exponents might be wrong to start with. Shadowing arguments may help [**11,12**], but *quantitative* shadowing computations are inherently more expensive than a straightforward numerical solution. The question is whether the extra information gained is worth the computational expense.

However, if we use a defect-controlled method from the outset, we can say that we have (cheaply) computed the exact orbit of a nearby system, and can (again cheaply) compute the (largest) Lyapunov exponent for that orbit of that system, correct to $O(\varepsilon)$ by the above analysis. With the pragmatic definition of chaos as the sensitivity of nearby systems to perturbations, this is as good as we could wish for.

A recent theorem in [**15**] allows us to quantitatively extend this result to the computation of all Lyapunov exponents, not just the largest one, if the 'gaps' between the Lyapunov exponents are not too small.

As another example, Aulbach's theorem can be used backwards—if the defect-controlled numerical solution (relative to x) approaches a fixed point at a rate α, then the original equation also has that fixed point and a similar Lyapunov exponent. The theorems of Taniguchi and Sell [**45,39**] extend this to allow small nonlinearities in x as well.

In addition, Lyapunov function arguments allow strong conclusions to be drawn about the geometry of such defect-controlled numerical solutions as compared to the geometry of the true solutions [**37**], as follows.

Definition (total stability, or stability under persistent disturbances) [**37**]. An equilibrium solution x^* of $\dot{x} = f(t, x)$ is called *totally stable* or *stable under persistent disturbances* if, for all $\eta > 0$ there exist $\delta_1, \delta_2 > 0$ such that for all t_0 and for all initial points $x_0 \in B_{\delta_1}(x^*)$ and for any $\varepsilon v(t, x)$ in (6.1) with $||v(t, x)|| \leq 1$ for all $x \in B_\eta(x^*)$ and $\varepsilon < \delta_2$, then $x(t)$ satisfying (6.1) starting with $x(t_0) = x_0$ will remain in $B_\eta(x^*)$. That is, all initial points starting sufficiently close to x^* will remain close to x^* for all time, though not necessarily settling down to any fixed point.

There are several theorems setting out conditions to ensure total stability of equilibria in [**37**]; we note here only a theorem of Malkin (1944) and Gorsin (1948) that states that if f is Lipshitz in x uniformly with respect to t and if the equilibrium point is uniformly asymptotically stable, then it is totally stable.

This theorem guarantees the existence of a user-tolerance ε which will allow the numerical solutions to successfully find stable equilibria. One could wish to know *a priori* just what that tolerance was, but it depends on the eigenvalues, as we will see in an example below. One could also wish for a converse to this theorem—if our numerical solution finds small persistent balls of radius which go to zero with the tolerance, does there exist a uniformly asymptotically stable fixed point in that ball? Unfortunately, the answer is *not necessarily* [**37, p. 84, Exercise 4.10**].

EXAMPLE. Suppose $dx/dt = \mu x - \nu y$ and $dy/dt = \nu x + \mu y$. Suppose further that $\mu < 0 < \nu$. Then a Lyapunov function for this problem is [**5, p. 158**] $V = x^2 + y^2$, for $dV/Dt = 2\mu V$ along trajectories. Now suppose that we solve this problem with a defect-controlled numerical method giving the exact solution to $dx/dt = \mu x - \nu y + \varepsilon v_1(t)$ and $dy/dt = \nu x + \mu y + \varepsilon v_2(t)$ where $||v_1||$ and $||v_2||$ are less than 1. Then $dV/dt = 2\mu V + 2\varepsilon(x v_1(t) + y v_2(t))$. Note that V is a Lipshitz function in, say, $-1 < x < 1$, $-1 < y < 1$, with Lipshitz constant $L = 2$. Then $dV/dt \leq 2\mu V + 4\varepsilon$ and we see that V must decrease exponentially as long as $V > -2\varepsilon/\mu$. Thus we can see that trajectories are eventually trapped inside a ball of radius $-2\varepsilon/\mu$. Note that the size of the (real part) of the eigenvalues, μ, plays an important part here. The fixed point at the origin will be detected by this numerical solution, but the required value of ε to make the radius of the trapping circle less than, say, δ, depends on μ and thus on the problem.

We now turn to the problem of uniformly attractive periodic orbits and more general attracting sets. The following theorem is not new—but it is more useful than it was before the implementation of defect-controlled codes. It is essentially a simplified version of Lemma 24.1 in [**50**], together with some additional results similar to those of [**32**]. We use the definitions and notation of [**32**] in what follows, which is an analogue of their Theorem 1.1. Similar results in the case of

Banach spaces, as in [**33**,**25**], are of interest for infinite-dimensional dynamical systems.

THEOREM. *Suppose that f is uniformly bounded in $\Re^n$ and that the differential equation $y' = f(y)$ has a compact, uniformly asymptotically stable set Λ. Then there is $\varepsilon_0 > 0$ such that for all $0 < \varepsilon < \varepsilon_0$, equation (6.1) has a compact, uniformly asymptotically stable set $\Lambda(\varepsilon)$ which contains Λ, and $\Lambda(\varepsilon) \to \Lambda$ in the Hausdorff metric.*

REMARK. As in [**32**] we talk about attracting sets and not attractors (which contain dense trajectories), in this theorem. We will discuss chaotic attractors by defect control shortly.

PROOF. The proof of this theorem is very similar to the proof in [**32**]. Some simplifications are possible here because equation (6.1) is easier to deal with than the local error criterion used in [**32**]. The results here apply equally well to multistep methods and exotic methods, so long as (6.1) is satisfied with $||v(t)|| < 1$.

As in [**32**] we note the existence of $R_0 > 0$ and a uniformly Lipshitzian (with Lipshitz constant L) Lyapunov function $V : S(\Lambda, R_0) \to [0, \infty)$, where $S(A, R_0) = \{x \in \Re^n : \text{dist}(x, A) < R_0\}$, with continuous strictly increasing functions $\alpha, \beta : [0, R_0) \to [0, \infty)$ with $\alpha(0) = \beta(0) = 0$, $\alpha(r) < \beta(r)$ and

$$\alpha\Big(\text{dist}(x, \Lambda))\Big) \leq V(x) \leq \beta\Big(\text{dist}(x, \Lambda)\Big) . \tag{6.7}$$

Further, there exists a constant $c > 0$ such that the upper-Dini derivative of V with respect to $x' = f(x)$ satisfies

$$D^+V(x) = \limsup_{h \to 0^+} \left(\frac{V(x + hf(x)) - V(x)}{h} \right) \leq -cV(x) . \tag{6.8}$$

It follows that $x(t; x_0) \in S(\Lambda, R_0)$ if x_0 is, and $V(x(t; x_0)) \leq \exp(-ct)V(x_0)$ for all $t > 0$ whenever $x_0 \in S(\Lambda, \beta^{-1}(\alpha(R_0))$.

Using (6.7) and the fact that V is Lipshitz, (6.1) implies

$$\limsup_{h \to 0^+} \left(\frac{V(x + hf(x) + h\varepsilon v(t, x)) - V(x)}{h} \right) \leq -cV(x) + L\varepsilon . \tag{6.9}$$

We have used a projection of the trajectory from the higher-dimensional space (which is t-dependent) onto the x-space. Essentially, we allow the vector field to be a little 'fuzzy', and so long as the fuzzy tangent vectors point inward on $V(x) =$ constant, trajectories will be driven towards Λ.

Now the theory of differential inequalities shows that

$$V(x(t; x_0)) \leq e^{-ct}V(x_0) + \frac{L\varepsilon}{c}\left(1 - e^{-ct}\right) \tag{6.10}$$

and this will enable us to prove the theorem.

Define $\eta(\varepsilon) = 2L\varepsilon/c$ and

$$\Lambda(\varepsilon) = \{x \in S(\Lambda, R_0) : V(x) \leq \eta(\varepsilon)\} . \tag{6.11}$$

$\Lambda(\varepsilon)$ is compact and contains Λ in its interior since V is continuous and $\eta > 0$ and Λ is the set where $V = 0$; moreover it is easy to see that $\Lambda(\varepsilon) \to \Lambda$ in the Hausdorff metric as $\varepsilon \to 0^+$, as follows. Since $\Lambda \subset \Lambda(\varepsilon)$, $H^*(\Lambda, \Lambda(\varepsilon)) = 0$ where $H^*(A, B) = \max(\mathrm{dist}(a, B) : a \in A)$. Thus the Hausdorff distance between Λ and $\Lambda(\varepsilon)$ is $H^*(\Lambda(\varepsilon), \Lambda)$ and is by the definition of α and η less than or equal to $\alpha^{-1}(\eta(\varepsilon))$ which tends to zero as $\varepsilon \to 0$ by the monotonicity of α.

$\Lambda(\varepsilon)$ is an invariant set for (6.1) since if $x_0 \in \Lambda(\varepsilon)$ we have

$$\begin{aligned} V(x(t; x_0)) \leq & \; e^{-ct}V(x_0) + \frac{L\varepsilon}{c}(1 - e^{-ct}) \\ \leq & \; e^{-ct}\eta(\varepsilon) + \tfrac{1}{2}\eta(\varepsilon)(1 - e^{-ct}) \\ \leq & \; \tfrac{1}{2}(1 + e^{-ct})\eta(\varepsilon) \\ < & \; \eta(\varepsilon) \end{aligned} \tag{6.12}$$

and so $x(t; x_0) \in \Lambda(\varepsilon)$. Now the attractivity of $\Lambda(\varepsilon)$ under (6.1) follows immediately from (6.10) and (6.7) and the theorem is proved.

REMARK. This theorem says nothing about the dynamics of (6.1) inside the attracting set, and for good reason. If $x' = f(x)$ possesses a chaotic attractor, it will also have an infinite number of unstable periodic orbits inside the attractor. By using only small perturbations on the dynamical system, one can stabilize any of these unstable periodic orbits (for initial conditions near enough) and thus control trajectories to stay near these periodic orbits [**40,44**]. Thus we need more information about the small perturbation $v(t, x)$ to say anything about the dynamics on the attractor. However, we note that if we can control our trajectory in such a fashion, choosing to go to any of several periodic solutions at will, then this is good evidence that there really is a chaotic attractor present (at least in a practical sense). There is no practical difference, of course, between a strange attractor and several very closely entwined 'slightly' stable periodic or quasi-periodic orbits in an attracting set; in the presence of real noise, the trajectories will divert from one orbit to the other in an unpredictable fashion, indistinguishably so from a 'truly chaotic' system.

I believe, however, that these attractors will be lower semicontinuous for *generic* perturbations $v(t)$—that is, it should take a very special, highly correlated, perturbation $v(t, x)$ to stabilize periodic orbits on the attractor; further, one expects that the Lyapunov spectrum will be stable under small *generic* perturbations $\varepsilon v(t)$. I am not aware of any proof of either of these conjectures.

We can return to our earlier examples, and consider what defect-controlled numerical methods do with $x' = x^2 - t$ and $x' = x^2$. In the first case it is easy to see by considering $x' = x^2 - t \pm \varepsilon$ that we get the correct asymptotic geometry, with a time-scale change of at most ε. Likewise in the second case we locate the (movable) singularity correct to $O(\varepsilon)$ if we use relative defect control. Thus in each case the geometry and the dynamics of the numerical solution are faithful to those of the true solution. We emphasize that the earlier numerical infidelity can be explained by the fact that the numerics were making a *large* perturbation on the equation, and of course this can change the qualitative behaviour.

7. An Experimental Approach

In considering the problem of correlations in the defect, which are sometimes observable even in the variable time-step case and may be relevant to the long-term dynamics of the system, one can take an experimental approach. Take for example the simple harmonic oscillator problem $y''+y=0$ or, in the Hamiltonian formulation used earlier, $H=\frac{1}{2}(p^2+q^2)$. Applying any non-symplectic or non-conservative numerical method to this problem leads to growth or decay in the orbits; likewise, rounding errors even in conservative methods lead to growth or decay, unless lattice methods are used [**16**]. One might like to *measure* the amount of artificial damping introduced by the method, instead of predicting it by, say, the method of modified equations (which can be used for variable time-step codes but requires knowledge of the error control used).

It would be advantageous to tie this experimental approach to a standard interpolant, to avoid questions of whether or not an appropriate interpolant was used by the code (or indeed any at all). Hence we consider the interpolant which minimizes the L_2-norm of the defect. On each time step, this gives the variational problem

$$\min I = \int_{t_k}^{t_{k+1}} \left(y''(\tau)+y(\tau)\right)^2 \, d\tau \,. \tag{7.1}$$

Since this integrand is convex, there exists a function $y(t)$ which minimizes I, defined on $t_k \le t \le t_{k+1}$ satisfying $y(t_k)=y_k$, $y(t_{k+1})=y_{k+1}$, and the derivative conditions given by requiring the interpolant to be C^1 at the meshpoints t_k (there are two degrees of freedom, one at either endpoint of integration). In fact, the solution is given by $y=A\cos(t)+B\sin(t)+Ct\cos(t)+Dt\sin(t)$ where A, B, C, and D are chosen to match the four conditions. The Euler-Lagrange equations determining this solution are $L^*Ly=0$, where $Ly=y''+y$ and, simply, $L^*=L$ here. In general, we can use classical variational calculus to find the minimum L_2-norm residual interpolant for the computed solution to $y'=f(y)$ by solving the the Euler-Lagrange equations for minimizing $I=\int(y'-f)^T(y'-f)dt$. The solution to these equations will not usually be continuous from time-step to time-step, but C^1 interpolants can be found (in theory) which arbitrarily approach the discontinuous solution and have residuals which are arbitrarily close to minimal.

For the special problem considered in this section, this minimum L_2-norm defect gives us a baseline—one can do at least this well without accounting for the 'structure' of the defect. However, most numerical solutions of $y''+y=0$ will introduce either spurious growth or spurious decay; one can then imagine trying to find the 'equivalent viscous damping' or dissipation of the implemented numerical method. That is, instead of trying to find the residual determined by (7.1), one tries to find the ζ which minimizes

$$\min\ I(\zeta) = \int_{t_k}^{t_{k+1}} \left(y'' + 2\zeta y' + y\right)^2 \, d\tau \tag{7.2}$$

and clearly this residual can only be less than the previous. By using a trick, we can estimate the optimal ζ without actually constructing the optimal interpolant: we expand the above, differentiate with respect to ζ, set the derivative to zero and solve for ζ, and then perform the integration. It turns out that the answers involve only the endpoints anyway, plus some integrals of the squares of the interpolant and its derivative; these remaining integrals can be computed numerically by integrating some extra differential equations as we go. One can also search for the frequency shift in this way, but we will use only the damping here. Of course, for other model problems, other parameters would be of interest and a similar analysis might help.

For this problem, we find that ζ is usually different from time-step to time-step, and thus it seems reasonable to fit a function of t to the results. [One could instead look at the 'total effective viscous damping' by integrating from t_0 to t_k instead of over just one step.] If we use the time-dependent approach to measure the 'step-wise viscous damping' of Matlab's ODE23, we find that $\zeta \sim 0.033\varepsilon t$, where ε is the input tolerance, and the residual in the modified equation is about half the size it is with $\zeta = 0$. That is, this modification accounts for half the residual, and moreover it is positive: this means that ODE23 adds some dissipation to its solution of this equation. Note that this quantity will be a characteristic of an implemented numerical code. That is, this experiment tells us something about the particular piece of Fortran (or C, or whatever) code used to solve the initial-value problem. When we perform this experiment using the defect-controlled method PAMETH [**19**] we see that it also adds dissipation to its solution, but Matlab's ODE45 (based on RKF45) has $\zeta = -0.03\varepsilon t$ and thus introduces exponential growth into the solution (albeit at a very slow rate). Trying to measure the effective damping or frequency shift of a symplectic method by this technique gives confusing results—the damping oscillates with a moderate amplitude around zero, while (in the more complicated model where both the damping and frequency are allowed to vary) the frequency shift also oscillates around 1. The residual is not much reduced, either; hence one concludes that this is not a good model for what symplectic methods do to this equation (the implicit midpoint rule was used here as a test case). Instead, the earlier technique of finding a modified Hamiltonian gives a better explanation. This 'experimental' technique will be pursued further in future work.

8. Limitations and Objections

Not everyone is happy with the idea of 'changing the problem'. Some are simply prejudiced—they are in the habit of thinking of forward error $x - x_{\text{true}}$ as the quantity they are interested in, and perhaps are even unaware of any alternatives. We see some of this resistance even in the area of polynomial rootfinding or the solution of linear systems, where the idea of backward error analysis has been used successfully for more than thirty years.

A similar objection is that defect control is somehow 'unethical'—a person who uses it 'isn't playing the game', where the 'game' is to get good solutions to the problems given to you (usually by someone else). This objection is unanswerable without questioning the validity of the fundamental assumption of the 'game', namely that the given problem *as written down* is sacred. I am not denying the utility of this 'game'—if one can cheaply and easily get good solutions to the specified problem, then one can indeed get good insight into the original problem. The point is that sometimes good solutions to the specified problem are much too expensive to compute, especially compared with a cheap alternative.

There are more rational objections: consider for example people who write *general-purpose* software. As such, they obviously cannot easily take into account the context of the problem being solved, which may not even exist at the time the code is being written. For such people, the model of trying for 'small global error' has proved very successful, for well-conditioned problems. It is unfortunate that the trajectories of chaotic problems are ill-conditioned, and in that case the goal of 'small global error' is unattainable, without an appeal to shadowing. It is true that interval arithmetic codes—enclosure methods—can compute solutions to chaotic initial-value problems with small global error, on finite time intervals. Unfortunately, the cost of such solutions necessarily grows exponentially with the length of the time interval. A much less expensive alternative is to attempt to computationally verify that shadowing has occurred, and to compute the shadowing distance. This will give you a global error bound on your computed solution, in the sense that you will have found a nearby true solution [**7**]. However, this process is itself inherently somewhat expensive, essentially requiring that interval rootfinding methods be used to prove the existence of and find the distance to the nearby true solution. A defect-controlled solution can be expected to be much cheaper, since it is essentially doing only a small part of the calculation above—one could graft such a computational shadowing procedure on top of a defect-controlled solution, in fact. The real issue is whether the extra effort is worth getting the global error bound. One suspects that this can only be decided on a problem-by-problem basis.

An interesting objection to the backward error idea is that some problems are 'forward-stable' but not 'backward-stable' [**13**]—that is, for such problems, there exist approximations which give good forward error, but no acceptable ones which give good backward error. Stiff problems (and perhaps DAE's) may be viewed as falling into this category: it is possible to have a solution (computed with a stiffly-stable method) with a large defect in the usual norms, but small global error. In fact, this may be a useful definition of stiffness. If this definition is accepted, we see a certain rough anti-symmetry with chaotic problems: a chaotic problem is one where a solution with good backward error may be easily computed with explicit methods while a solution with good forward error is too expensive; on the other hand, a stiff problem is one where a solution with good forward error is easily computed using implicit methods, while a solution with good backward error is too expensive.

This anti-symmetry is only rough: it may be that there are problems which partake of both properties, having some very large and negative Lyapunov exponents and some small and positive ones: while the trajectory is moving towards an attractor, it might be that some degree of implicitness will be useful to efficiently solve the system, to locate and stay on the attractor—in which case one might call the problem stiff—but once on the attractor care must be taken to get good backward error, and this is likely to be an important restriction on the stepsize.

Further objections to defect control include the fact that we have added a time-dependent perturbation to a (perhaps) initially autonomous problem, and thus made it more complicated—putting the solution in a higher-dimensional space, in effect. This is indeed a complication, but one must remember that physical perturbations of autonomous problems can be time-dependent, and so this objection is not necessarily insuperable. Another objection is that the numerical solution usually merely goes on for a very long time—one wants an infinite time solution, which numerically is difficult unless one is near an equilibrium or limit cycle. But here again the use of infinite time is only a mathematical model of physical reality—one is usually only interested in infinite time as a 'first approximation' or simple model for 'very long times', and so whether one simulates for a billion years instead of forever usually makes little practical difference. It is different, of course, if the infinite-time behaviour is easily available, through asymptotics or other approximations. Then, infinite time results provide useful approximations for the lengthy times we are usually interested in practically. See [**17**] for a discussion of available infinite-time backward error results for dynamical systems.

Finally, the choice of norm appropriate for the defect is clearly problem-dependent. This requires more complicated user interfaces, so the user can tell the computer code what norm to use, and concomitantly requires the user to be sophisticated enough to tell the code to use the right one. For differential-algebraic equations (supposing the approach to be useful at all for such problems, and supposing that appropriate norms have been identified) the norm will have to take into account the greater rigidity of algebraic constraints, for example, and raises the question of whether or not the chosen norm of the defect can be controlled at all by choosing the stepsize, order, or method. By showing that defect control is asymptotically equivalent to local error per unit step, one can recover existing results on the non-vanishing of the stepsize for stable problems, but with other norms this is no longer so clear. It is possible that an inappropriate underlying scheme might be detected by the non-controllability of the defect, and this may open a fruitful line of investigation for numerical analysts.

9. Concluding Remarks

We have seen that backward error analysis can be useful for analysis of long-time numerical simulations of chaotic dynamical systems. The method of modified equations on its own is not sufficient to get long-time results, and a simple defect-control analysis, while normally very useful, may sometimes lead to vexing questions about correlations. These objections can be overcome by combining the two methods.

Existing perturbation results are very useful for determining the relation between the long-time behaviour of the numerical simulation and the asymptotic behaviour of the dynamical system. The results presented here include the fact that attracting sets are preserved by numerical methods, though nothing (beyond conjecture) has been said about attractors. One could wish for a more complete list of such perturbation results, including results for the infinite-dimensional case. Interpolants are widely used in codes for solving delay equations, for example, and it is possible that some of these ideas will carry directly over, and similarly for partial differential equations.

It is likely that the concept of 'structured backward stability', currently of great interest in the linear algebra community [**27, 13**], will be of importance in this context. One would prefer a backward stability result for numerical solutions of the Navier-Stokes equations that took proper account of the preservation of conservation laws—that is, the requisite norm used to measure how close the actual dynamical system solved was to the Navier-Stokes equations should not allow violation of important conservation laws. It is not even clear if this will be possible at all, though the success of the lattice gas approach (which exactly solves a physical problem for a different sort of fluid) suggests that it will be.

The remaining pure mathematical challenges include a complete characterization of the stability of attractors under generic perturbations of the types (6.1)–(6.3), and extension of those results to infinite-dimensional systems.

Acknowledgements.

Cleve Moler provided the graphs for Figures 5.1 and 5.2, using Matlab 4.0. Discussions with Andrew Stuart and with Wayne Enright were very helpful. J. M. Sanz-Serna was kind enough to explain the method of modified equations to me, with special attention to the case of Hamiltonian systems. Others of my colleagues, too numerous to mention, also contributed to the perspective of this paper, for which they have my thanks again.

References

1. E. Adams, W. F. Ames, W. Kühn, W. Rufeger, and H. Spreuer, *Computational Chaos May Be Due to a Single Local Error*, J. Comp. Phys. **104** (1993), 241–250.
2. Bernd Aulbach, *On Linearly Perturbed Linear Systems*, J. Math. Anal. and Appl **112** (1985), 317–327.
3. Carl M. Bender and Steven A. Orzag, *Advanced Mathematical Methods for Scientists and Engineers*, McGraw-Hill, 1978.
4. Wolf-Jürgen Beyn, *Numerical Methods for Dynamical Systems*, Advances in Numerical Analysis (Will Light, ed.), vol. I, Oxford Science Publications, 1991, pp. 175–236.
5. Garret Birkhoff and Gian-Carlo Rota, *Ordinary Differential Equations*, 4th ed., Wiley, 1989.
6. Bruce W. Char, Keith O. Geddes, Gaston H. Gonnet, Benton L. Leong, Michael B. Monagan, and Stephen M. Watt, *The Maple V Language Reference Manual*, Springer-Verlag, New York, 1992.
7. Shui-Nee Chow and Erik S. Van Vleck, *A Shadowing Lemma Approach to Global Error Analysis for Initial Value ODEs*, SIAM J. Sci. Comp. (to appear).
8. Robert M. Corless, *What good are numerical simulations of chaotic dynamical systems?*, Computers in Mathematics with Applications (1993) (to appear).
9. ———, *Defect-controlled numerical methods and shadowing for chaotic differential equations*, Physica D **60** (1992), 323–334.
10. Robert M. Corless and G. F. Corliss, *Rationale for Guaranteed ODE Defect Control*, Proceedings of SCAN 1991: International Symposium on Computer Arithmetic and Scientific Computing (Oldenburg, October 1–4, 1991) (J. Herzberger, ed.), IMACS Annals on Computing and Applied Mathematics, 1991.
11. Robert M. Corless and S. Yu. Pilyugin, *Approximate and Real Trajectories for Generic Dynamical Systems*, J. Math. Anal. and Applic. (1994) (to appear).
12. ———, *Evaluation of Upper Lyapunov Exponents on Hyperbolic Sets*, J. Math. Anal. and Applic. (1994) (to appear).
13. James Demmel, *private communication* (1991).
14. Robert L. Devaney, *An Introduction to Chaotic Dynamical Systems*, 2nd ed., Addison-Wesley, 1989.
15. Luca Dieci, Robert D. Russell, and Erik S. Van Vleck, *On the Computation of Lyapunov Exponents for Continuous Dynamical Systems*, Math: 040893-006, Georgia Tech. Report (1993).
16. David J. D. Earn and Scott Tremaine, *Exact Numerical Studies of Hamiltonian maps: Iterating without roundoff error*, Physica D **56** (1992), 1–22.
17. Timo Eirola, *Aspects of backward error analysis of numerical ODEs*, J. Comp. Appl. Maths. **45** (1993), 65–73.
18. W. H. Enright, *A New Error-Control for Initial Value Solvers*, Appl. Math. Comput. **31** (1989), 288–301.
19. ———, *PAMETH: fortran subroutine for defect-controlled Runge-Kutta solution of ODE's*, *private communication* (1990).
20. Gene Golub and Charles Van Loan, *Matrix Computations*, Johns Hopkins University Press, Baltimore, 1983.
21. D. F. Griffiths and J. M. Sanz-Serna, *On the scope of the Method of Modified Equations*, No 3, SIAM J. Sci. Stat. Comput. **7** (1986), 994–1008.
22. D. F. Griffiths, *The dynamics of some linear multistep methods with step-size control*, Numerical Analysis 1987 (D. F. Griffiths and G. A. Watson, eds.), Longman Scientific and Technical, 1987, pp. 115–134.
23. Jack K. Hale, *private communication.*
24. ———, *Ordinary Differential Equations*, 2nd. ed., Kreiger, 1978.
25. ———, *Asymptotic Behavior of Dissipative Systems*, Mathematical Surveys and Monographs, No. 25, American Mathematical Society, 1988.
26. Peter Henrici, *Applied and Computational Complex Analysis*, vol. 2, Wiley-Interscience, 1977.

27. Desmond J. Higham, *Defect Estimation in Adams PECE Codes*, No. 5, SIAM J. Sci. Stat. Comput. **10** (1989), 964–976.
28. J. Hubbard and B. West, *Differential Equations, a dynamical systems approach, Part I*, Springer-Verlag, 1991.
29. Bruno Salvy, *private communication* (1993).
30. N. J. A. Sloane, *Handbook of Integer Sequences*, 2nd. ed., Academic Press, 1993.
31. A. R. Humphries and A. M. Stuart, *Runge-Kutta Methods for Dissipative and Gradient Dynamical Systems*, NA-92-17, Computer Science Department, Stanford (1992).
32. P. E. Kloeden and J. Lorenz, *Stable Attracting Sets in Dynamical Systems and in their One-Step Discretizations*, No. 5, SIAM J. Numer. Anal. **23** (1986), 986–995.
33. ———, *Lyapunov Stability and Attractors Under Discretization*, Differential Equations: Proceedings of the EQUADIFF Conference (C. M. Dafermos, G. Ladas, and G. Papanicolaou, eds.), Marcel Dekker Inc., 1989, pp. 361–367.
34. G. Labelle, *Sur l'Inversion et l'Itération Continue des Séries Formelles*, Europ. J. Combinatorics **1** (1980), 113–138.
35. Francis C. Moon, *Chaotic Vibrations*, Wiley, New York, 1987.
36. Sergei Yu. Pilyugin, *Introduction to Structurally Stable Systems of Differential Equations*, Birkhaüser, 1992.
37. N. Rouche, P. Habets, and M. Laloy, *Stability Theory by Liapunov's Direct Method*, AMS **22**, Springer-Verlag, 1977.
38. J. M. Sanz-serna, *private communication.*.
39. G. R. Sell, *Some perturbation problems in ordinary differential equations*, Funkcial. Ekvac. **10** (1967), 1–13.
40. Troy Shinbrot, Edward Ott, Celso Grebogi, and James A. Yorke, *Using Chaos to Direct Trajectories to Targets*, No. 26,, Phys. Rev. Lett. **65** (1990), 3215–3218.
41. A. M. Stuart and A. R. Humphries, *Model Problems in Numerical Stability Theory for Initial Value Problems*, NA-92-16, Computer Science Department, Stanford (1992).
42. ———, *An Analysis of Local Error Control for Dissipative, Contractive, and Gradient Dynamical Systems*, NA-92-18, Computer Science Department, Stanford (1992).
43. A. M. Stuart, *private communication.*
44. S. Talwar and N. Sri Namachchivaya, *Control of Chaotic Systems: Application to the Lorenz Equations*, Nonlinear Vibrations (R. A. Ibrahim, N. S. Namachchivaya, and A. K. Bajaj, eds.), DE-Vol. 50, AMD-Vol. 144, Proceedings of the Winter Annual Meeting of the ASME, Anaheim, CA, 1992, pp. 47–58.
45. Takeshi Taniguchi, *On the Estimate of Solutions of Perturbed Linear Differential Equations*, J. Math. Anal. and Appl. **153** (1990), 288–300.
46. M. Vidyasagar, A. Boyarsky, and A. Vannelli, *On the Stability Properties of Perturbed Linear Nonstationary Systems*, J. Math. Anal. and Appl. **88** (1982), 245–256.
47. E. Hairer, S. P. Nørsett, and G. Wanner, *Solving Ordinary Differential Equations I*, Springer Series in Computational Mathematics 8, Springer-Verlag, Berlin, 1980.
48. Ernst Joachim Weniger, *Nonlinear Sequence Transformations for the Acceleration of Convergence and the Summation of Divergent Series*, Computer Physics Reports **10**, North-Holland, 1989.
49. James H. Wilkinson, *The Perfidious Polynomial*, Studies in Numerical Analysis (Gene H. Golub, ed.), MAA Studies in Mathematics, Vol 24, Mathematical Association of America, 1984, pp. 1–28.
50. T. Yoshizawa, *Stability Theory by Liapunov's Second Method*, vol. 9, Math. Soc. Japan, 1966.

Appendix A: Maple code for evaluation of $B(v)$

The following procedures evaluate $B(v)$ from (3.4).

```
c:=proc(n)
local i;
option remember;
 -1/(n-1)*convert([seq(binomial(n-i+1,i+1)*c(n-i),i=1..n-1)],`+`)
end;
c(1):=1:  # Remember the base of the recurrence.
Bseries := v -> evalf(Sum('c(n)*v**(n-1)',n=1..infinity));
B := proc(v) local p,v0,u0;
 if not type(v,numeric) then 'B(v)' else
  if v < -1 then
   (1.+v)**2/(1.+2*v)*B(v*(v+1.))
  elif v=-1 then
   0.
  elif -1 < v and v < 0 then
   # Forward use of the recurrence.  (Ignore v=-1/2).
   v0 := v; p := 1.;
   while v0 < -0.1 do
    p := p*(1.+v0)**2/(1+2.*v0);
    v0 := v0*(v0+1.);
   od:
   p*Bseries(v0)
  elif v = 0 then
   1.
  else
   # Backward use of the recurrence.
   u0 := v; p := 1.;
   while u0 > 0.1 do
    u0 := 2.*u0/( 1.+(1.+4.*u0)**(1/2) );
    p := p*(1.+2.*u0)/(1+u0)**2;
   od:
   p*Bseries(u0)
  fi
 fi
end:
```

Department of Applied Mathematics, University of Western Ontario, London, CANADA N6A 5B7

E-mail address: rcorless@uwo.ca

Contemporary Mathematics
Volume **172**, 1994

Modified equations for ODEs

M. P. CALVO, A. MURUA AND J. M. SANZ-SERNA

ABSTRACT. We study the method of modified equations for the analysis of discretizations of ordinary differential equations. We show how to systematically construct modified system of any order. Some applications are presented.

1. Introduction

Modified equations [**28**], [**13**] are a means for the analysis of numerical methods for differential equations. Modified equations are strongly related to the idea of *backward error analysis,* explained in all numerical analysis textbooks. Given a problem $\mathcal{P}$ with true solution $\mathcal{S}$ and given an approximate solution $\tilde{\mathcal{S}}$, *forward error analysis* consists of estimating the distance between $\tilde{\mathcal{S}}$ and $\mathcal{S}$. Backward error analysis consists of showing that $\tilde{\mathcal{S}}$ exactly solves a problem $\tilde{P}$ which is close to $\mathcal{P}$. While backward error analysis has played a role of paramount importance in areas like numerical linear algebra, error analysis of numerical methods for evolutionary problems has essentially been of the forward variety (see nevertheless [**4**], [**23**], [**10**]).

However, there are many instances where forward error analysis of numerical simulations of evolutionary problems is doomed to fail. In regimes where true orbits of the system quickly diverge from each other, including chaotic dynamics, any numerical method in realistic circumstances will produce an answer $\tilde{\mathcal{S}}$ very different from the true $\mathcal{S}$. Hence the outcome of forward analysis would be that any method performs badly, a conclusion at odds with the fact that numerical simulations have been helpful in ascertaining the behaviour of the systems involved. Similarly, classical error bounds are meaningless in long-time simulations performed in order to find the qualitative behaviour of most nontrivial dynamical systems. Therefore there is a clear need for analyses that depart from

1991 *Mathematics Subject Classification.* Primary 65L05; Secondary 70H15, 05C05.

This research has been supported by project DGICYT PB89-0351.

This paper is in final form and no version of it will be submitted for publication elsewhere.

the classical (forward) error bounds for numerical integrators of evolutionary problem.

An idea that has become quite prominent is that of *shadowing*. Originally introduced in a dynamical systems context [**2**], [**5**], it has recently gained popularity in numerical analysis applications, see e.g. [**1**], [**3**], [**8**], [**9**], [**12**], [**17**], [**18**], [**19**], [**20**], [**21**], [**22**], [**26**], [**27**]. In a shadowing approach, the numerically computed orbit with initial value u_0 is compared not with the true orbit from u_0, but with the shadowing orbit, the exact orbit of the system being simulated corresponding to a slightly perturbed $\tilde{u}_0$. Typically it is shown that the distance between the numerical and shadowing orbits is small in some sense, while the distance between the numerical orbit and the true orbit (i.e. the classical error) is unacceptable. The similarity with backward error analysis is evident.

In the shadowing approach, the initial condition is allowed to be changed, while keeping the same evolutionary system. In the modified equation technique the numerical solution is compared with a solution of a perturbed system. The idea of modified equations has been around for some time (see e.g. [**28**]), mainly in the study of dissipation and dispersion properties of numerical schemes for partial differential equations. However it is only recently that the method has been applied to ordinary differential equations (ODEs), specially to investigate symplectic methods for Hamiltonian systems [**24**], [**25**], [**14**].

In this paper, the attention is restricted to one-step numerical methods for initial value problems for D-dimensional systems of ODEs:

$$\frac{du}{dt} = f(u), \qquad u(0) = u_0. \tag{1}$$

For simplicity, we assume that the vector field f is defined in the whole of $\mathcal{R}^D$ and of class $\mathcal{C}^\infty$. Of course the system in (1) may arise from the discretization in space of a system of time-dependent partial differential equations.

Simple examples of numerical methods are Euler's rule

$$u_{n+1} = u_n + hf(u_n), \tag{2}$$

the implicit midpoint rule

$$u_{n+1} = u_n + hf(\frac{1}{2}(u_n + u_{n+1})), \tag{3}$$

and Runge's second order method

$$u_{n+1} = u_n + hf(u_n + \frac{h}{2}f(u_n)). \tag{4}$$

Here h denotes the time step and u_n is the numerical solution at time $t_n = nh$.

A numerical method is consistent of order $p \geq 1$ if, for all u in $\mathcal{R}^D$,

$$\psi_{h,f}(u) - \phi_{h,f}(u) = O(h^{p+1}), \qquad h \to 0, \tag{5}$$

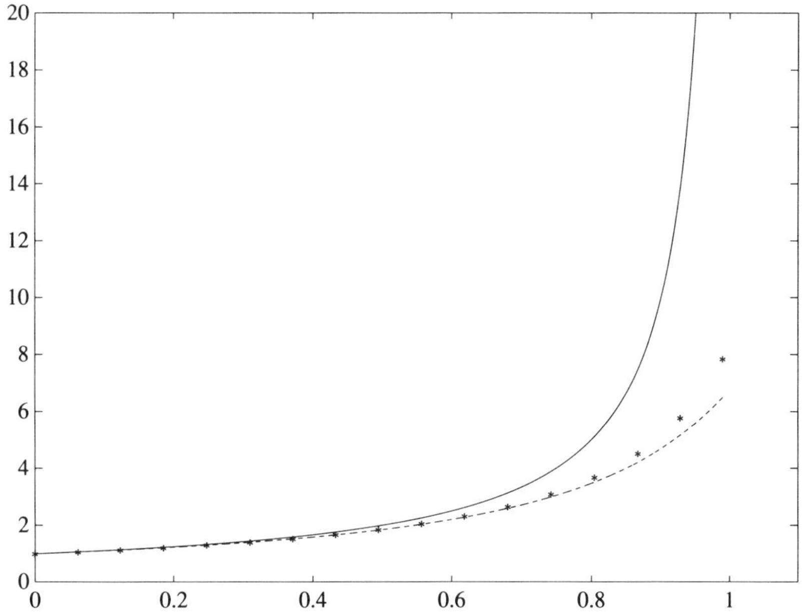

FIGURE 1. True solution $u = u(t)$ (solid), Euler solution (stars) and modified solution (dashes) for $du/dt = u^2$.

where $\psi_{h,f}(u)$ and $\phi_{h,f}(u)$ respectively denote the numerical and theoretical solutions after one step of length h taken from the initial condition u (so that $\phi_{h,f}$ is the flow of (1)). From the local error estimate (5), it follows [**6**], [**15**] that, as $h \to 0$, the global errors $u_n - u(t_n)$ are $O(h^p)$, uniformly in bounded time intervals $0 \leq t_n \leq t_{max}$ contained in the interval of existence of the true solution $u(t)$ (convergence of order p). The order is 2 for (3) and (4) and only 1 for (2).

We now present a simple example of the application of the method of modified equations. We integrate in the interval $0 \leq t \leq T = 0.99$ the equation $du/dt = f(u) = u^2$ with initial condition $u(0) = 1$ (solution $u(t) = 1/(1-t)$). The solid line in Figure 1 represents the true solution, while the stars depict the numerical solution for Euler's rule with $h = T/16$. We see that there is little agreement between the behaviour of the numerical and theoretical solutions. Can we find a differential equation whose solution with initial condition $u_0 = 1$ behaves as the numerically computed points? To be more precise, we try to find a modified equation $du/dt = \tilde{f}(u)$ such that, for all u,

$$\psi_{h,f}(u) - \phi_{h,\tilde{f}}(u) = O(h^3),$$

i.e. such that Euler's rule, consistent of the first order with the problem being integrated, is consistent of the *second* order with the modified equation. By going from local to global errors, the computed points u_n will lie at distance $O(h^2)$ from the solution of the modified equation with initial value u_0. To find

$\tilde{f}$ we start with an ansatz $\tilde{f}(u) = u^2 + hF(u)$, where F is a function of u to be determined. Note that $\tilde{f}$ depends on h. Expanding in powers of h the flow $\phi_{h,\tilde{f}}(u)$, it is found that it differs from $\psi_{h,f}(u) = u + hu^2$ in $O(h^3)$ terms if $F = -u^3$, which leads to the modified $\tilde{f} = u^2 - hu^3$. The solution of the modified equation is shown in Figure 1 by a dotted line. It is clear that the numerical solution is better described by the modified equation than by the original equation being solved. It is now possible to look for an even better $\tilde{f}$, $\tilde{f}_2(u) = u^2 - hu^3 + h^2F_2(u)$, to have consistency of the third order and more generally for vector fields $\tilde{f}_N(u) = u^2 - hu^3 + h^2F_2(u) + \cdots + h^{N-1}F_{N-1}(u)$ leading to consistency of order N. We then say that $du/dt = \tilde{f}_N(u)$ is a modified equation of order N.

In §2 of this paper we show how to systematically construct modified systems of any order for one-step methods. The formulae we present are due to Hairer [**14**]; however our methodology for the derivation of those formulae is different from and easier than that presented in the original paper. An example of the application of modified equation techniques is given in §3.

2. Constructing modified equations

It is well known [**6**], [**15**] that (rooted) trees are an important tool in the analysis of one-step methods. The trees with four of fewer nodes are depicted in Figure 2. The symbol τ_1 denotes the only tree with one node. It is common to denote by $[\tau^1, \tau^2, \ldots, \tau^m]$ the tree that consists of the root and m leaving edges to which the trees $\tau^1, \tau^2, \ldots, \tau^m$ are attached. Thus in Figure 2, $\tau_2 = [\tau_1]$, $\tau_{31} = [\tau_1, \tau_1]$, $\tau_{32} = [\tau_2]$, etc. For each tree τ, the integers $\rho(\tau)$ and $\alpha(\tau)$ respectively denote its order (number of nodes) and number of monotonic labellings. These functions can be computed recursively by the formulae $\rho(\tau_1) = \alpha(\tau_1) = 1$ and, for $\tau = [\tau^1, \ldots, \tau^m]$,

$$\begin{aligned} \rho(\tau) &= 1 + \rho(\tau^1) + \cdots + \rho(\tau^m), \\ \alpha(\tau) &= \frac{(\rho(\tau) - 1)!}{\rho(\tau^1)! \cdots \rho(\tau^m)!} \alpha(\tau^1) \ldots \alpha(\tau^m) \frac{1}{\mu_1! \mu_2! \cdots}. \end{aligned}$$

The integers μ_i count the number of equal trees among $\tau^1, \ldots, \tau^m$. Finally, in connection with the system in (1), an $\mathcal{R}^D$-valued function $F(\tau)(u)$ (elementary differential) is associated with each tree τ. The recursive definition of the $F(\tau)(u)$'s is $F(\tau_1)(u) = f(u)$ and for $\tau = [\tau^1, \ldots, \tau^m]$

$$F(\tau)(u) = f^{(m)}(u)(F(\tau^1)(u), \ldots, F(\tau^m)(u)),$$

where $f^{(m)}(u)$ represents the m-th Frechet derivative of f evaluated at u.

With these notations, the formal Taylor expansion of the flow $\phi_{h,f}$ in powers of h is given by

$$\phi_{h,f}(u) = u + \sum_{\tau \in T} \frac{h^{\rho(\tau)}}{\rho(\tau)!} \alpha(\tau) F(\tau)(u),$$

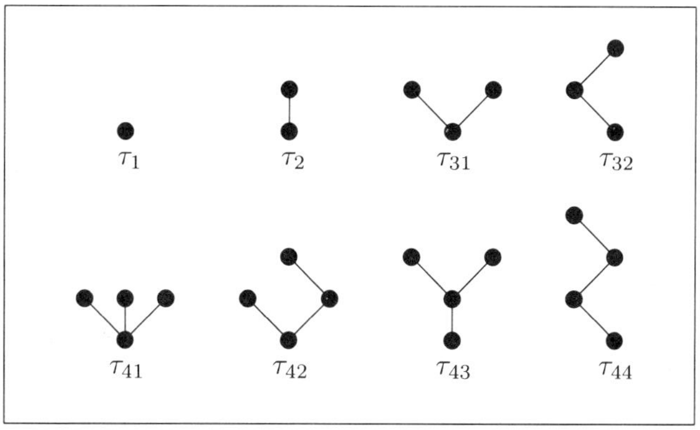

FIGURE 2. Trees of order ≤ 4.

where T denotes the set of all trees.

Numerical methods can be Taylor expanded in a similar way. Methods (2), (3) and (4) are particular instances of Runge-Kutta methods. A general Runge-Kutta method is specified by an integer $s \geq 1$ (the number of stages) and real coefficients a_{ij}, b_i, $i, j = 1, \ldots, s$. Its application to (1) results in the formulae

$$\begin{aligned} U_i &= u_n + h \sum_{j=1}^{s} a_{ij} f(U_j), \\ u_{n+1} &= u_n + h \sum_{i=1}^{s} b_i f(U_i). \end{aligned}$$

For Euler's rule $s = 1$, $a_{11} = 0$, $b_1 = 1$; for the midpoint rule $s = 1$, $a_{11} = 1/2$, $b_1 = 1$; for the method (4), $s = 2$, $a_{21} = 1/2$, $b_2 = 1$ and the remaining coefficients are 0. The Taylor expansion of a Runge-Kutta method is

$$\psi_{h,f}(u) = u + \sum_{\tau \in T} \frac{h^{\rho(\tau)}}{\rho(\tau)!} \alpha(\tau) \left(\gamma(\tau) \sum_{i=1}^{s} b_i \Phi_i(\tau) \right) F(\tau)(u),$$

where the recursive definitions of γ and Φ_i are $\gamma(\tau_1) = 1$, $\Phi_i(\tau_1) = 1$ and

$$\begin{aligned} \gamma(t) &= \rho(\tau)\gamma(\tau^1) \ldots \gamma(\tau^m), \\ \Phi_i(\tau) &= \sum_{j_1, \ldots, j_m} a_{ij_1} \Phi_{j_1}(\tau^1) \ldots a_{ij_m} \Phi_{j_m}(\tau^m). \end{aligned}$$

In view of the Taylor expansions above, Hairer and Wanner [**16**] introduced the notion of a B-series. Given a real valued mapping a defined in the union of T and the set $\{\emptyset\}$, a B-series $B(a, u)$ is a formal power series

$$a(\emptyset)u + \sum_{\tau \in T} \frac{h^{\rho(\tau)}}{\rho(\tau)!} \alpha(\tau) a(\tau) F(\tau)(u).$$

Thus the true flow $\phi_{h,f}$ corresponds to $a \equiv 1$, while for a Runge-Kutta method $a(\emptyset) = 1$ and

$$a(\tau) = \gamma(\tau) \sum_{i=1}^{s} b_i \Phi_i(\tau).$$

These ideas are not confined to Runge-Kutta methods. The Taylor expansion of most one-step methods used in practice is also a B-series. In the remainder of the section we assume that we are dealing with a method $\psi_{h,f}(u)$ corresponding to a suitable B-series $B(a, u)$, without specifying the exact nature of the method. We suppose that the method is at least of order 1, i.e.

$$a(\emptyset) = 1, \qquad a(\tau_1) = 1. \tag{6}$$

Our aim is to construct a formal power series $\tilde{f}$

$$\sum_{\tau \in T} \frac{h^{\rho(\tau)-1}}{\rho(\tau)!} \alpha(\tau) b(\tau) F(\tau)(u) \tag{7}$$

so that for each integer $N \geq 1$

$$\frac{du}{dt} = \tilde{f}_N(u) = \sum_{1 \leq \rho(\tau) \leq N} \frac{h^{\rho(\tau)-1}}{\rho(\tau)!} \alpha(\tau) b(\tau) F(\tau)(u) \tag{8}$$

provides a modified equation of order N.

An essential tool for our purposes is the formula for composition of B-series, see Theorem 11.6 in [**15**]. If a and b are B-series coefficients with $a(\emptyset) = 1$ then the composition $B(b, B(a, y))$ is a again a B-series $B(ab, y)$ whose coefficients $ab(\tau)$ can be found in a systematic way from the a's and b's. The formulae for the first $ab(\tau)$ are

$$\begin{aligned}
ab(\emptyset) &= b(\emptyset), && (9)\\
ab(\tau_1) &= b(\emptyset)a(\tau_1) + b(\tau_1), && (10)\\
ab(\tau_2) &= b(\emptyset)a(\tau_2) + 2b(\tau_1)a(\tau_1) + b(\tau_2), && (11)\\
ab(\tau_{31}) &= b(\emptyset)a(\tau_{31}) + 3b(\tau_1)a(\tau_1)^2 + 3b(\tau_2)a(\tau) + b(\tau_{31}), && (12)\\
ab(\tau_{32}) &= b(\emptyset)a(\tau_{32}) + 3b(\tau_1)a(\tau_2) + 3b(\tau_2)a(\tau) + b(\tau_{32}). && (13)
\end{aligned}$$

We introduce a real parameter λ and write the flow of the vector field in (7) as a B-series

$$\phi_{\lambda h, \tilde{f}}(u) = u + \sum_{\tau \in T} \frac{h^{\rho(\tau)}}{\rho(\tau)!} \alpha(\tau) a_\lambda(\tau) F(\tau)(u).$$

Next we substitute this series into the equation

$$\frac{d}{dt} \phi_{t,\tilde{f}} = \tilde{f}(\phi_{t,\tilde{f}});$$

in doing so the B-series of the right hand side is computed by the formula for composing B-series. In this way we find that the $a_\lambda(\tau)$ satisfy, for each tree τ,

$$\frac{d}{d\lambda}a_\lambda(\tau) = (a_\lambda b)(\tau). \tag{14}$$

Furthermore at $\lambda = 0$, $\phi_{0,\tilde{f}}(u) = u$ and hence, for each τ,

$$a_0(\tau) = 0. \tag{15}$$

The relations (14)–(15) allow the computation of the $a_\lambda(\tau)$'s in terms of the $b(\tau)$'s when the latter are known. In our setting, the b coefficients are determined to ensure that, for each τ, at $\lambda = 1$

$$a_1(\tau) = a(\tau), \tag{16}$$

to impose that, as formal power series, $\phi_{h,\tilde{f}}$ and $\psi_{h,f}$ coincide.

The relations (14)–(16) make it possible to recursively compute the b coefficients. Let us illustrate this. For τ_1 we obtain from (14) and (10), $(d/d\lambda)a_\lambda(\tau_1) = b(\tau_1)$, so that, according to (15), $a_\lambda(\tau_1) = \lambda b(\tau_1)$. If we now impose (16), we obtain the relation $b(\tau_1) = a(\tau_1)$. We conclude, from the consistency assumption (6), that $b(\tau_1) = 1$, and therefore, as expected, $\tilde{f}$ differs from f in $O(h)$ terms.

If we now go through the same steps for the next tree τ_2, we succesively obtain

$$\begin{aligned}
\frac{d}{d\lambda}a_\lambda(\tau_2) &= 2b(\tau_1)a_\lambda(\tau_1) + b(\tau_2) \\
&= 2\lambda b(\tau_1)^2 + b(\tau_2), \\
a_\lambda(\tau_2) &= \lambda^2 b(\tau_1)^2 + \lambda b(\tau_2), \\
a(\tau_2) &= b(\tau_1)^2 + b(\tau_2).
\end{aligned}$$

The last equation yields $b(\tau_2)$. Note that for a method of order ≥ 2, $a(\tau_2) = 1$, which, in tandem with $b(\tau_1) = 1$, leads to $b(\tau_2) = 0$ and $\tilde{f}$ and f differ in $O(h^2)$ terms.

Similarly the equations for finding $b(\tau_{32})$ and $b(\tau_{31})$ turn out to be

$$\begin{aligned}
a(\tau_{31}) &= b(\tau_1)^3 + \frac{3}{2}b(\tau_2)b(\tau_1) + b(\tau_{31}), \\
a(\tau_{32}) &= b(\tau_1)^3 + 3b(\tau_2)b(\tau_1) + b(\tau_{32}).
\end{aligned}$$

From here $b(\tau_{31}) = b(\tau_{32}) = 0$ for methods of order ≥ 3.

We summarize our findings in the following theorem, due to Hairer [**14**].

THEOREM 1. *Assume that an order p, $p \geq 1$, one-step method can be formally Taylor expanded into a B-series $B(a, u)$. There is a unique B-series (7), differing from $f(u)$ in $O(h^p)$ terms, such that, for each integer $N \geq 1$, (8) provides a modified system of order N. The coefficients b can be recursively found as functions of the coefficients a.*

We emphasize that the formal power series (7) in general does not converge. Lack of space prevents us from discussing further this point and the interested reader is referred to [**24**] and to Chapter 10 in [**25**].

3. An application

We now illustrate the use of modified equations in ODEs. We consider the pendulum system, that we write in terms of the components p and q of u as

$$\frac{dp}{dt} = -\sin q, \qquad \frac{dq}{dt} = p.$$

This is a Hamiltonian problem [**24**], [**25**] with Hamiltonian function (energy) $H = (1/2)p^2 + 1 - \cos q$. Let (p_0, q_0) be an initial condition with energy H_0, $0 < H_0 < 2$, leading to a periodic solution. In phase plane the trajectory corresponds to the level set $H = H_0$; the period T_0 of the solution is an increasing function of H_0. Furthermore, we respectively denote by f_0 and g_0 the vector field f evaluated at (p_0, q_0) and the energy gradient at (p_0, q_0). The vectors f_0 and g_0 are mutually orthogonal by conservation of energy.

This initial value problem is integrated by a one-step method of order p with step length h, that for simplicity we assume to be of the form $h = T_0/\nu$, with ν a positive integer. Let $e_M(h)$ be the global error $u_n - u(t_n)$ after $n = M\nu$ steps, i.e. after simulating M periods of the solution. Then it is not too difficult to show (see [**7**]) that

$$e_M(h) = Me_1(h) + \frac{1}{2}(M^2 - M)\Big(g_0, e_1(h)\Big)\,\delta_0 f_0 + O(h^{2p}), \qquad h \to 0, \tag{17}$$

where $(\cdot, \cdot)$ means inner product and δ_0 denotes the derivative of the period T with respect to the energy H evaluated at the initial condition. Therefore, ignoring the $O(h^{2p})$ remainder, the error $e_M(h)$ grows quadratically with M. The leading M^2 growth is in the direction of f_0, i.e. tangent to the solution at the initial point, thus corresponding to a *phase error*. However linear error growth with M is possible: if $(g_0, e_1(h)) = O(h^{2p})$ (i.e the error after one period is almost orthogonal to the energy gradient), then

$$e_M(h) = Me_1(h) + O(h^{2p}), \qquad h \to 0.$$

To sum up, the way global errors build up is determined by the *direction* of the error $e_1(h)$. This is not suprising: if after one period the error $e_1(h)$ has a significant component in the direction of g_0, then the numerical solution has jumped in phase plane to a neighbouring trajectory corresponding to a different (say larger) value of the energy. Thereafter, the method, when evaluating the vector field f, picks up wrong information as to the solution period and is lead to believe that the motion is faster than it really is. As the integration proceeds the numerical solution keeps jumping to higher and higher energy levels and getting unduly speeded up. This is the mechanism leading to quadratic growth in the *phase* error. On the other hand, if $e_1(h)$ is essentially in the direction of

f_0, then there is no energy error: the method is basically describing the right trajectory with a slightly distorted average velocity and errors grow linearly. These considerations apply to all nonlinear oscillators with one degree of freedom [**7**] and even to some partial differential equations [**11**].

We now use the method of modified equations to investigate the direction of $e_1(h)$. We begin with the midpoint rule (3). The modified system with $N = 4$ is found to be

$$\begin{aligned}\frac{dp}{dt} &= -\sin q + \frac{h^2}{24}(\sin 2q - p^2 \sin q),\\ \frac{dq}{dt} &= p - \frac{h^2}{12} p \cos q.\end{aligned}$$

There are no $O(h^3)$ terms: the b coefficients corresponding to trees of order 4 vanish, a consequence of the symmetry of the midpoint rule [**14**]. The modified system is the Hamiltonian system with Hamiltonian function

$$\tilde{H} = \frac{1}{2}p^2 + (1 - \cos q) + \frac{h^2}{48}(-2p^2 \cos q + \cos 2q - 1).$$

The Hamiltonian character of the modifed system is linked to the symplecticness of the midpoint rule [**24**], [**25**], [**14**]. The modified solution $\tilde{u}$ conserves $\tilde{H}$ exactly and hence, Taylor expanding,

$$\begin{aligned}0 = \tilde{H}(\tilde{u}(T_0)) - \tilde{H}(u_0) &= (\tilde{g}_0, \tilde{u}(T_0) - u_0) + O(|\tilde{u}(T_0) - u_0|^2)\\ &= (\tilde{g}_0, \tilde{u}(T_0) - u_0) + O(h^4). \qquad (18)\end{aligned}$$

Here $\tilde{g}_0$ is the gradient of $\tilde{H}$ at the initial point u_0 and we have used that

$$\tilde{u}(T_0) - u_0 = \tilde{u}(T_0) - u(T_0) = O(h^2), \qquad (19)$$

due to the periodicity of the true solution and to the fact that the true and modified vector fields differ in $O(h^2)$ terms. From (18)–(19), along with $\tilde{g}_0 - g_0 = O(h^2)$, we obtain

$$(g_0, \tilde{u}(T_0) - u_0) = O(h^4), \qquad (20)$$

and finally, since $\tilde{u}$ and the numerical solution differ in $O(h^4)$ terms,

$$(g_0, e_1(h)) = O(h^4). \qquad (21)$$

We take this to (17) and conclude that for the midpoint rule

$$e_M(h) = c_0 M h^2 f_0 + O(h^4), \quad h \to 0,$$

where c_0 is a constant depending on the initial condition, but independent of M and h. This is illustrated in Figure 3, where the initial condition is $p_0 = 0$, $q_0 = \pi/3$ and $h = T_0/400$. The dash-dot line gives the Euclidean norm of the error as a function of M. The linear growth is clear.

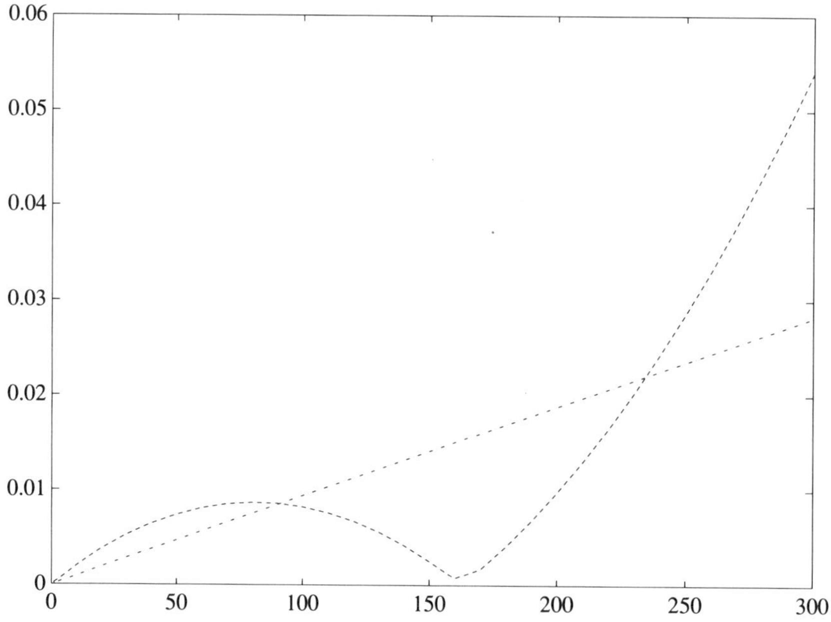

FIGURE 3. Euclidean norm of the error against time measured in periods. The dash-dot line corresponds to the midpoint rule and the dash line to Runge's method.

We next use the method (4). The modified equations with $N = 3$ turn out to be

$$\begin{aligned}\frac{dp}{dt} &= -\sin q - \frac{h^2}{24}(p^2 \sin q + 2\sin 2q),\\ \frac{dq}{dt} &= p + \frac{h^2}{6} p\cos q,\end{aligned}$$

and for $N = 4$ we find

$$\begin{aligned}\frac{dp}{dt} &= -\sin q - \frac{h^2}{24}(p^2 \sin q + 2\sin 2q) + \frac{h^3}{8} p\cos^2 q,\\ \frac{dq}{dt} &= p + \frac{h^2}{6} p\cos q + \frac{h^3}{16}(p^2 \sin q + \sin 2q).\end{aligned}$$

The modified system of order $N = 3$ is not Hamiltonian but has the reversibility property of being invariant under the change of p into $-p$ and t into $-t$. It has the following invariant of motion

$$I = \frac{p^2}{2\,(1 + (h^2/6)\cos q)^{\frac{1}{2}}} + \int_0^q \frac{\sin\xi + (h^2/3)\sin 2\xi}{(1 + (h^2/6)\cos\xi)^{\frac{3}{2}}}\,d\xi.$$

This quantity plays now the role played by $\tilde{H}$ in the midpoint rule analysis, so that (20) still holds. However since we are dealing with a modified system of order 3 we only conclude that

$$(g_0, e_1(h)) = O(h^3), \tag{22}$$

(rather than (21)), and (17) implies

$$e_M(h) = C_0 M f_0 h^2 + O(h^3), \quad h \to 0.$$

To expliciy obtain the h^3 term in the asymptotic expansion of $e_M(h)$ we have to resort to the modified problem with $N = 4$. Now, this is a system with negative dissipation. Straightforward differentiation reveals that along its solutions the $O(h^3)$ quantity dI/dt remains positive for $|q| \leq \pi/2$, leading to $O(h^3)$ outward spiraling. From here we see that the inner product in (22) is actually of size $O(h^3)$ and no better. Then

$$e_M(h) = C_0 M f_0 h^2 + (D_0 M^2 + E_0 M) f_0 h^3 + F_0 M g_0 h^3 + O(h^4), \quad h \to 0,$$

where C_0, D_0, E_0 and F_0 are real constants independent of h and M, with C_0, D_0 and F_0 different from 0.

The dotted line in Figure 3 gives the actual error norm for the value $h = T_0/400$ used before. For M large, $M^2 h^3$ dominates over Mh^2 and what we see is the quadratic growth of the $O(h^3)$ terms in the expansion. For M small (say less than 20) $M^2 h^3$ is negligible relatively to the leading Mh^2 term and we see linear growth. We infer from the figure that for M near 160 the Mh^2 and $M^2 h^3$ contributions are of equal size and cancel each other.

References

1. F. Alouges and A. Debussche, *On the qualitative behavior of the orbits of a parabolic differential equation and its discretization in the neighborhood of a hyperbolic fixed point,* Numer. Funct. Anal and Optimiz. **12** (1991), 253–269.
2. D. V. Anosov, *Geodesic flows and closed Riemannian manifolds with negative curvature,* Proc. Stelov Inst. Math. **90** (1967).
3. W.-J. Beyn, *On the numerical approximation of phase portraits near stationary points,* SIAM J. Numer. Anal. **24** (1987), 1095–1113.
4. ______, *Numerical methods for dynamical systems,* Advances in Numerical Analysis, Vol. I (W. Light ed.), Clarendon, Oxford, 1991, pp. 175–236.
5. R. Bowen, *ω-limit sets for axiom A-diffeomorphisms,* J. Diff. Eq. **18** (1975), 333–339.
6. J. C. Butcher, *The Numerical Analysis of Ordinary Differential Equations,* "John Wiley", Chichester, 1987.
7. M. P. Calvo and J. M. Sanz-Serna, *The development of variable-step symplectic integrators, with application to the two-body problem,* SIAM J. Sci. Comput. **14** (1993), 936–952.
8. S. N. Chow and K. J. Palmer, *On the numerical computation of orbits of dynamical systems: the one-dimensional case,* Dynamics and Diff. Eqn. **3** (1991), 361–380.
9. S. N. Chow and E. S. Van Vleck, *A shadowing lemma approach to global error analysis for initial value ODEs,* Report No. 92-12, Department of Mathematics and Statistics, Simon Fraser University.
10. T. Eirola, *Aspects of backward error analysis of numerical ODEs,* J. Comput. Appl. Math. **45** (1993), 65–73.

11. J. de Frutos and J. M. Sanz-Serna, *Error growth and invariant quantities in numerical methods: a case study,* Report 1993/4, May 1993, Applied Mathematics and Computation Reports, Universidad de Valladolid.
12. C. Grebogi, S. M. Hammel, J. A. Yorke and T. Sauer, *Shadowing of physical trajectories in chaotic dynamics: containment and refinement,* Phys. Rev. Lett. **65** (1990), 1527–1530.
13. D. F. Griffiths and J. M. Sanz-Serna, *On the scope of the method of modified equations,* SIAM J. Sci. Comput. **7** (1986), 994–1008.
14. E. Hairer, *Backward analysis of numerical integrators and symplectic methods* (submitted).
15. E. Hairer, S. P. Nørsett and G. Wanner, *Solving Ordinary Differential Equations I, Nonstiff Problems,* "Springer", Berlin, 1987.
16. E. Hairer and G. Wanner, *On the Butcher group and general multivalue methods,* Computing **13** (1974), 1–15
17. J. K. Hale, *Dynamics and numerics,* The Dynamics of Numerics and the Numerics of Dynamics (D. S. Broomhead and A. Iserles, eds.), Clarendon, Oxford, 1992, pp. 243–253.
18. S. M. Hammel, J. A. Yorke and C. Grebogi, *Do numerical orbits of chaotic dynamical processes represent true orbits?,* J. Complexity **3** (1987), 136–145.
19. ———, *Numerical orbits of chaotic processes represent true orbits,* Bull. Am. Math. Soc. **19** (1988), 465–469.
20. S. Larsson and J. M. Sanz-Serna, *The behavior of finite element solutions of semilinear parabolic problems near stationary points,* SIAM J. Numer. Anal. (to appear).
21. Ch. Lubich, K. Nipp and D. Stoffer, *Runge-Kutta solutions of stiff differential equations near stationary points,* Report No. 93-01, April 1993, Seminar für Angewandte Mathematik, Eidgenössische Technische Hochschule, Zürich.
22. H. E. Nusse and J. A. Yorke, *Is every approximate trajectory of some process near an exact trajectory of a nearby process?,* Commun. Math. Phys. **114** (1988), 381–347.
23. J. M. Sanz-Serna, *Numerical ordinary differential equations vs. dynamical systems,* The Dynamics of Numerics and the Numerics of Dynamics (D. S. Broomhead and A. Iserles, eds.), Clarendon, Oxford, 1992, pp. 81–106.
24. ———, *Symplectic integrators for Hamiltonian problems: an overview,* Acta Numerica **1** (1992), 243–286.
25. J. M. Sanz-Serna and M. P. Calvo, *Numerical Hamiltonian problems,* "Chapman and Hall", London, 1993.
26. J. M. Sanz-Serna and S. Larsson, *Shadows, chaos and saddles,* Appl. Numer. Math. (to appear).
27. T. Sauer and J. A. Yorke, *Rigorous verification of trajectories for the computer simulation of dynamical systems,* Nonlinearity **4** (1991), 961–979.
28. R. F. Warming and B. J. Hyett, *The modified equation approach to the stability and accuracy of finite difference methods,* J. Comput. Phys. **14** (1974), 159–179.

Departamento de Matemática Aplicada y Computación, Universidad de Valladolid, Valladolid, Spain
E-mail address: maripaz@cpd.uva.es

Facultad de Informática, Apartado 649, 20080 San Sebastián, Spain
E-mail address: ccpmuura@sisb00.si.ehu.es

Departamento de Matemática Aplicada y Computación, Universidad de Valladolid, Valladolid, Spain
E-mail address: sanzserna@cpd.uva.es

Contemporary Mathematics
Volume **172**, 1994

THE DYNAMICS OF SOME ITERATIVE IMPLICIT SCHEMES

H.C. Yee
NASA Ames Research Center, Moffett Field, CA 94035, USA
and
P.K. Sweby
University of Reading, Whiteknights, Reading RG6 2AX, England

Abstract

The global asymptotic nonlinear behavior of some standard iterative procedures in solving nonlinear systems of algebraic equations arising from four implicit linear multistep methods (LMMs) in discretizing 2×2 systems of first-order autonomous nonlinear ordinary differential equations is analyzed using the theory of dynamical systems. With the aid of parallel Connection Machines (CM-2 and CM-5), the associated bifurcation diagrams as a function of the time step, and the complex behavior of the associated "numerical basins of attraction" of these iterative implicit schemes are revealed and compared. Studies showed that all of the four implicit LMMs exhibit a drastic distortion and segmentation but less shrinkage of the basin of attraction of the true solution than standard explicit methods. The numerical basins of attraction of a noniterative implicit procedure mimic more closely the basins of attraction of the differential equations than the iterative implicit procedures for the four implicit LMMs.

1. Relevance and Objective

It has been shown recently by the authors, and others [1-11] that the dynamics of the numerical discretizations of nonlinear differential equations (DEs) can differ significantly from that of the original DEs themselves. For example, the discretizations can possess

1991 *Mathematics Subject Classification.* Primary 65D30, 39A11, 76N99.

spurious steady-state solutions and spurious asymptotes which do not satisfy the original DEs. These spurious numerical solutions may be stable or unstable and may occur both below and above the linearized stability limit of the numerical scheme (on the time step for the equilibrium or asymptote of the DE). In [2,3,8], we showed how "numerical" basins of attraction can complement the bifurcation diagrams in gaining more detailed global asymptotic numerical solution behaviors for nonlinear DEs. We showed how, in the presence of spurious asymptotes the basins of the true stable steady states can be segmented by the basins of the spurious stable and unstable asymptotes. One major consequence of this phenomena which is not commonly known is that this spurious behavior can result in a dramatic distortion, and in most cases, a dramatic shrinkage and segmentation of the basin of attraction of the true solution for finite time steps. Such distortion, shrinkage and segmentation of the numerical basins of attraction will occur regardless of the stability of the spurious asymptotes.

We use the term "spurious asymptotic numerical solutions" to mean asymptotic solutions that satisfy the discretized counterparts but do not satisfy the underlying ordinary differential equations (ODEs) or partial differential equations (PDEs). Asymptotic solutions here include steady-state solutions, periodic solutions, limit cycles, chaos and strange attractors. Here the basin of attraction is a domain of a set of initial conditions whose solution curves (trajectories) all approach the same asymptotic state. Also we use the term "exact" and "numerical" basins of attraction to distinguish "basins of attraction of the underlying DEs" and "basins of attraction of the discretized counterparts".

Studies in [1-11] are particularly important for computational fluid dynamics (CFD), since it is a common practice in CFD computations to use a time-dependent approach to obtain steady-state numerical solutions of complicated steady fluid flows which often consist of stiff nonlinear PDEs of mixed type. When a time-dependent approach is used to obtain steady-state numerical solutions of a fluid flow or a steady PDE, a boundary value problem is transformed into an initial-boundary value problem with unknown initial data. If the steady PDE is strongly nonlinear and/or contains stiff nonlinear source terms, phenomena such as slow convergence, non-convergence or spurious steady-state numerical solutions commonly occur even though the time step is well below the linearized stability limit and the initial data are physically relevant.

It is also a common practice in CFD to use implicit methods to solve stiff problems. Such schemes, however, introduce the added difficulty of solving the implicit equations (nonlinear algebraic equations) in order to obtain the solution at the next time level. Various options such as linearization or iteration are available for this purpose. In [2] we included a study on the dynamics of a noniterative linearized version of the implicit Euler and trapezoidal methods. In this work we generalize our earlier work [2] to include iterative solution procedures of the implicit discretized equations, namely simple iteration and full and modified Newton iteration. In addition we also study the 3-level backward differentiation formula (BDF) and a midpoint implicit method (one-legged cousin of the trapezoidal method) with these four methods (noniterative vs iterative) of solving the resulting nonlinear algebraic equations applied to it. Comparison of the above combination of methods with the "straight" Newton method in solving the steady part of the equations is also performed. We use the term "straight Newton" to distinguish it from the combination of "implicit LMM + Newton" type of solution procedure. Some study on variable time step control in avoiding spurious asymptotes will also be addressed.

Since all four implicit methods under consideration are linear multistep methods

(LMMs) they will not exhibit spurious steady states (fixed points of order one). However, as discussed in [2] and as we shall see in later sections, some implicit LMMs have the property of increasing the stability range for the stable fixed points of the ODE [11], accompanied in some instances by the stabilization of unstable fixed points of the ODE [2]. One consequence of the stabilization of unstable fixed points is a distortion, shrinkage and/or segmentation of the resulting numerical basin of attraction. In addition the method of solution of the implicit equations generated by these schemes can itself contribute to the dynamics of the discretization since different numerical methods and/or solution procedures result in entirely different nonlinear discrete maps. Iserles [10] and Dieci and Estep [11] were the first to examine some of the stability issues. Our attempt here is to address issues that were not investigated in [10,11]. Our main purpose is to study the global asymptotic behavior in terms of bifurcation diagrams and numerical basins of attraction of these four procedures for solving nonlinear systems of algebraic equations arising from implicit LMM discretizations.

2. Model Nonlinear First-Order Autonomous ODEs

Consider a 2×2 system of first-order autonomous nonlinear ODEs of the form

$$\dot{U} = S(U), \tag{2.1}$$

where U and S are vector functions of dimension 2, and $S(U)$ is nonlinear in U. Three of the four 2×2 systems of nonlinear first-order autonomous model ODEs considered in [2] are considered here. As before, we do not treat any system parameter present in the DEs as a bifurcation parameter, but instead keep it constant throughout each numerical calculation so that only the discretization parameters come into play. The systems considered with $U^T = (u, v)$ or $z = u + iv$ are a

(1) **Dissipative complex** model

$$\dot{z} = z(i + \epsilon - |z|^2), \tag{2.2}$$

(2) **Predator-Prey** model

$$\begin{aligned} \dot{u} &= -3u + 4u^2 - 0.5uv - u^3 \\ \dot{v} &= -2.1v + uv, \end{aligned} \tag{2.3}$$

(3) **Perturbed Hamiltonian system** model

$$\begin{aligned} \dot{u} &= \epsilon(1 - 3u) + 3\left[1 - 2u + u^2 - 2v(1 - u)\right]/4 \\ \dot{v} &= \epsilon(1 - 3v) - 3\left[1 - 2v + v^2 - 2u(1 - v)\right]/4. \end{aligned} \tag{2.4}$$

Here ϵ is the system parameter for (2.2) and (2.4).

The perturbed Hamiltonian model can be related to the numerical solution of the viscous Burgers' equation with no source term

$$(2.5) \qquad \frac{\partial u}{\partial t} + \frac{1}{2}\frac{\partial (u^2)}{\partial x} = \beta \frac{\partial^2 u}{\partial x^2} \qquad \beta > 0.$$

Let $u_j(t)$ represent an approximation to $u(x_j, t)$ of (2.5) where $x_j = j\Delta x$, $j = 1, ..., J$, with Δx the uniform grid spacing. Consider the three-point central difference spatial discretization with periodic condition $u_{J+j} = u_j$, and assume $\sum_{j=1}^{J} u_j =$ constant, which implies that $\sum_{j=1}^{J} \frac{du_j}{dt} = 0$. If we take $J = 3$ and $\Delta x = 1/3$, then with $\epsilon = 9\beta$ this system can be reduced to the 2×2 system of first-order nonlinear autonomous ODEs (2.4) with $U^T = (u_1, u_2) = (u, v)$. In this case the nonlinear convection term is contributing to the nonlinearity of the ODE system (2.4).

Fixed Points of (2.2)-(2.4): The dissipative complex model has a unique fixed point at (u, v)=(0,0) for $\epsilon \leq 0$. The fixed point is a stable spiral if $\epsilon < 0$ and a center if $\epsilon = 0$. For $\epsilon > 0$, the fixed point (0,0) becomes unstable with the birth of a stable limit cycle with radius equal to $\sqrt{\epsilon}$ centered at (0,0). Here the entire (u, v) plane belongs to the basins of attraction of the stable fixed point (0,0) if $\epsilon < 0$. On the other hand, if $\epsilon > 0$, the entire (u, v) plane except the unstable fixed point (0,0) belongs to the basins of attraction of the stable limit cycle. The predator-prey model (2.4) has four fixed points of which (0,0) is a stable node, (2.1,1.98) is a stable spiral, and (1,0) and (3,0) are saddles. The perturbed Hamiltonian (semi discrete system of the viscous Burgers' equation with three-point central difference in space) has four steady-state solutions of which three are saddles and one is a stable spiral at (1/3, 1/3) for $\epsilon > 0$. For $\epsilon = 0$ the stable spiral becomes a center.

3. Numerical Methods

The four LMMs for (2.1) with the time step Δt considered are

(1) **Implicit Euler method**

$$(3.1) \qquad U^{n+1} = U^n + \Delta t S^{n+1},$$

(2) **Trapezoidal method**

$$(3.2) \qquad U^{n+1} = U^n + \frac{1}{2}\Delta t(S^n + S^{n+1}),$$

(3) **3-level backward differentiation formula (BDF)**

$$(3.3) \qquad U^{n+1} = U^n + \frac{2}{3}\Delta t S^{n+1} + \frac{1}{3}(U^n - U^{n-1}),$$

(4) **Mid-point implicit method (one-legged trapezoidal)**

$$(3.4) \qquad U^{n+1} = U^n + \Delta t S\left[\frac{1}{2}(U^{n+1} + U^n)\right].$$

Performing standard perturbation analysis on the above equations at the fixed points $\bar{U}$ of the ODE system by writing $U^n = \bar{U} + \delta^n$ and discarding terms of $O(\delta^2)$ yields

$$\delta^{n+1} = K(\bar{U})\delta^n \tag{3.5}$$

where the matrix $K(\bar{U})$ is defined implicitly for the 3-level BDF method. The stability of the perturbation is governed by the eigenvalues μ of the matrix $K(\bar{U})$. For the implicit Euler method

$$K(\bar{U}) = \left[I - \Delta t J(\bar{U})\right]^{-1}, \tag{3.6}$$

where $J(\bar{U})$ is the Jacobian dS/dU evaluated at the fixed point, while for the trapezoidal and midpoint implicit methods

$$K(\bar{U}) = \left[I - \frac{1}{2}\Delta t J(\bar{U})\right]^{-1}\left[I + \frac{1}{2}\Delta t J(\bar{U})\right], \tag{3.7}$$

and for the 3-level BDF method we have the relationship

$$\left[I - \frac{2}{3}\Delta t J(\bar{U})\right] K(\bar{U})^2 - \frac{4}{3}K(\bar{U}) + \frac{1}{3}I = 0. \tag{3.8}$$

Thus, if the eigenvalues of $J(\bar{U})$ are λ, we have μ of $K(\bar{U})$ as follows

(3.9a) Implicit Euler $\quad \mu = \dfrac{1}{1 - \Delta t\lambda}$

(3.9b) Trapezoidal & Midpoint Implicit $\quad \mu = \dfrac{2 + \Delta t\lambda}{2 - \Delta t\lambda}$

(3.9c) 3-level BDF $\quad \mu = \dfrac{1}{2 \mp \sqrt{1 + 2\Delta t\lambda}}$

with both matrices sharing the same eigenvectors. (Note that since the expression (3.8) is a quadratic in $K(\bar{U})$ there are four possible modes corresponding to the different eigenvalues λ and the two possible signs of the square root. However, it is the modes taking the negative square root which have larger modulus and therefore govern stability of the perturbation.)

The stability of the corresponding fixed points based on the eigenvalues of $K(\bar{U})$ can be determined exactly and were used to check our numerical computations later. When numerically computing the full discretized equations there are various options which can be used to solve the equations (3.1), (3.2) or (3.3). We consider here linearization, simple iteration, Newton iteration and modified Newton iteration.

Linearization (a noniterative procedure [2]) is achieved by expanding S^{n+1} as $S^n + J(U^n)(U^{n+1} - U^n)$. **Simple Iteration** is the process in which where, given a scheme of the form $U^{n+1} = G(U^n, U^{n+1})$, we perform the iteration

$$U^{n+1}_{(\nu+1)} = G(U^n, U^{n+1}_{(\nu)}) \tag{3.10}$$

where $U_0^{n+1} = U^n$ and "(ν)" indicates the iteration index. The iteration is continued either until some tolerance between iterates is achieved or a limiting number of iterations has been performed. In all of our computations the tolerance "tol" is set as $||U_{(\nu)}^{n+1} - U_{(\nu-1)}^{n+1}|| \le \text{tol}$ and the maximum number of iterations is 15. The major drawback with simple iteration is that for guaranteed convergence the iteration must be a contraction, i.e.

$$||G(U^n, V) - G(U^n, W)|| \le \alpha ||V - W|| \tag{3.11}$$

where $\alpha < 1$. Whether or not the iteration is a contraction at the fixed points will influence the stability of that fixed point, over riding the stability of the implicit scheme. Away from the fixed points the influence will be on the basins of attraction. For the four LMMs **at the fixed points** this translates as follows:

(3.12a) Implicit Euler $\Delta t||J(\bar{U})|| \le \alpha < 1$

(3.12b) Trapezoidal & Midpoint Implicit $\frac{1}{2}\Delta t||J(\bar{U})|| \le \alpha < 1$

(3.12c) 3-level BDF $\frac{2}{3}\Delta t||J(\bar{U})|| \le \alpha < 1$

As we shall see later, our numerical results illustrate this limitation well.

Newton Iteration for the implicit schemes is of the form

$$U_{(\nu+1)}^{n+1} = U_{(\nu)}^{n+1} - F'(U^n, U_{(\nu)}^{n+1})^{-1} F(U^n, U_{(\nu)}^{n+1}) \tag{3.13}$$

where $U_0^{n+1} = U^n$. The differentiation is with respect to the second argument and the scheme has been written in the form (for two-level schemes) $F(U^n, U^{n+1}) = 0$. **Modified Newton iteration** is the same as (3.13) except it uses a frozen Jacobian $F'(U^n, U^n)$. The same tolerance and maximum number of iterations used for the simple iteration are also used for the Newton and modified Newton iterations. In all of the computations, the starting scheme for the 3-level BDF is the linearized implicit Euler.

We also considered two variable time step control methods. The first one is "implicit Euler + Newton iteration with local truncation error control" [11]

(3.14a)

$$U_{(0)}^{n+1} = U^n$$

$$U_{(\nu+1)}^{n+1} = U_{(\nu)}^{n+1} + \left[I - \Delta t^n J(U_{(\nu)}^{n+1})\right]^{-1} \left[U_{(\nu)}^{n+1} - U^n - \Delta t^n S(U_{(\nu)}^{n+1})\right] \quad \nu = 1, \ldots$$

with

$$\Delta t^n = 0.9 \Delta t^{n-1} \{\text{tol}_1 / ||U^n - U^{n-1} - \Delta t^{n-1} S(U^n)||\}^{1/2} \tag{3.14b}$$

where the $(n+1)$th step is rejected if $||U^n - U^{n-1} - \Delta t^{n-1} S(U^n)|| > 2\text{tol}_1$ In this case, we set $\Delta t^{n-1} = \Delta t^n$. The value "$\text{tol}_1$" is a prescribed tolerance and the norm is an infinity norm. The second one is the popular "ode23" method

$$k_1 = S(U^n)$$
$$k_2 = S(U^n + \Delta t^n k_1)$$
$$k_3 = S(U^n + \Delta t^n (k_1 + k_2)/4)$$
$$U^{n+1} = U^n + \Delta t^n (k_1 + k_2 + 4k_3)/6$$
$$\Delta U^{n+1} = \Delta t^n (k_1 + k_2 - 2k_3)/3 \tag{3.15a}$$

with

$$\Delta t^n = 0.9 \Delta t^{n-1} \sqrt{\frac{\mathrm{tol}_1}{||\Delta U^n||}} \tag{3.15b}$$

where the $(n+1)$th step is rejected if $||\Delta U^{n+1}|| > \mathrm{tol}_1 \max\{1, ||U^{n+1}||\}$. In that case, we set $\Delta t^{n-1} = \Delta t^n$. Again, "$\mathrm{tol}_1$" is a prescribed tolerance and the norms are infinity norms. We also employ Newton's method in solving the solutions of $S(U) = 0$ which is the one-step Newton iteration of the implicit Euler method of (3.13).

4. Numerical Results

Although we purposely selected the model equations with known analytical solutions, depending on the scheme, the dynamics of their discretized counterparts are very difficult and might not be possible to analyze analytically. Only some analysis is possible for the lower order schemes. Part of the global asymptotic numerical solution behavior can be obtained by the pseudo arclength continuation method devised by Keller [12], a standard numerical method for obtaining bifurcation curves in bifurcation analysis. Besides not being able to provide the numerical basins of attraction, one deficiency of the pseudo arclength continuation method is that for problems with complicated bifurcation patterns it cannot provide the complete bifurcation diagram without a known solution for each of the main bifurcation branches. For spurious asymptotes it is usually not easy to locate even just one solution on each of these branches. For the majority of the cases where rigorous analysis is impractical, we utilized numerical experiments. See Sweby and Yee [6] for some of the analysis. Also, analytical representations (except in isolated cases) for numerical basins of attraction rarely exist for nonlinear DEs. Methods such as generalized cell mapping [13] can provide an efficient approach to locating these basins, but might not be exact. Here our aim is to numerically compute the basins of attraction as accurately as possible and in the most straightforward way in order to illustrate the key points.

Due to the complicated nature of the these discrete maps, analysis without a supercomputer is nearly impossible. The nature of our calculations requires thousands of iterations of the same equation with different ranges of initial data on a preselected (u, v) domain and ranges of the discretized parameter space Δt. The NASA Ames CM-2 and CM-5 allow vast numbers (typically 65,536) of calculations to be performed in parallel. Thus each processor could represent a single initial datum and thereby all the computations can be done in parallel to produce detailed global stability behavior and the resulting basins of attraction. With the aid of the CM-2 and CM-5, we were able to detect a wealth of the detailed nonlinear behavior of these schemes which would have been overlooked had isolated initial data been chosen on the Cray-YMP or other serial or vector machine.

Two different representations of the numerical basins of attraction are computed on the NASA Ames CM-2 and CM-5. One is bifurcation diagrams as a function of Δt with numerical basins of attraction superimposed on a constant $v-$ or u-plane. The other is the numerical basins of attraction with the stable asymptotes superimposed on the phase plane (u, v) with selected values of Δt.

To obtain a bifurcation diagram with numerical basins of attraction superimposed, the preselected domain of initial data on a constant $v-$ or u-plane and the preselected range of the Δt parameter are divided into 512 equal increments. For the bifurcation part of the computations, with each initial datum and Δt, the discretized equations are preiterated 5,000 - 9,000 steps before the next 6,000 iterations (more or less depending on the problem and scheme) are plotted. The preiterations are necessary in order for the solutions to settle to their asymptotic value. A high number of iterations are overlaid on the same plot in order to detect periodic orbits or invariant sets. The reader is reminded that with this method of computing the bifurcation diagrams, only the stable branches are obtained. While computing the bifurcation diagrams it is possible to overlay basins of attraction for each value of Δt used. For the basins of attraction part of the computations with each value of Δt used, we keep track of where each initial datum asymptotically approaches and color code them (appearing as a vertical strip) according to the individual asymptotes. While efforts were made to match color coding of adjacent strips on the bifurcation diagram, it was not always practical or possible. Care must therefore be taken when interpreting these overlays.

For the basins of attraction on the phase plane (u, v) with selected values of Δt and the stable asymptotes superimposed, the (u, v) domain is divided into 512×512 points of initial datum. With each initial datum and Δt, we preiterate the respective discretized equation 5,000 - 9,000 steps and plot the next 6,000 steps to produce the asymptotes (fixed points of various order and limit cycles). Again, for the basins of attraction part of the computations, for each value of Δt used, we keep track of where each initial datum asymptotically approaches and color code them according to the individual asymptotes. Details of the techniques used for detection of asymptotes and basins of attraction are given in the appendix of Sweby and Yee [6]. Note that in all of the plots, if color printing is not available, the different shades of grey represent different colors.

Due to space limitation, only selected results for the dissipative complex equation (2.2) for both representations of numerical basins of attraction are shown in Figs. 1 - 10. An expanded version of the paper which includes models (2.3) and (2.4) is reported in [15]. In the plots, $r = \Delta t$. The "r=aDt" label denotes a scaling parameter "a" (set to unity for calculations presented here) times the time step Δt. White dots and white curves on the basins of attraction with bifurcation diagrams superimposed represent the bifurcation curves. White dots and white closed curves on the basins of attraction with the numerical asymptotes superimposed represent the stable fixed points, stable periodic solutions or stable limit cycles. The black regions represent divergent solutions.

Note that the streaks on some of plots are either due to the non-settling of the solutions within the prescribed number of iterations or the existence of small isolated spurious asymptotes. Due to the high cost of computation, no further attempts were made to refine their detailed behavior since our purpose was to show how, in general, the different numerical methods behave in the context of nonlinear dynamics. From our numerical studies, the midpoint implicit method (linearized or iterative methods) behaves the same as or very similarly to the trapezoidal method for the studied model. Thus, no figures will be shown of the midpoint implicit method.

4.1. Numerical Results for the Dissipative Complex Equation

Figures 1 - 10 show selected results on the two representations of numerical basins of attraction diagrams for model (2.2) for $\epsilon = 1$. Recall that the exact solution for (2.2) is a stable limit cycle with radius equal to 1 centered at (0,0). The entire (u, v) plane except the unstable fixed point (0,0) belongs to the basins of attraction of the stable limit cycle. The four LMMs using the four solution procedures exhibit spurious stable and unstable asymptotes except spurious steady states. It is fascinating to see the dramatic difference in shapes and sizes of numerical basins of attraction for the different methods and solution procedure combinations compared with the exact basin of attraction. The evolution of the numerical basin of attraction as Δt changes is very traumatic for all four LMMs and for the same LMM with different solution procedures. For larger Δt, the linearized implicit Euler, "implicit Euler + Newton", and "straight Newton" give the same numerical basins of attraction. Simple iteration behaves similarly to typical explicit methods (in terms of stability and the size of numerical basin of attraction) for all of the four studied implicit methods and for models (2.3) and (2.4) as well. The main advantage of the simple iteration procedure over standard explicit methods is that spurious steady states cannot occur.

Figures 1 - 3 show selected results of the bifurcation diagram as a function of Δt with numerical basins of attraction superimposed. They compare the noniterative (linearized) vs. iterative methods in solving the nonlinear algebraic equations using implicit Euler, trapezoidal and 3-level BDF methods. Figures 4 - 10 show selected results of the stable numerical asymptotes with basins of attraction superimposed using four different Δt by these three implicit LMMs. The red regions are the numerical basins of attraction for the stable limit cycle except in Fig. 5, $\Delta t = 2.65$, in Fig. 6, $\Delta t = 1.5, 2, 4$, in Fig. 8, $\Delta t = 1.5$, in Fig. 9, $\Delta t = 2, 2.5, 4$, and in Fig. 10, $\Delta t = 1.515, 2.5$. The green regions shown in Figs. 1, 3-5 are the numerical basins of attraction for the stabilized fixed point (0,0). Note how the implicit method turns the unstable fixed point (0,0) of the ODE system into a stable one for $\Delta t \geq 1$. For $\Delta t \geq 1$, the structure of the numerical basins of attraction are so different for the same LMM with different solution procedures.

Figures 1, 3-6, 8-10 illustrate the situation where unconditionally stable LMM schemes such as the implicit Euler and 3-level BDF methods can converge to a wrong solution if one picks the initial data inside the green region which are valid physical initial data for the ODE. Thus even though LMMs preserved the same number of fixed points as the underlying ODE, these fixed points can change type and stability. This phenomenon is related to the "non-robustness" of implicit methods sometimes experienced in CFD computations. In these types of computations where the initial data are not known, the highest probability of avoiding spurious asymptotes is achieved when a fraction of the allowable linearized stability limit of Δt is employed.

To aid the understanding of some of the results shown on Figs. 8-10, the following gives some explanation on how to interpret the basins of attraction diagrams with the stable numerical asymptotes superimposed. All of the selected time steps Δt shown in Figs. 4-10 are based on Figs. 1-3 where the bifurcation diagrams with the basins of attraction are superimposed. These time steps were chosen to illustrate selected features of the different bifurcation phenomena on the (u, v) plane.

For example, it is easier to understand Fig. 8, $\Delta t = 1.5$ using "trapezoidal + Modified

Newton'' if we look at the fourth plot in Fig. 2. Fig. 2 shows that the original limit cycle bifurcates into a ''period 2 type'' limit cycle near $\Delta t = 1.25$. Figure 2 shows distinctively that both rings share the same basin (only one distinct solid red basin of attraction). The lack of a solid basin of attraction in the third plot of Fig. 8 is due to the coloring algorithm, which requires a repetition of the limit cycle within the time of integration in order to distinguish basins of attraction. When this repetition is not present the resulting coloring gives a crude indication of trajectories. Additional preiteration steps (many more than 9,000) would likely alleviate the problem. Note that we need all of the trajectories corresponding to the 512×512 initial data to settled to within the prescribed preiterations before a solid basin results. The fourth plot of Fig. 8 together with Fig. 2 hints at a rapid period doubling transition to instability.

Fig. 9 for $\Delta t = 2$ illustrates failure of the coloring algorithm to detect the basin of attraction due to an insufficient number of preiterations. The result is again a crude indication of the trajectories. Even though not correctly colored the non-black region gives the size of the basin. Also the third plot of Fig. 3 gives a clear indication of the size of the corresponding numerical basin of attraction for $\Delta t = 2$.

Figure 10 for $\Delta t = 1.515$ illustrates the surprising presence of an embedded region of instability within the basin of attraction of the limit cycle (see also Fig. 3). The lack of a distinct (single red colored) basin is again an artifact of the coloring algorithm. Additional preiteration steps would likely alleviate this.

Figures 1-10 also illustrate the unreliability of trying to compute a true limit cycle with any sizable Δt. This should not be surprising since the scheme only gives an $O(\Delta t^p)$ approximation to the solution trajectories. In addition, since the limit cycle is not a fixed point, we would expect inaccuracies to be introduced. However, inaccuracies are not easy to detect in practice, especially when a numerical solution produces the qualitative features expected. Overall, the trapezoidal method and the midpoint implicit method give a more accurate solution for the limit cycle.

Our studies indicate that variable time step control implicit and explicit methods (3.14) and (3.15) can alleviate spurious dynamics most of the time. However, the allowable time step, determined by (3.14) or (3.15), is too small for practical usage, especially for the explicit method (3.15). For the implicit method (3.14) the allowable time step, determined by (3.14), is slightly larger than the explicit method (3.15) but it is still impractical to use. With the wrong combination of starting time step, initial data and tolerance value, spurious dynamics could occasionally be produced.

4.2. General Discussion of Numerical Results

Studies showed that all of the four implicit LMMs exhibit a drastic distortion but less shrinkage of the basin of attraction of the true solution than standard explicit methods studied in [2]. The numerical basins of attraction of a noniterative implicit procedure mimics more closely the basins of attraction of the DEs than the iterative implicit procedures for the four implicit LMMs. In general the numerical basins of attraction bear no resemblance to the exact basins of attraction. The size can increase, decrease and/or fragmented depending on the time step. Also the possible existence of the largest numerical basin of attraction that is larger than the exact one does not occur when the time step is the smallest. The dynamics of numerics of the implicit methods differ significantly from each other, and the different methods of solving the resulting nonlinear algebraic equations are very different from each other since different numerical methods and solution procedures result in entirely different nonlinear discrete maps. Although unconditionally stable implicit methods allow theoretically large Δt, the numerical basins of attraction (allowable initial data) for large Δt sometimes are so fragmented and/or so small that the safe (or practical) choice of Δt is slightly larger or comparable to the stability limit of standard explicit methods. In general, if one uses a Δt that is a fraction of the stability limit, one has a higher chance of convergence to the correct asymptote.

One of the causes of the above behavior is the existence of stable and unstable spurious asymptotes other than steady states which have a similar detrimental (in terms of robustness) effect as explicit methods. Another cause of the observed behavior is due to the fact that an unstable fixed point can become a stable fixed point and can change type e.g., from a saddle to a stable or unstable node. One consequence of this behavior is that the flow pattern can change topology as the discretized parameter is varied. Thus even though LMMs preserve the same number (but not the same types) of fixed points as the underlying DEs, the numerical basins of attraction of LMMs do not coincide with the exact basins of attraction of the DEs even for small Δt. Some of the dynamics of the LMMs observed in our study can be used to explain the root of why one cannot achieve the theoretical linearized stability limit of the typical implicit LMMs in practice when solving strongly nonlinear DEs e.g., in CFD.

Straight Newton vs. Other Studied Methods: Studies indicated that contrary to popular belief, the initial data using the straight Newton method may not have to be close to the exact solution for convergence. Straight Newton also exhibit stable and unstable spurious asymptotes. Initial data can be reasonably removed from the asymptotic values and still be in the basin of attraction. However the basins can be fragmented even though the corresponding exact basins of attraction are single closed domains. The cause of non-convergence may just as readily be due to the fact that the numerical basins of attraction are fragmented. In many cases, the results obtained are better than those obtained by the trapezoidal and 3-level BDF methods (regardless of the three iterative procedures). If one use a time step slightly bigger than the stability limit of standard explicit methods for the four LMMs, straight Newton might have similar or better performance. In fact, using a large Δt by the linearized implicit Euler method or the implicit Euler + Newton procedure has the same chance of obtaining the correct steady state as the Newton method if the initial data are not known or arbitrary initial data is taken.

References

1. H.C. Yee, P.K. Sweby and D.F. Griffiths, "Dynamical Approach Study of Spurious Steady-State Numerical Solutions for Nonlinear Differential Equations, Part I: The Dynamics of Time Discretizations and Its Implications for Algorithm Development in Computational Fluid Dynamics," NASA TM-102820, April 1990, also J. Comput. Phys., Vol. 97, 1991, pp. 249-310.

2. H.C. Yee and P.K. Sweby, "Dynamical Approach Study of Spurious Steady-State Numerical Solutions for Nonlinear Differential Equations, Part II, Global Asymptotic Behavior of Time Discretizations RNR Technical Report RNR-92-008, March 1992, to appear, Intern. J. CFD.

3. A. Lafon and H.C. Yee, "Dynamical Approach Study of Spurious Steady-State

Numerical Solutions for Nonlinear Differential Equations, Part III: The Effects of Nonlinear Source Terms and Boundary Conditions in Reaction-Convection Equations,'' NASA TM-103877, July 1991.

4. A. Lafon and H.C. Yee, ''Dynamical Approach Study of Spurious Steady-State Numerical Solutions of Nonlinear Differential Equations, Part IV: Stability vs. Numerical Treatment of Nonlinear Source Terms,'' ONERA-CERT Technical Report DERAT 45/5005.38, Feb. 1992.

5. D.F. Griffiths, P.K. Sweby, P.K. and H.C. Yee, ''On Spurious Asymptotic Numerical Solutions of Explicit Runge-Kutta Methods,'' IMA J. in Numerical Analysis, Vol. 12, 1992, pp. 319-338.

6. P.K. Sweby and H.C. Yee, ''On Spurious Asymptotic Numerical Solutions of 2×2 Systems of ODEs,'' Numerical Analysis Report 7/91, October 1991, University of Reading, England.

7. Griffiths, D.F, Stuart, A.M., Yee, H.C.: ''Numerical Wave Propagation in Hyperbolic Problems with Nonlinear Source Terms,'' SIAM J. of Numerical Analysis, 1992.

8. H.C. Yee, P.K. Sweby, P.K. and A. Lafon, ''Basins of Attraction and the Time-Dependent Approach to Obtaining Steady-State Numerical Solutions,'' Proceedings of the ICFD Conference on Numerical Methods for Fluid Dynamics, April 7-10, 1992, Reading, England.

9. A.R. Mitchell and D.F. Griffiths, ''Beyond the Linearized Stability Limit in Non Linear Problems,'' Report NA/88 July 1985, Department of Mathematical Sciences, University of Dundee, Scotland U.K.

10. A. Iserles, ''Stability and Dynamics of Numerical Methods for Nonlinear Ordinary differential Equations,'' DAMTP NA1, 1988, University of Cambridge, Cambridge England.

11. L. Dieci and D. Estep, ''Some Stability Aspects of Schemes for the Adaptive Integration of Stiff Initial Value Problems,'' School of Math. Report, 1990, Georgia Institute of Technology.

12. H.B. Keller, ''Numerical Solution of Bifurcation and Nonlinear Eigenvalue Problems,'' in *Applications of Bifurcation Theory*, P.H. Rabinowitz, ed., Academic Press, 1977, pp. 359-384.

13. H. Flashner and R.S. Guttalu, ''A Computational Approach for Studying Domains of Attraction for Non-Linear Systems,'' Int. J. Non-Linear Mechanics, Vol. 23, No. 4, 1988, pp. 279-295.

14. J. Guckenheimer and P. Holmes, *Nonlinear Oscillations, Dynamical Systems, and Bifurcations of Vector Fields*, Springer-Verlag, New York, 1983.

15. H.C. Yee and P.K. Sweby, ''Global Asymptotic Behavior of Iterative Implicit Schemes,'' RIACS Report, Nov. 1993.

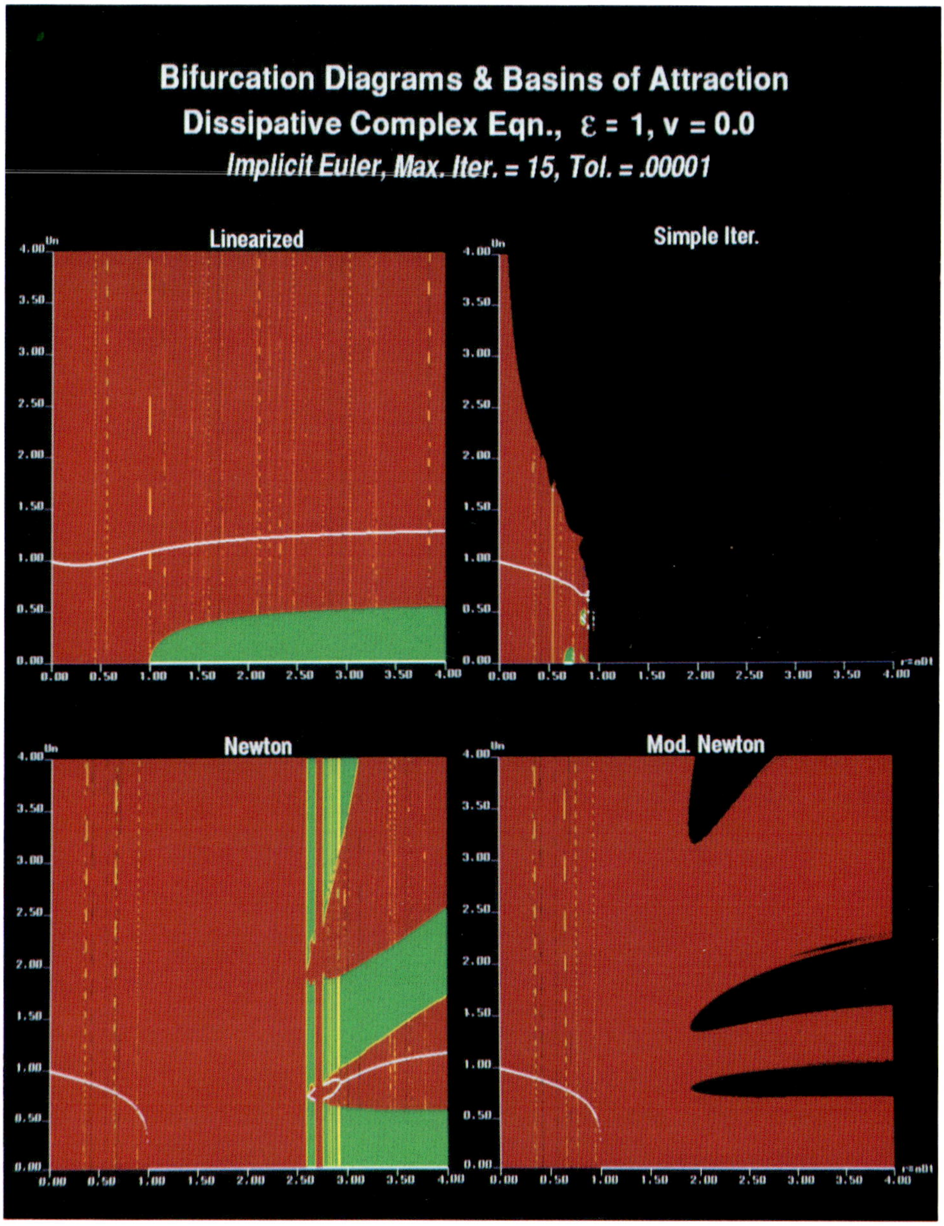
Bifurcation Diagrams & Basins of Attraction
Dissipative Complex Eqn., ε = 1, v = 0.0
Implicit Euler, Max. Iter. = 15, Tol. = .00001
Linearized
Simple Iter.
Newton
Mod. Newton

Figure 1.

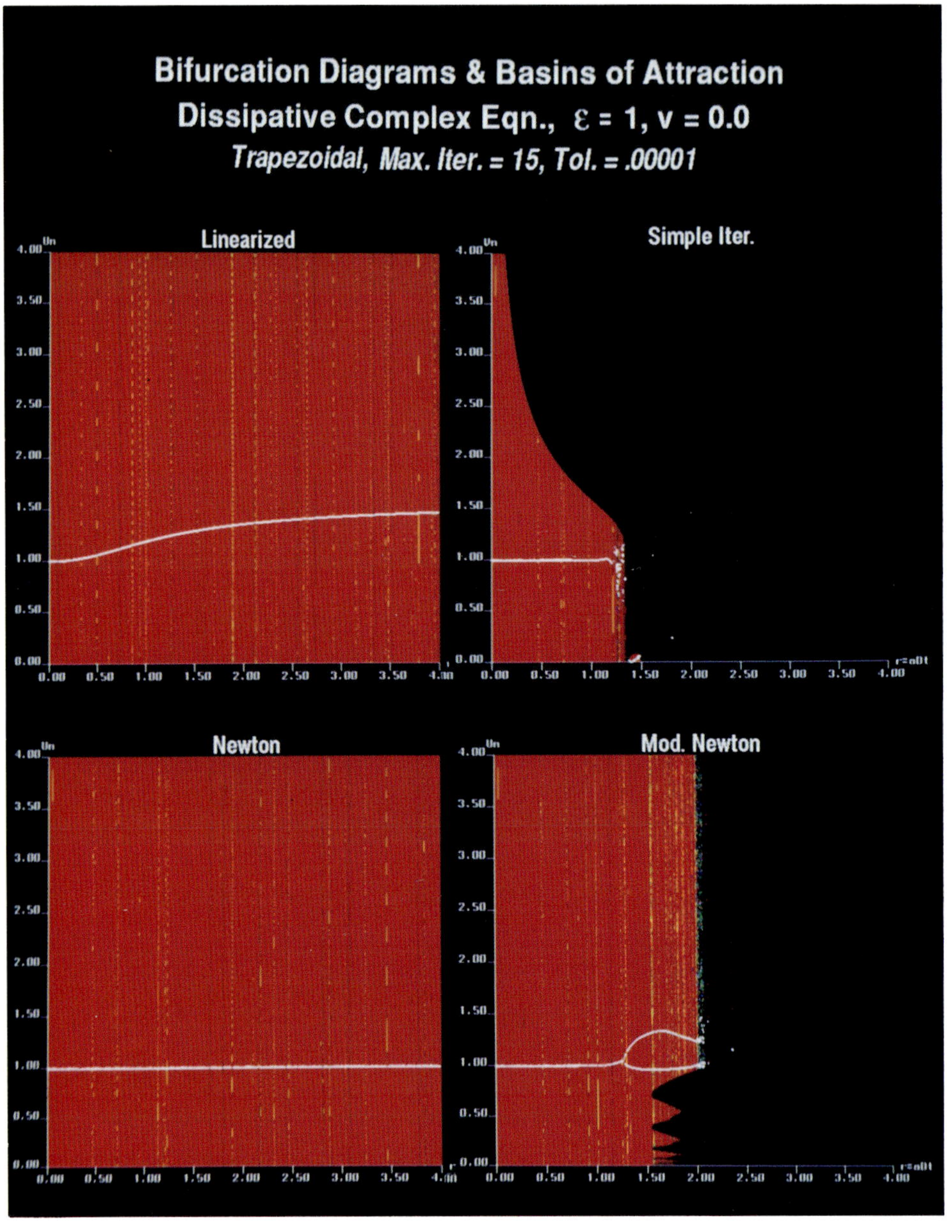
Bifurcation Diagrams & Basins of Attraction
Dissipative Complex Eqn., ε = 1, v = 0.0
Trapezoidal, Max. Iter. = 15, Tol. = .00001
Linearized
Simple Iter.
Newton
Mod. Newton

Figure 2.

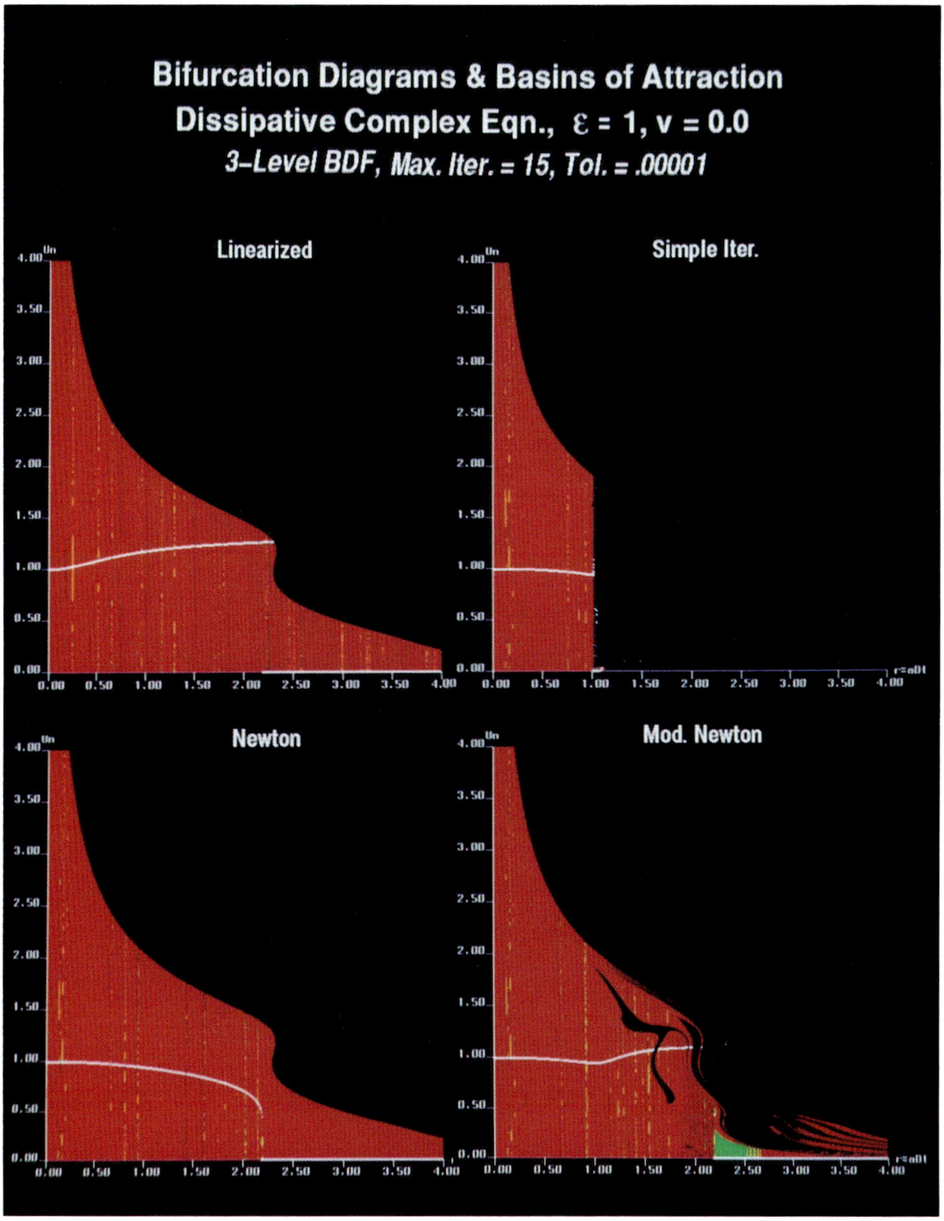
Bifurcation Diagrams & Basins of Attraction
Dissipative Complex Eqn., ε = 1, ν = 0.0
3-Level BDF, Max. Iter. = 15, Tol. = .00001
Linearized
Simple Iter.
Newton
Mod. Newton
Un
4.00
3.50
3.00
2.50
2.00
1.50
1.00
0.50
0.00
0.00 0.50 1.00 1.50 2.00 2.50 3.00 3.50 4.00

Figure 3.

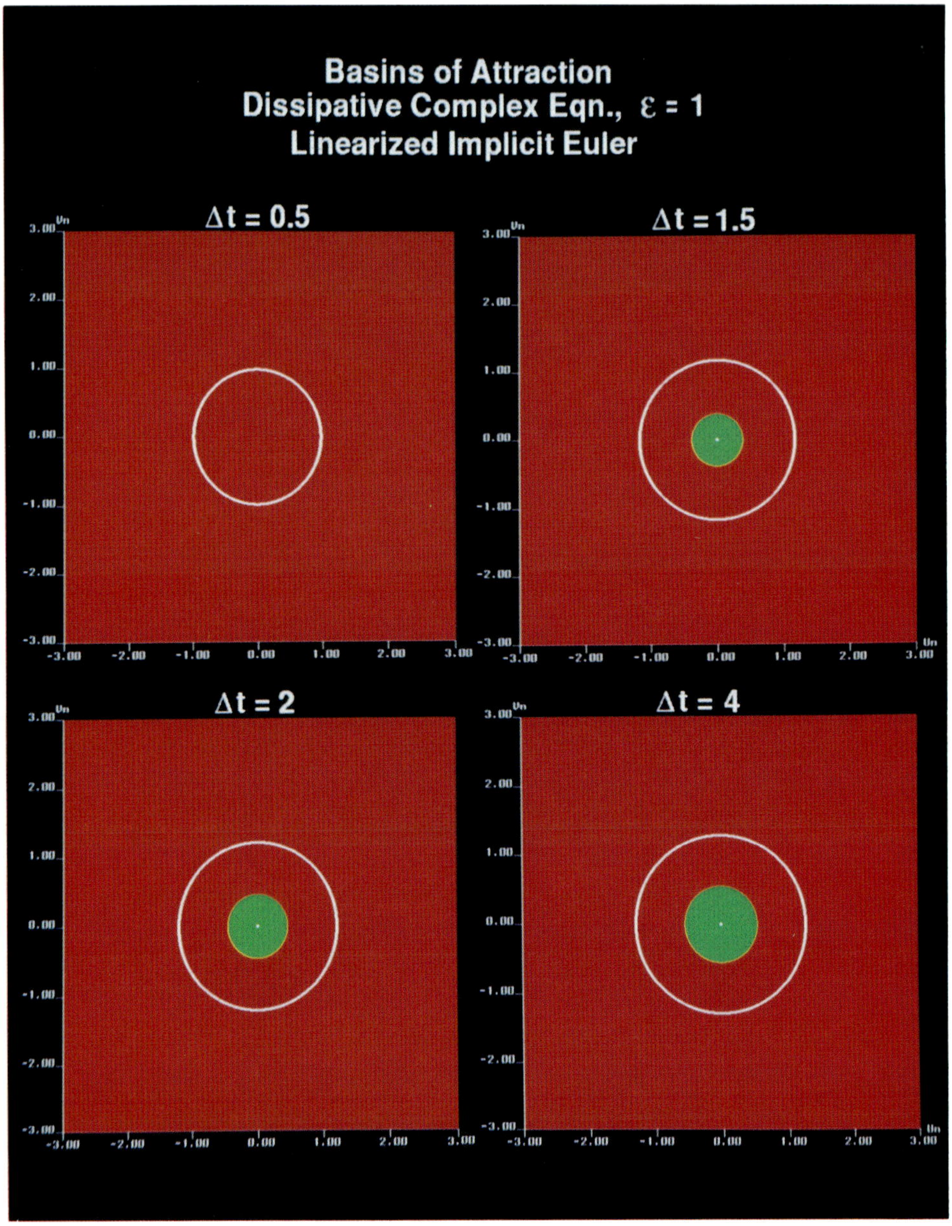
Basins of Attraction
Dissipative Complex Eqn., ε = 1
Linearized Implicit Euler
Δt = 0.5
Δt = 1.5
Δt = 2
Δt = 4

Figure 4.

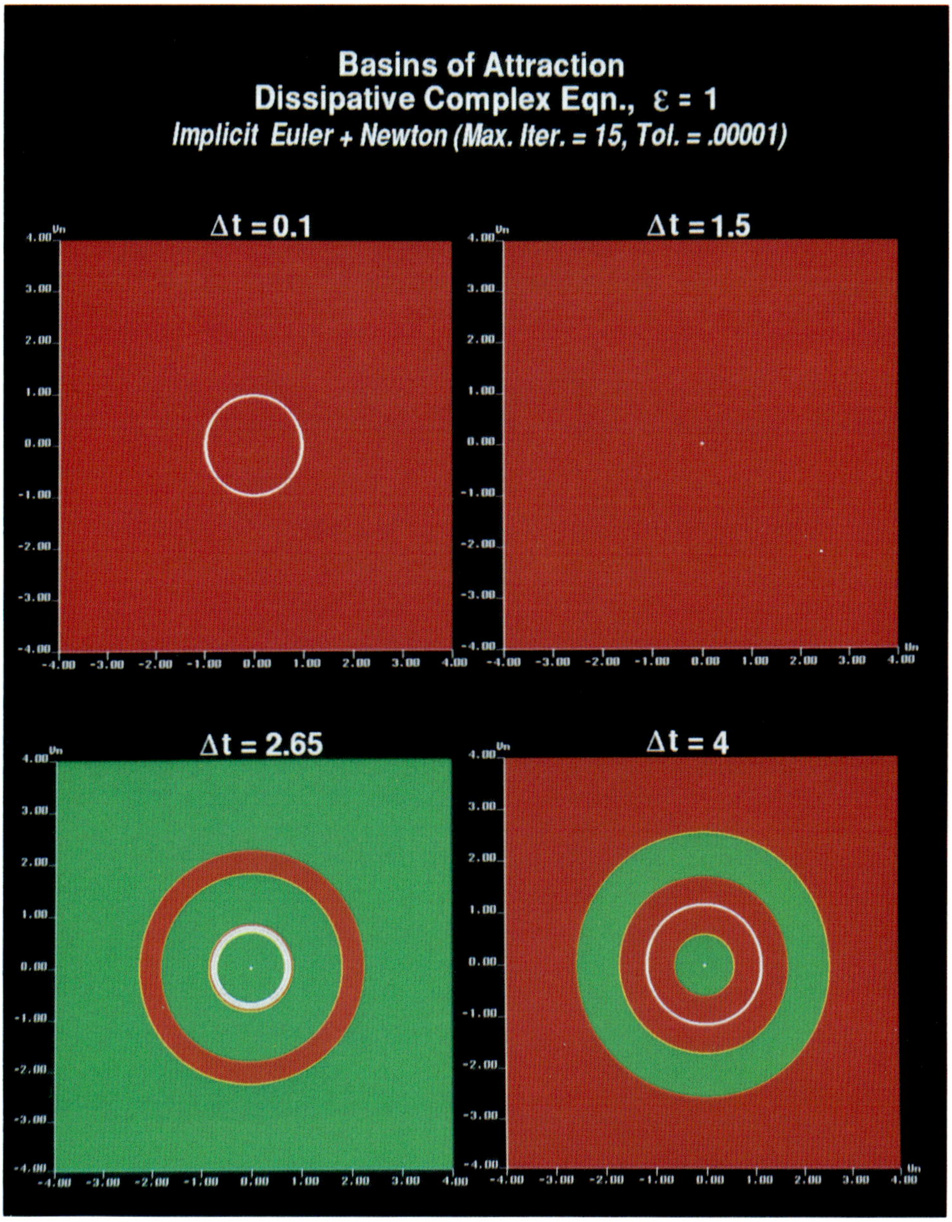
Basins of Attraction
Dissipative Complex Eqn., ε = 1
Implicit Euler + Newton (Max. Iter. = 15, Tol. = .00001)
Δt = 0.1
Δt = 1.5
Δt = 2.65
Δt = 4

Figure 5.

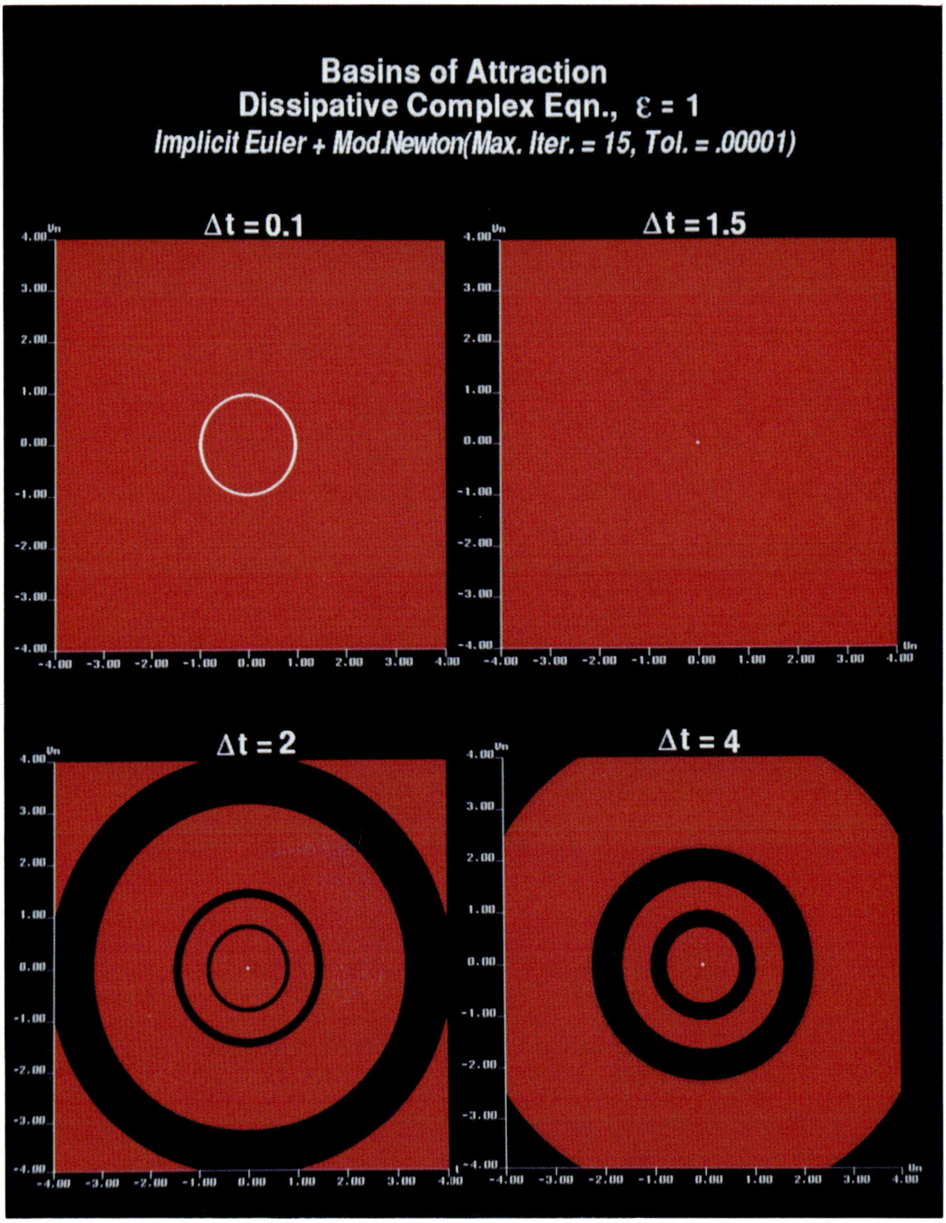
Basins of Attraction
Dissipative Complex Eqn., ε = 1
Implicit Euler + Mod.Newton(Max. Iter. = 15, Tol. = .00001)
Δt = 0.1
Δt = 1.5
Δt = 2
Δt = 4

Figure 6.

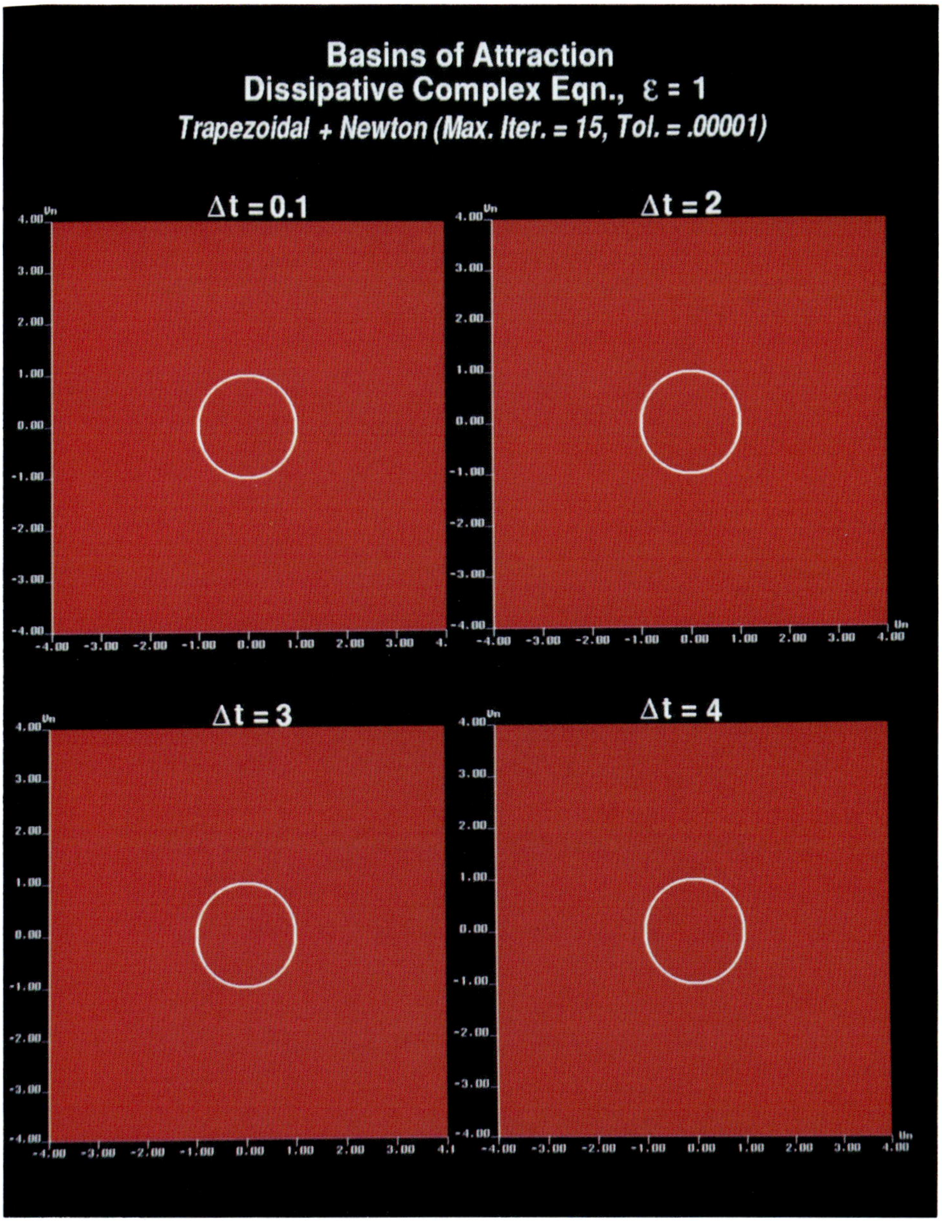
Basins of Attraction
Dissipative Complex Eqn., ε = 1
Trapezoidal + Newton (Max. Iter. = 15, Tol. = .00001)
Δt = 0.1
Δt = 2
Δt = 3
Δt = 4

Figure 7.

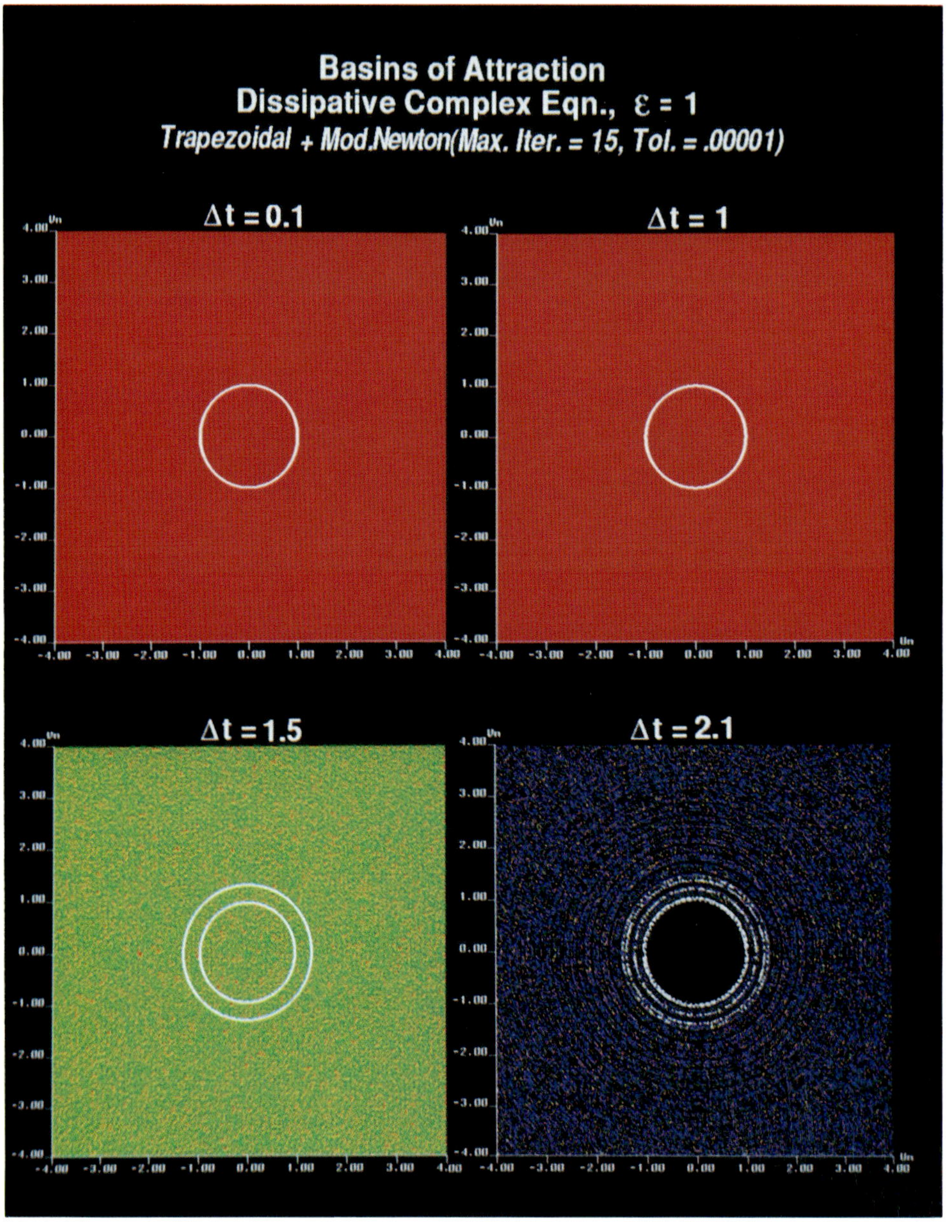
Basins of Attraction
Dissipative Complex Eqn., ε = 1
Trapezoidal + Mod.Newton(Max. Iter. = 15, Tol. = .00001)
Δt = 0.1
Δt = 1
Δt = 1.5
Δt = 2.1
4.00
3.00
2.00
1.00
0.00
-1.00
-2.00
-3.00
-4.00

Figure 8.

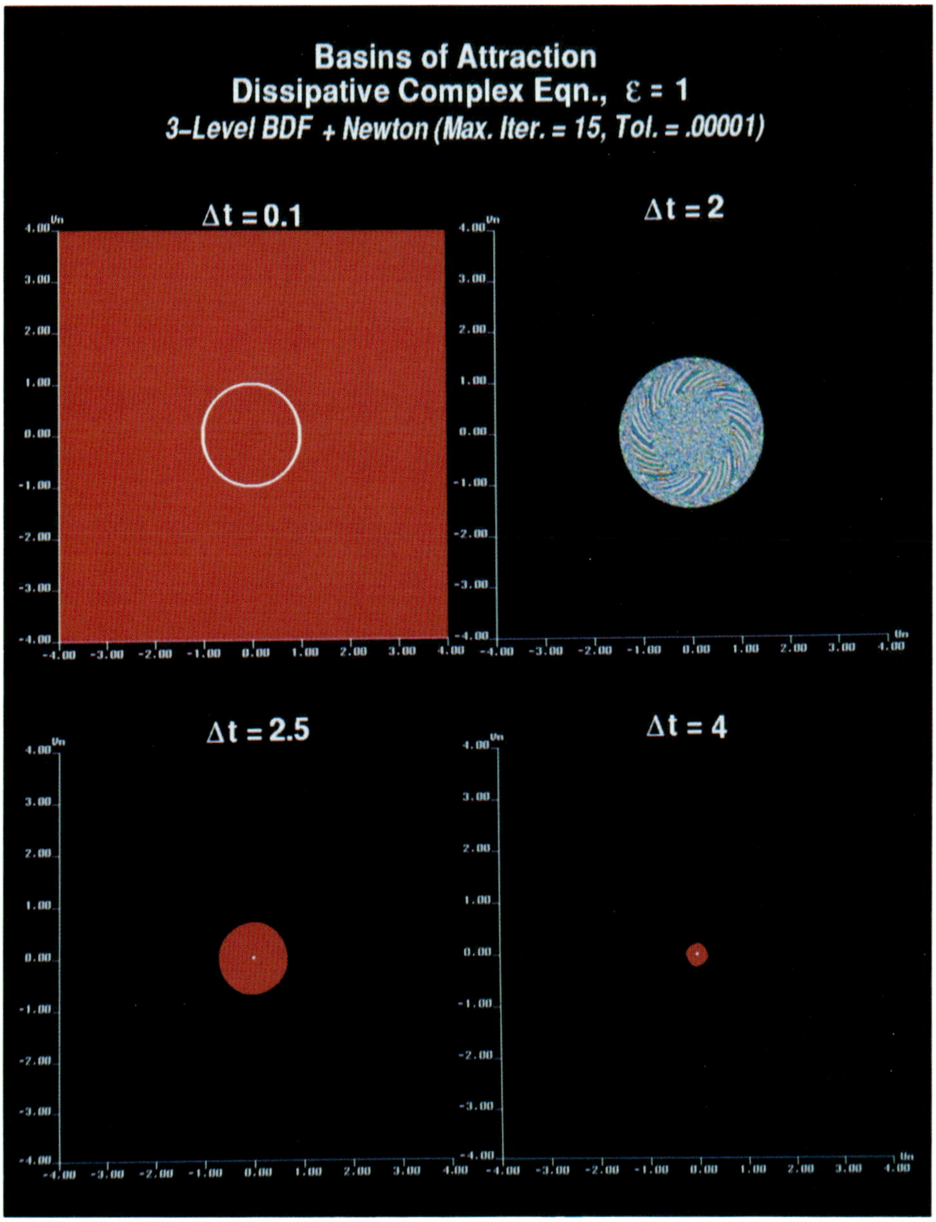
Basins of Attraction
Dissipative Complex Eqn., ε = 1
3-Level BDF + Newton (Max. Iter. = 15, Tol. = .00001)
Δt = 0.1
Δt = 2
Δt = 2.5
Δt = 4

Figure 9.

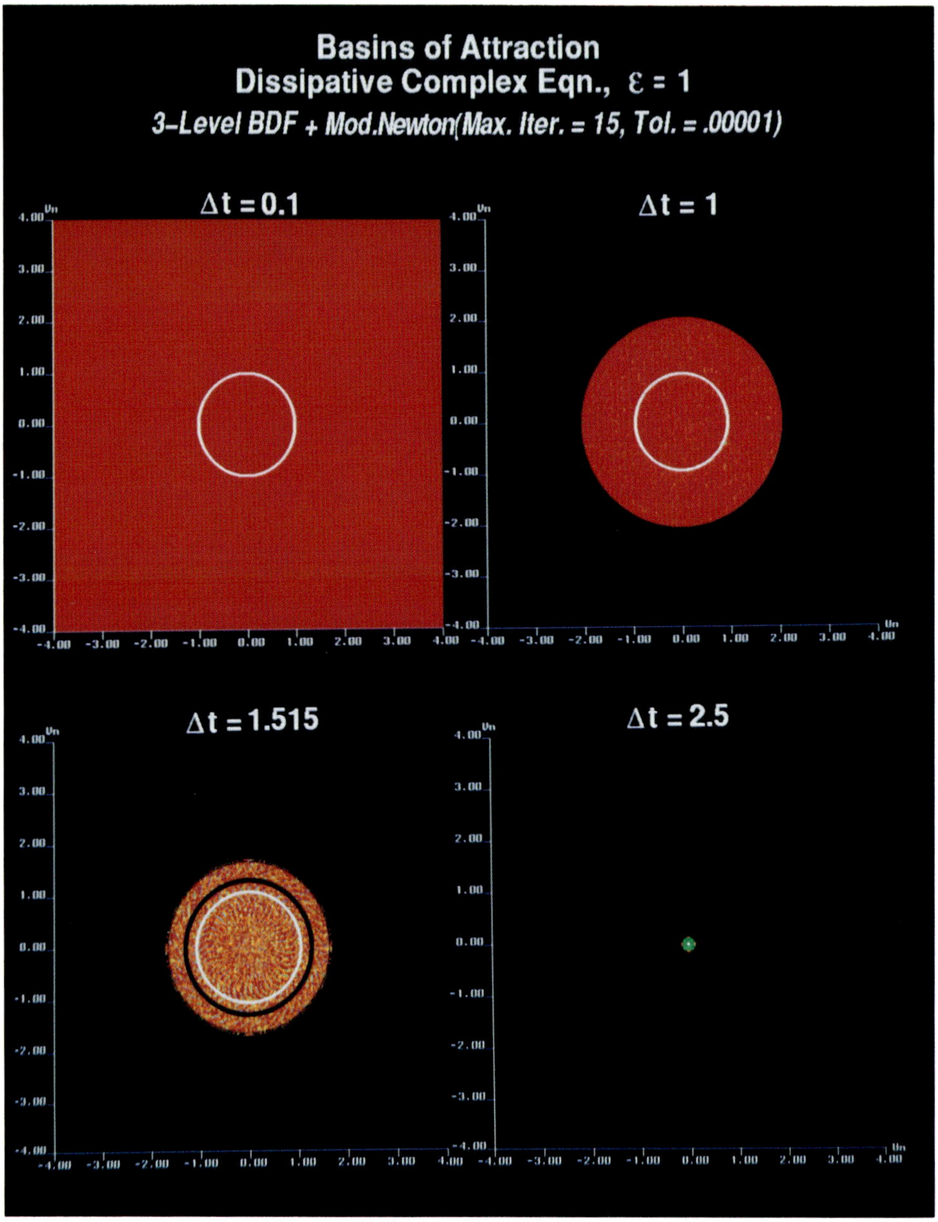
Basins of Attraction
Dissipative Complex Eqn., ε = 1
3-Level BDF + Mod.Newton(Max. Iter. = 15, Tol. = .00001)
Δt = 0.1
Δt = 1
Δt = 1.515
Δt = 2.5

Figure 10.

Contemporary Mathematics
Volume **172**, 1994

Shadowing of Lattice Maps

SHUI-NEE CHOW AND ERIK S. VAN VLECK

ABSTRACT. The shadowing lemma provides an alternative means of characterizing global errors. In this paper we study the shadowing properties of lattice maps, typically discretizations of partial differential equations. Theoretical results are presented that emphasize the relationship between exponential dichotomy and the shadowing property. A simple algorithm for determining the shadowing distance numerically is applied to the logistic map, the Henon map, a discretization of Burgers' equation and a discretization of the Korteweg-de Vries equation.

1. Introduction

Modeling of physical phenomena in space-time is important in the study of nonlinear behavior. Lattice maps and lattice ODEs have been proposed as models in gas dynamics, fluid dynamics, solid-state physics, optics, chemical reaction with diffusion, and biology. Lattice maps that correspond to Navier-Stokes equations have been proposed in [**17**]. These maps allow for convective coupling, diffusion-type spatial average, and cut off for high velocity. A lattice ODE has been proposed in [**12**] that is the discrete counterpart of the continuum model known as the Cahn-Hilliard equation. This lattice ODE is a solid-solution model for inhomogeneous systems based on a correction to Fick's second law for chemical diffusion. The Navier-Stokes and Cahn-Hilliard equations are characteristic of partial differential equations whose discretization exhibit nearest neighbor type interactions. In this paper we consider the time evolution of one-dimensional lattice maps of the form

$$x_{n+1}(i) = f_n(x_n(i+k), ..., x_n(i), ..., x_n(i-k))$$

1991 *Mathematics Subject Classification.* Primary 58F13, 58F15; Secondary 65G05, 65L70.

This work was supported in part by a grant from Darpa/NIST. The work of S.N.C. was supported in part under NSF Grant #DMS-9005420. The work of E.S.V.V. was supported in part under NSERC Grant #OGP0121873.

This paper is in final form and no version of it will be submitted for publication elsewhere.

with appropriate boundary conditions.

The errors associated with computing orbits of lattice maps are due to round off error as opposed to discretization error. A small local error may multiply and incur a large global error so that the computed orbit corresponds to an extraneous or ghost solution. It is important to determine if the computed solution is a spurious solution or if it corresponds to some true solution that could be computed given an infinite precision computer. It is possible that the computed orbit is uniformly "close" to some true orbit with slightly different initial data. To determine if the computed orbit corresponds to a true orbit we employ the concept of shadowing.

Shadowing is a technique for providing a qualitative global error analysis of computer generated orbits. If every δ-pseudo orbit $x = \{x_n\}_0^T$ of a sequence of maps $\{f_n\}_0^{T-1}$ has a true orbit $y = \{y_n\}_0^T$ with $||x - y|| \leq \epsilon$ for some $\epsilon > 0$, then $\{f_n\}_0^{T-1}$ is said to have the ϵ-shadowing property; i.e. given an orbit x generated by a sequence of maps with a local error uniformly less then δ, then the sequence of maps is said to have the ϵ-shadowing property, if there exists an orbit y generated by the same sequence of maps uniformly "close" to x. Shadowing has been applied to finding the distance from a computed orbit to some true orbit for explicitly defined maps in one, two, and three dimensions (see [**7,8**] and [**15,16**]). Theoretical shadowing for mappings appear in [**1**], [**3**], [**5,6**] and [**9**] while theoretical shadowing results for differential equations appear in [**2**], [**4**], [**5**], [**10**] and [**14**]. Numerical shadowing results have been obtained for ordinary differential equations in the papers [**2**], [**10**], [**20**] and [**21**]. In this paper we apply the shadowing concept to discretizations of partial differential equation in higher dimensions.

Specifically, we consider discretizations of Burgers' equation with Dirichlet boundary conditions, and the Korteweg-de Vries equation with periodic boundary conditions. Burgers' Equation

$$v_t = a(x,t)v_{xx} + b(x,t)vv_x \tag{1.1}$$

is a streamline diffusion method to a model problem for compressible fluid flow. For Burgers' equation we consider functions such that $a(x,t) > 0$, and $b(x,t) < 0$. The Korteweg-de Vries equation is a third-order nonlinear equation

$$v_t = a(x,t)v_{xxx} + b(x,t)vv_x \tag{1.2}$$

first encountered in the study of water waves with $a(x,t) < 0$ and $b(x,t) > 0$.

This paper is organized as follows. In section 2 a proof of the shadowing lemma is presented. Section 3 contains a perturbation theorem for exponential dichotomy and some applications to lattice maps and is independent of what follows in sections 4-6. An introduction to iterative boundary value problems with separated boundary conditions and a solution method via decoupling using the QR decomposition is presented in section 4. Section 5 contains an algorithm for finding the shadowing distance ϵ, and in section 6 numerical results are presented. Section 7 contains our conclusions, and sections 8-10 contain a review of error analysis for standard vector/matrix operations, the computation of orbits and jacobians, and the QR decomposition, respectively.

2. The Shadowing Lemma

Let $G \equiv \mathbb{R}^{N(T+1)}$ and $H \equiv \mathbb{R}^{NT}$ where N and T are positive integers. Define the functional equation $F : G \to H$ by

$$(F(x))_n = x_{n+1} - f_n(x_n)$$

for $n = 0, ..., T-1$. If $x \in G$, then x is a δ-pseudo orbit of a sequence of maps $\{f_n\}_0^{T-1}$, $f_n : \mathbb{R}^N \to \mathbb{R}^N$, if and only if

$$||F(x)|| \leq \delta$$

and $y \in G$ is a true orbit of $\{f_n\}_0^{T-1}$ if and only if

$$||F(y)|| = 0$$

where $||F(z)|| = \sup_n ||(F(z))_n||_\infty$. If the maps $\{f_n\}_0^{T-1}$ are C^1 then F is C^1 and for $u \in G$ we have

$$(DF(x)u)_n = u_{n+1} - Df_n(x_n)u_n \tag{2.1}$$

for $n = 0, ..., T-1$. Since $DF(x)$ is onto, it has a right inverse. Assume $DF(x)^\dagger$ is a right inverse of $DF(x)$. To find a true orbit in a neighborhood of a pseudo orbit we employ an approximate implicit function theorem.

THEOREM 2.1. *Let G, H be Banach spaces and $F : G \to H$ a C^1 map. Let x be a point in G such that $DF(x)^\dagger$ exists and let $\epsilon_0 > 0$ be chosen so that*

$$||DF(x) - DF(y)|| \leq 1/(2||DF(x)^\dagger||) \tag{2.2}$$

for $||x - y|| \leq \epsilon_0$. If $||F(x)|| \leq \delta$ for some $\delta > 0$ and

$$\epsilon \equiv 2||DF(x)^\dagger||\delta \tag{2.3}$$

is such that $0 < \epsilon \leq \epsilon_0$, then the equation $F(y) = 0$ has a solution y such that $||x - y|| \leq \epsilon$.

PROOF. (see [**5**]) $\square$

To find the shadowing distance ϵ we must determine $||DF(x)^\dagger||$. If, for each n, $J_n = Df_n(x_n)$ is invertible, then the linear difference equation

$$u_{n+1} = J_n u_n$$

is said to have an *exponential dichotomy* if there exist projection-valued functions P_n and constants, $K \geq 1$, $1 > \lambda \geq 0$, such that

$$P_{n+1}J_n = J_nP_n, \forall n,$$
$$||({}^{n-1}J^m)P_m|| \leq K\lambda^{(n-m)}, n > m,$$
$$||({}^{m-1}J^n)^{-1}(I - P_m)|| \leq K\lambda^{(m-n)}, m > n.$$

where $({}^iJ^j) = J_i \cdots J_j$ for $i \geq j$. The following lemma illustrates how the dichotomy constants K, λ can be used to bound $||DF(x)^\dagger||$.

LEMMA 2.2. *Suppose $DF(x)$ has an exponential dichotomy with constants K, λ. Then $DF(x)$ has a right inverse such that*

$$||DF(x)^\dagger|| \leq 2K(1 - \lambda)^{-1}.$$

PROOF. Let $J_n = Df_n(x_n)$ and let $h = \{h_n\}_0^\infty \in l^\infty(\mathbb{N})$. Set

$$u_n = \sum_{m=1}^{n-1}({}^{n-1}J^m)P_mh_{m-1} + P_nh_{n-1} - \sum_{m=n+1}^{\infty}({}^{n-1}J^m)^{-1}(I - P_m)h_{m-1}$$

Then for $n - 1 \geq m$ we have,

$$||({}^{n-1}J^m)P_mh_m|| \leq K\lambda^{(n-m)}||h||,$$

and for $m - 1 \geq n$ we have,

$$||({}^{m-1}J^n)^{-1}(I - P_m)h_m|| \leq K\lambda^{(m-n)}||h||.$$

Therefore,

$$\begin{aligned} ||u_n|| &\leq ||h||K\left\{\sum_{m=0}^{n-1}\lambda^{n-m} + 1 + \sum_{m=n+1}^{T}\lambda^{m-n}\right\} \\ &= ||h||2K\sum_{m=0}^{T}\lambda^m \\ &\leq ||h||2K(1 - \lambda)^{-1}. \end{aligned}$$

Thus, $u = \{u_n\}_0^\infty \in l^\infty(\mathbb{N})$ and $DF(x)u = h$ so that $DF(x)$ is onto and has a right inverse $DF(x)^\dagger$ with $||DF(x)^\dagger|| \leq 2K(1 - \lambda)^{-1}$. □

3. Exponential Dichotomy and Lattice Maps

In this section we consider lattice maps that occur as discretizations of (1.1) or (1.2). Given a lattice map that has exponential dichotomy with either $a(x, t) = 0$ or $b(x, t) = 0$ the results in this section can be used to determine bounds on the coefficients $b(x, t)$ or $a(x, t)$, respectively, such that the perturbed system has exponential dichotomy. The results in this section are similar to arguments used in [**5**] to prove the shadowing lemma.

The following is a perturbation theorem for exponential dichotomy in linear difference equations (see also Proposition 1, [**11**, pg.42]).

THEOREM 3.1. *(Perturbation Theorem* [**19**]*) Suppose that for each i in an index set I and for each integer n, $C_n^{(i)}$ is an invertible $N \times N$ matrix satisfying*

$$||C_n^{(i)}||, ||(C_n^{(i)})^{-1}|| \leq L,$$

such that the linear difference equation

$$u_{n+1} = C_n^{(i)} u_n$$

has an exponential dichotomy with constants K, λ with the rank of the projections all independent of i. Then there exists $M > 0$, $\Delta > 0$, both depending only on L, K, λ with the following property:

Let B_n be a sequence of $N \times N$ matrices such that, for each integer m, there exists $i_m \in I$, for which

$$||B_n - C_n^{(i_m)}|| < \Delta$$

for $m \leq n \leq m + M$. Then the equation

$$u_{n+1} = B_n u_n$$

has an exponential dichotomy with constants $2K^4$, $\lambda - \eta$ and projections Q_n such that for all m, there exists $i \in I$ with

$$||Q_{n+m} - P_n^{(i)}|| < \eta$$

for $|n| < M$ where $0 < \eta < \min(\frac{1}{12}, \frac{\lambda}{2})$ is given.

PROOF. (see [**19**] or [**9**]) □

Consider a difference equation of the form

$$x_{n+1} = J_n x_n + B_n h_n(x_n)$$

where $\{J_n\}_0^\infty$ is a sequence of invertible matrices and $\{B_n\}_0^\infty$ is a sequence of $N \times N$ matrices and $\{h_n\}_0^\infty$ is a sequence of C^1 functions on $\mathbb{R}^N$. We now state a corollary of the Perturbation Theorem.

COROLLARY 3.2. *Suppose the linear difference equation*

$$u_{n+1} = J_n u_n$$

has an exponential dichotomy with constants K, λ, and let $\eta > 0$ be given. Given a sequence $\{x_n\}_0^\infty$ such that $Dh_n(x_n)$ is bounded for all n, there exists a constant $\delta > 0$ depending on $||J|| \equiv \sup_n ||J_n||, K, \lambda, \{h_n\}_0^\infty$ such that for $||B_n|| \leq \delta$ the linear difference equation

$$u_{n+1} = J_n u_n + B_n Dh_n(x_n) u_n$$

has an exponential dichotomy with constants $2K^4$, $\lambda - \eta$.

PROOF. Apply the Perturbation Theorem. □

Corollary 3.2 applies to discretizations of Burgers' equation (1.1) and the KdV equation (1.2) with appropriate boundary conditions. In particular, it applies to uniform central difference in space/Forward Euler in time discretizations when

$a(x,t) \equiv a(t)$ is slowly varying (see Proposition 1 [**11**, pg. 50]) and $|b(x,t)|$ is sufficiently small for all x, t.

We now consider a difference equation of the form

$$x_{n+1} = A_n J_n x_n + h_n(x_n) \equiv g_n(x_n). \tag{3.1}$$

where $\{A_n\}_0^\infty$ is a sequence of $N \times N$ matrices.

The following theorem is an application of the Perturbation Theorem when the terms corresponding to the nonlinearity have exponential dichotomy. It applies to discretizations of (1.1) and (1.2) for appropriately chosen $b(x,t)$ and $|a(x,t)|$ sufficiently small for all x, t.

A set $S \in \mathbb{R}^N$ is said to be a *hyperbolic invariant set* for a sequence $\{h_n\}_0^\infty$ of C^1 diffeomorphisms on $\mathbb{R}^N$ if:

(i) $h_n(S) \subset S$ for all n,
(ii) for all $x_0 \in S$ the variational equation

$$u_{n+1} = Dh_n(x_n)u_n \tag{3.2}$$

has exponential dichotomy with both the constants K, λ and the rank of the projection $P_n(x_0)$ independent of x_0.

THEOREM 3.3. *Assume that the linear difference equation (3.2) has exponential dichotomy with constants K, λ and suppose*

$$||J_n|| \leq L,$$

and let $\eta > 0$ be given. Suppose also that $\{h_n(x)\}_0^\infty$ and $\{Dh_n(x)\}_0^\infty$ are uniformly bounded and uniformly equicontinuous in a closed σ-neighborhood O of the hyperbolic invariant set S. Then there exists a constant $\delta > 0$ depending on $K, \lambda, L, \{h_n\}_0^\infty$ such that for $||A_n|| \leq \delta$ the linear difference equation

$$u_{n+1} = A_n J_n u_n + Dh_n(y_n)u_n \equiv Dg_n(y_n)u_n \tag{3.3}$$

has an exponential dichotomy with constants $2K^4$, $\lambda - \eta$.

PROOF. Let

$$||Dh_n|| = \sup\{||Dh_n(x)|| : x \in O\}$$

and

$$||Dh|| = \sup_n\{||Dh_n||\}.$$

Let

$$||h_n|| = \sup\{||h_n(x)|| : x \in O\}$$

and

$$||h|| = \sup_n\{||h_n||\}.$$

For $\delta \geq 0$ set

$$\omega_n(\delta) = \sup\{||Dh_n(y) - Dh_n(x)|| : x \in S, y \in \mathbb{R}^N, ||y - x|| \leq \delta\}$$

and define

$$\omega(\delta) = \sup_n \{w_n(\delta)\}.$$

By assumption $||h||, ||Dh|| < \infty$ and $\omega(\delta) \downarrow 0$ as $\delta \downarrow 0$. Let M, Δ be the numbers in the Perturbation Theorem corresponding to $||Dh||, K, \lambda$. Then let $\delta > 0$ be the largest number satisfying

$$\begin{aligned}\Delta \geq \delta L + \omega(\delta L((\delta L + ||h||)^M + (\delta L + ||h||)^{M-1}||Dh|| + \\ \cdots + (\delta L + ||h||)||Dh||^{M-1} + ||Dh||^M)),\\ \sigma \geq \delta L((\delta L + ||h||)^M + (\delta L + ||h||)^{M-1}||Dh|| + \\ \cdots + (\delta L + ||h||)||Dh||^{M-1} + ||Dh||^M).\end{aligned}$$

We want to use the Perturbation Theorem to show that the linear difference equation (3.3) has an exponential dichotomy for $||A_n|| \leq \delta$. First we claim that for all integers m,

$$||({}^n g^m)(y_m) - ({}^n h^m)(y_m)|| \leq \delta L((\delta L + ||h||)^{n-m} + \cdots + ||Dh||^{n-m})$$

for $m \leq n \leq m + M$. We prove this by induction. By assumption it holds for $n = m$. Assuming it is true for some n with $m \leq n \leq m + M$, we prove it for $n + 1$ as follows:

$$\begin{aligned}&||({}^{n+1} g^m)(y_m) - ({}^{n+1} h^m)(y_m)||\\ =&||A_{n+1} J_{n+1}({}^n g^m(y_m)) + h_{n+1}({}^n g^m(y_m)) - h_{n+1}({}^n h^m(y_m))||\\ \leq&||A_{n+1} J_{n+1}({}^n g^m(y_m))|| + ||h_{n+1}({}^n g^m(y_m)) - h_{n+1}({}^n h^m(y_m))||\\ \leq&\delta L(\delta L + ||h||)^{n-m+1} + ||Dh||||({}^n g^m(y_m)) - ({}^n h^m(y_m))||\\ =&\delta L((\delta L + ||h||)^{n-m+1} + \cdots + ||Dh||^{n-m+1}).\end{aligned}$$

So the induction step is verified. Now, if $m \leq n \leq m + M$,

$$\begin{aligned}&||A_{n+1} J_{n+1} + Dh_{n+1}({}^n g^m(y_m)) - Dh_{n+1}({}^n h^m(y_m))||\\ \leq&||A_{n+1} J_{n+1}|| + ||Dh_{n+1}({}^n g^m(y_m)) - Dh_{n+1}({}^n h^m(y_m))||\\ \leq&\delta L + \omega(||({}^n g^m(y_m)) - ({}^n h^m(y_m))||)\\ \leq&\delta L + \omega(\delta L((\delta L + ||h||)^{n-m} + \cdots + ||Dh||^{n-m}))\\ \leq&\Delta.\end{aligned}$$

By the Perturbation Theorem the difference equation (3.3) has an exponential dichotomy with constants $2K^4$ and $\lambda - \eta$. □

4. The QR Decomposition and Decoupling

When theoretical estimates of the dichotomy constants are not possible or are difficult to obtain we must find other techniques for estimating $||DF(x)^\dagger||$. With this in mind consider the following iterative boundary value problem (IBVP) with separated boundary conditions

$$(4.1)\qquad \begin{cases} u_{n+1} = J_n u_n + g_n \\ B_0 u_0 = 0, B_T u_T = 0 \end{cases}$$

where $u_n, g_n \in \mathbb{R}^N$ and $J_n, B_0, B_T \in \mathbb{R}^{N\times N}$. We are free to choose the boundary conditions defined by the matrices B_0 and B_T. Then we can compute $DF(x)^\dagger$ using (4.1) so that

$$||DF(x)^\dagger|| = \sup_{g\neq 0} \frac{||DF(x)^\dagger g||}{||g||} = \sup_{g\neq 0} \frac{||u||}{||g||}.$$

The difference equation (4.1) will have a global dichotomy since the number of iterates is finite, but to minimize the norm of u the boundary conditions must be chosen to have the correct rank. The optimal choice (see [**13**]) is to choose B_0 and B_T such that $rank(B_0) = rank(P_0)$ and $rank(B_T) = N - rank(P_0)$ where P_0 is the projection associated with the exponential dichotomy of the difference equation. Once the boundary conditions are determined all that remains is to solve the (IBVP).

To solve the (IBVP) we perform decoupling transformations. In this paper we consider decoupling transformations based on the QR decomposition. To decouple the (IBVP) the decoupling transformations must satisfy the discrete Lyapunov equation

$$S_{n+1}R_n - J_n S_n = 0$$

where S_n is the nth decoupling transformation and R_n is at least block upper triangular.

We perform the QR decomposition as follows:

$$Q_{n+1}R_n = J_n Q_n, \quad n = 0, \dots, T-1$$

where R_n is upper triangular, Q_0 is given and the Q_n are orthogonal. The QR decomposition is used for decoupling in the popular boundary value problem solver in [**18**]. The QR decomposition tends to order the spectrum of the upper triangular matrix along the diagonal. The corresponding decoupled IBVP is

$$\begin{cases} w_{n+1} = R_n w_n + h_n \\ C_0 w_0 = 0, C_T w_T = 0 \end{cases}$$

where

$$S_n w_n = u_n$$

and $S_n = Q_n$ for all n.

5. The Algorithm

To find $||DF(x)^\dagger||$ we find the solution of a decoupled IBVP such that

$$C_0 = \begin{pmatrix} 0 & 0 \\ 0 & I_p \end{pmatrix}$$

and

$$C_T = \begin{pmatrix} I_q & 0 \\ 0 & 0 \end{pmatrix}$$

where $p + q = N$.

To bound $||DF(x)^\dagger||_\infty$ we choose $h_n^j \in \{+1, -1\}$, where h_n^j is the jth component of the vector h_n, to maximize the norm of w, so that

$$\begin{aligned} ||DF(x)^\dagger||_\infty &= \sup_{g\neq 0} \frac{||DF(x)^\dagger g||_\infty}{||g||_\infty} = \sup_{g\neq 0} \frac{||u||_\infty}{||g||_\infty} \\ &\leq \sup_n \{||S_n||_\infty ||S_n^{-1}||_\infty\} \cdot ||w||_\infty \end{aligned}$$

To find $||w||_\infty$, let $w_n = \begin{pmatrix} w_n^{(1)} \\ w_n^{(2)} \end{pmatrix}$, $h_n = \begin{pmatrix} h_n^{(1)} \\ h_n^{(2)} \end{pmatrix}$, and suppose

$$R_n = \begin{pmatrix} R_n^{(11)} & R_n^{(12)} \\ 0 & R_n^{(22)} \end{pmatrix}$$

where $w_n^{(2)}$ denotes a real vector of length p and $R_n^{(12)}$ denotes a real matrix of dimension $q \times p$. Now we must solve the linear difference equation

$$\begin{pmatrix} w_{n+1}^{(1)} \\ w_{n+1}^{(2)} \end{pmatrix} = \begin{pmatrix} R_n^{(11)} & R_n^{(12)} \\ 0 & R_n^{(22)} \end{pmatrix} \begin{pmatrix} w_n^{(1)} \\ w_n^{(2)} \end{pmatrix} + \begin{pmatrix} h_n^{(1)} \\ h_n^{(2)} \end{pmatrix}$$

so that for a given integer $s \geq 0$ we have

$$w_{n+1}^{(2)} = \begin{cases} 0, & n = -1 \\ h_n^{(2)}, & n = 0 \\ R_n^{(22)} \cdots R_{n-r}^{(22)} w_{n-r}^{(2)} & \\ +(\sum_{k=1}^{r} R_n^{(22)} \cdots R_{n-k+1}^{(22)} h_{n-k}^{(2)}) + h_n^{(2)}, & n = 1, ..., T-1 \end{cases}$$

where $r = \min\{s, n-1\}$ and we define $(A_n \cdots A_n) = A_n$. Similarly,

$$w_n^{(1)} = \begin{cases} 0, & n = T \\ (R_{n+r}^{(11)} \cdots R_n^{(11)})^{-1}(w_{n+r+1}^{(1)} - g_{n+r}) & \\ -\sum_{k=1}^{r} (R_{n+r-k}^{(11)} \cdots R_n^{(11)})^{-1} g_{n+r-k}, & n = 0, ..., T-1 \end{cases}$$

where $g_n = (R_n^{(12)} w_n^{(2)} + h_n^{(1)})$ and $r = \min\{s, T-1-n\}$.

To perform the forward sweep defined by the computation of $w_n^{(2)}$ we compute

$$\tilde{w}_{n+1}^{(2)} = \begin{cases} 0, & n=-1 \\ h_n^{(2)}, & n=0 \\ |R_n^{(22)} \cdots R_{n-r}^{(22)}|\tilde{w}_{n-r}^{(2)} & \\ +(\sum_{k=1}^{r} |R_n^{(22)} \cdots R_{n-k+1}^{(22)}|h_{n-k}^{(2)}) + h_n^{(2)}, & n=1,...,T-1 \end{cases}$$

where $r = \min\{s, n-1\}$, $|A| = (|a_{ij}|)$ for $A = (a_{ij})$ and we set $h_n^{(2)} = (1,...,1)^T$ for $n = 0,...,T$.

Similarly, we set

$$\tilde{w}_n^{(1)} = \begin{cases} 0, & n=T \\ |(R_{n+r}^{(11)} \cdots R_n^{(11)})^{-1}|(\tilde{w}_{n+r+1}^{(1)} + \tilde{g}_{n+r}) & \\ +\sum_{k=1}^{r} |(R_{n+r-k}^{(11)} \cdots R_n^{(11)})^{-1}|\tilde{g}_{n+r-k}, & n=0,...,T-1 \end{cases}$$

where $\tilde{g}_n = (|R_n^{(12)}|\tilde{w}_n^{(2)} + h_n^{(1)})$ and $r = \min\{s, T-1-n\}$ and $h_n^{(1)} = (1,...,1)^T$ for $n = 0,...,T$.

REMARKS.

(i) We have $||\tilde{w}_n||_\infty \geq ||w_n||_\infty$ for all sequences $\{h_n\}$ with $\sup_n ||h_n||_\infty \leq 1$ (see [**10**]).

(ii) We replace the inequality (2.2) with $L_{Df}\epsilon \leq \frac{1}{2c}$ where $||DF(x)^\dagger||_\infty \leq c$ and L_{Df} is the Lipschitz constant for Df in an ϵ-neighborhood of the computed sequence $\{x_n\}$. This implies that $L_{Df}\epsilon^2 \leq \delta$ (note conjecture in [**15,16**]) if we set $\epsilon := 2\delta c$.

(iii) We determine the number of stable and unstable modes by assuming initially that there are N stable modes and then monitoring the norm of each component of w. If $L_{Df}(2\delta N||w(i)||_\infty)^2 > \delta$ for some $i, 1 \leq i \leq N$ (see Remark (ii) above), then we decrease the number of stable modes accordingly. Here we set $||w(i)|| = \sup_n |w(i)_n|$. The value used for δ can be either an a priori bound or the current computed bound for δ.

(iv) The algorithm presented here is similar to that presented in [**8**] in that it combines the use of a discrete Gronwall's inequality and a fixed point argument. The algorithm above is based upon the supremum norm as the local norm while the Euclidean norm was used in [**8**]. The idea of determining the number of stable nodes dynamically appears to be new.

In summary our algorithm consists of the following steps.

(1) Generate orbit, find δ, determine the number of stable modes, compute $||\tilde{w}||_\infty$ and set $c := N \cdot ||\tilde{w}||_\infty$.

(2) If $L_{Df}\epsilon \leq \frac{1}{2c}$ for $\epsilon := 2\delta \cdot c$, then set the shadowing distance to ϵ.

Up to now we have not discussed the numerical errors that can occur while computing the jacobian, J_n, decomposing the jacobian, and finding the next iterate of $w_n^{(2)}$ and $w_n^{(1)}$ by matrix vector multiplication and back substitution, respectively. Our philosophy is to correct at each iterate by increasing the magnitude of the components of $w_n^{(2)}$ and $w_n^{(1)}$ in accordance with the error estimates that appear in the appendices. The error bounds for computing the jacobian of the logistic map, the Henon map, and the discretizations of Burgers' equation, and Korteweg-de Vries equation can be found in Appendix 2. Bounds for decomposing the jacobians can be found in Appendix 3, while Appendix 1 contains an error analysis of standard vector/matrix operations. Bounds for δ are in Appendix 2.

6. Numerical Results

In this section numerical results are presented in which shadowing distances are obtained using the algorithm presented in section 4. The method is applied to the logistic map, the Henon map and discretizations of Burgers' equation, and the Korteweg-de Vries equation. Our numerical experiments were performed on a Silicon Graphics Indigo with machine precision $\epsilon_M \approx 2.2e - 16$.

EXAMPLE 1. The first example we consider is the logistic map

$$x_{n+1} = ax_n(1 - x_n).$$

In Tables 1 and 2 we compute with $a = 3.8$ and initial condition $x_0 = 0.3$ using 4-digit and 10-digit arithmetic and compare with exact computations done using rational numbers.

Table 1. 4-digit arithmetic

Iterate	Exact	Computed
x_0	.3	.3
x_{10}	.9336	.9340
x_{20}	.9485	.9076
x_{30}	.9497	.3187

Table 2. 10-digit arithmetic

Iterate	Exact	Computed
x_0	.3	.3
x_{20}	.9484616646	.9484616661
x_{50}	.7722739427	.7726487778
x_{70}	.6913512585	.8464781103

Note that there is not an actual solution that stays "close" to the numerically computed solution if the same initial condition is used. This is a typical behavior for this parameter value because orbits are expanding; i.e. they have a positive

Lyapunov exponent. In general a forward error interpretation is not possible for a mapping with an expanding component.

Figure 1 illustrates the "sublinear" growth of the global shadowing error as a function of the number of iterates. To compute the shadowing distance ϵ note that for this example we may set $L_{Df} = 2$ and use the bound on δ that appears the appendix. We have set $s = 0$ to obtain these results.

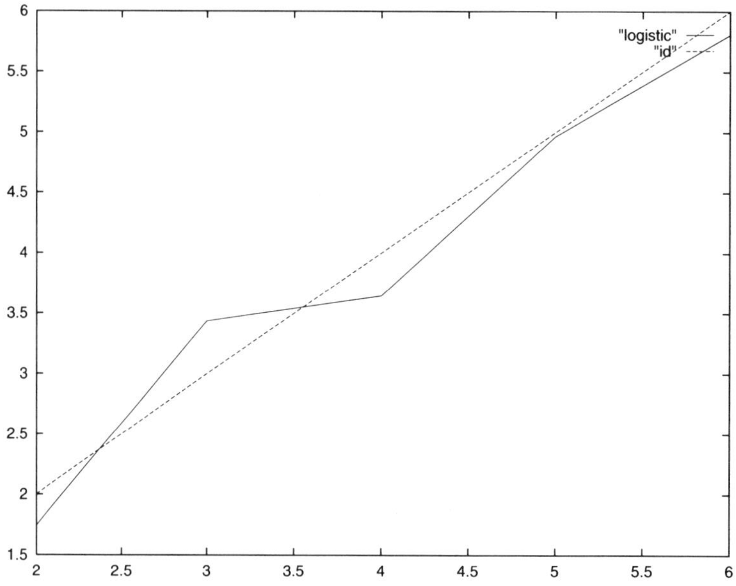

FIGURE 1. $\log_{10}(n)$ vs. $\log_{10}(c)$ for Logistic Map

EXAMPLE 2. We consider now the Henon map

$$x_{n+1} = f(x_n) \equiv f(x_n(1), x_n(2)) \equiv \begin{pmatrix} 1 + x_n(2) - a x_n^2(1) \\ b x_n(1) \end{pmatrix}$$

with the floating point representations of the standard parameter values $a = 1.4$ and $b = 0.3$. The parameter s is set to zero since the number of stable modes is one and the number of unstable modes is one. In this case any positive value of s will give the same results but will be computationally more expensive. For this example we set $L_{Df} = 2.8$. Note in Figure 2 that the growth in the global shadowing error is "sublinear" for small n but becomes "superlinear" for n large with our algorithm.

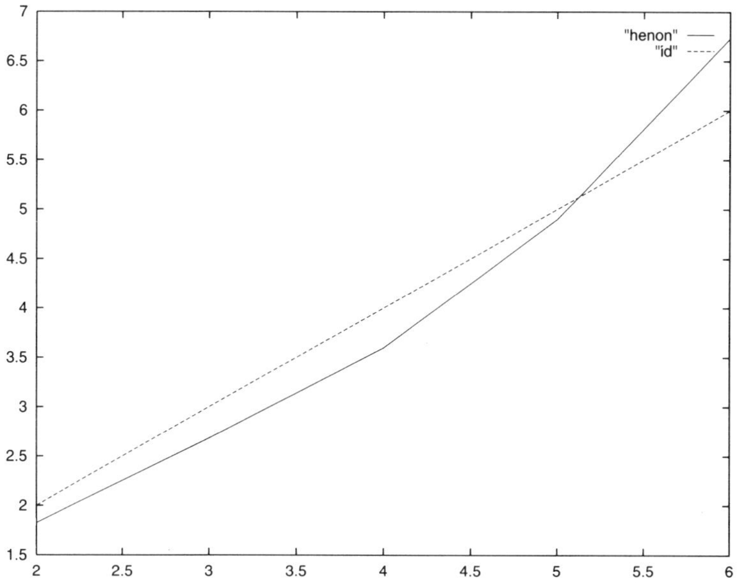

FIGURE 2. $\log_{10}(n)$ vs. $\log_{10}(c)$ for Henon Map

EXAMPLE 3. Next consider the following discretization of Burgers' Equation using central differences in space

$$x_{n+1}(i) = x_n(i) + a_n(i)(x_n(i+1) - 2x_n(i) + x_n(i-1)) + b_n(i)(x_n^2(i+1) - x_n^2(i-1))$$

with Dirichlet boundary conditions. The parameter values for Burgers' equation satisfy $a_n(i) > 0$ and $b_n(i) < 0$. We discretize the initial data $\sin(2\pi x) + \frac{1}{2}\sin(\pi x)$ on the interval $[0, 1]$. We have $L_{Df} = \sup_{n,i} |b_n(i)|$ for this example. Some typical results are summarized in Table 3. To read Tables 3 and 4 note that "N" is the dimension of the problem, "T" is the number of iterates that were taken, "s" is the parameter introduced in Section 4, "p" is the number of stable modes as determined by the algorithm, "c" is a bound on $||DF(x)^\dagger||_\infty$, and "ϵ" is the computed value for the shadowing distance. In all of the these examples the local error δ was approximately $1.0 \cdot 10^{-14}$.

Table 3. (Burgers' Equation $N = 9$)

$a_n(i)$	$b_n(i)$	T	s	p	c	ϵ
10^{-3}	-10^{-2}	100	0	9	$1.63 \cdot 10^2$	$1.93 \cdot 10^{-11}$
10^{-3}	-10^{-2}	1000	0	9	$1.10 \cdot 10^4$	$1.98 \cdot 10^{-9}$
10^{-3}	-10^{-2}	1000	100	9	$8.30 \cdot 10^3$	$1.50 \cdot 10^{-9}$
10^{-3}	-10^{-2}	10,000	0	9	$2.40 \cdot 10^4$	$4.31 \cdot 10^{-9}$
10^{-3}	-10^{-2}	10,000	100	9	$1.97 \cdot 10^4$	$3.56 \cdot 10^{-9}$

EXAMPLE 4. We consider the following discretization of the Korteweg-de Vries equation:

$$x_{n+1}(i) = x_n(i)+a_n(i)(x_n(i+2) - 2x_n(i+1) + 2x_n(i-1) - x_n(i-2)) \\ +b_n(i)(x_n^2(i+1) - x_n^2(i-1))$$

with $a_n(i)_n < 0$ and $b_n(i) > 0$ and periodic boundary conditions. The soliton initial data $-2sech^2(x+10)$ discretized on the computational domain $[-20, 20]$ is employed. We set $L_{Df} = \sup_{n,i} |b_n(i)|$ and summarize our results in Table 4.

Table 4. (Korteweg-de Vries' Equation $N = 9$)

$a_n(i)$	$b_n(i)$	T	s	p	c	ϵ
-10^{-4}	10^{-2}	100	0	9	$4.76 \cdot 10^2$	$8.56 \cdot 10^{-11}$
-10^{-4}	10^{-2}	100	20	9	$4.44 \cdot 10^2$	$8.01 \cdot 10^{-11}$
-10^{-4}	10^{-2}	1000	0	7	$1.79 \cdot 10^6$	$3.21 \cdot 10^{-7}$
-10^{-4}	10^{-2}	1000	50	7	$1.13 \cdot 10^6$	$2.05 \cdot 10^{-7}$
-10^{-4}	10^{-2}	1000	100	7	$7.59 \cdot 10^5$	$1.37 \cdot 10^{-7}$

7. Conclusions

In our algorithm we decoupled at each step, but this is not essential. The algorithms could have been performed by decoupling at every kth step. The problem with such an approach is that if k is large then the error from performing the matrix multiplications may be unmanageable, so k must be chosen carefully.

Other approaches for finding the norm of the right inverse are possible. One approach is to use the pseudo inverse as the right inverse. In this way there is no need to choose boundary conditions (i.e. determine the number of stable and unstable modes), but then finding an estimate of the norm of the pseudo inverse may be computationally expensive (see [**10**]). Another approach that does not appear to have been tried is to compute the dichotomy constants directly and then apply Lemma 2.2.

Perhaps more appropriate as discretizations for Burgers' equation and the KdV equation are the leap frog or upwinding scheme and the nondissipative implicit midpoint scheme, respectively. To determine the shadowing distance from actual solutions of the PDE or to solutions of a semidiscretization one must understand the error in our approximation of the norm of the right inverse of the mapping as compared with the norm of a right inverse that is defined in terms of local solution operators of the PDE or semidiscretization (see [**10**] for results for the spatial discretization of a reaction diffusion equation).

8. Appendix 1 - Error Analysis of Standard Operations

Here we state an analysis of rounding errors for basic vector, matrix operations following Wilkinson ([**23**]). We assume the ∞-norm unless stated otherwise and assume that ϵ_M is the machine precision.

Scalar Operations: If $a, b \in \mathbb{R}$, then

$$|fl(a \bullet b) - a \bullet b| \leq |a \bullet b|\epsilon_M. \tag{8.1}$$

where $\bullet = \times, \div, +$ or $-$.

Vector inner product: Let $x, y \in \mathbb{R}^N$., then

$$|fl(x^T y) - x^T y| \leq (1.06)N^2||x|| \cdot ||y|| \cdot \epsilon_M \tag{8.2}$$

if $N \cdot \epsilon_M < 0.1$ (see [**23**, pg. 114]).

Matrix times matrix: For $A, B \in \mathbb{R}^{N\times N}$ we have

$$||fl(AB) - AB|| \leq (1.06)N||A|| \cdot ||B|| \cdot \epsilon_M$$

if $N \cdot \epsilon_M < 0.1$ (see [**23**, pg. 115]). Thus, for $A_1, A_2, ..., A_m \in \mathbb{R}^{N\times N}$ we have

$$\begin{aligned} &||fl(A_1 \cdots A_m) - A_1 \cdots A_m|| \\ \leq &||A_1||\{(1.06)N \cdot \epsilon_M||fl(A_2 \cdots A_m)|| + ||fl(A_2 \cdots A_m) - A_2 \cdots A_m||\} \\ \leq &((1.06)N \cdot \epsilon_M)\{\sum_{i=0}^{M-2}(1 + (1.06)N \cdot \epsilon_M)^i\}||A_1|| \cdots ||A_m|| \end{aligned} \tag{8.3}$$

9. Appendix 2 - Error Analysis of Orbit and Jacobian

To determine δ we need an upper bound on the local error in computing the orbit. Let $\{\delta_n\}_0^T$ be a sequence of numbers denoting an upper bound of computing the nth iterate and let $\delta = \sup_n\{\delta_n\}$. For the logistic map

$$\delta_n \leq a \cdot (1 + \epsilon_M) \cdot \epsilon_M$$

and for the Henon map

$$\delta_n \leq \max\{3 \cdot (1 + |x_n(2)| + ax_n^2(1)(1 + \epsilon_M)) \cdot \epsilon_M, b|x_n(1)| \cdot \epsilon_M\}.$$

The local error in computing a single step of the discrete Burgers' equation and the discrete Korteweg-de Vries equation is bounded by

$$\begin{aligned} \delta_n \leq 2\cdot &\max\{C_1||a_n|| \cdot ||x_n|| \cdot (1 + \epsilon_M), ||b_n||\}\cdot \\ &\max\{1, \sup_{1\leq i\leq N}\{|(x_n^2(i+1) - x_n^2(i-1))|(1 + \epsilon_M)\}\} \cdot \epsilon_M. \end{aligned}$$

where $C_1 = 6$ for Burgers' equation and $C_1 = 8$ for Korteweg-de Vries equation.

To determine the error in computing the jacobian we consider again the ∞-norm and denote the error in computing the nth jacobian, J_n, by E_n. For the logistic map

$$||E_n|| \leq a|1 - 2x_n|(1 + \epsilon_M) \cdot \epsilon_M$$

while for the Henon map

$$||E_n|| \leq 2(1 + 2a|x_n(1)|(1 + \epsilon_M)) \cdot \epsilon_M.$$

The error in computing the jacobian of the discrete Burgers' equation and the discrete Korteweg-de Vries equation is bounded by

$$||E_n|| \leq 3 \cdot \max\{1, ||a_n||\} \cdot ||b_n|| \sup_{1 \leq i \leq N} \{|x_n^2(i+1) - x_n^2(i-1)|(1+\epsilon_M)\} \cdot \epsilon_M.$$

10. Appendix 3 - Error Analysis for the QR Decomposition

Following Stewart [**22**, pg. 516] we employ the following error bounds for the QR decomposition. If $J_n = \tilde{J}_n + E_n$, $J_n = Q_n R_n$ and $\tilde{J}_n = \tilde{Q}_n \tilde{R}_n$, then under appropriate assumptions (see [**22**, Thm. 3.1, pg. 516]) $\tilde{Q}_n = Q_n + W_n$ and $\tilde{R}_n = R_n + F_n$ where

$$||W_n|| \leq \frac{3||J_n^{-1}|| \cdot ||E_n||}{1 - 2||J_n^{-1}|| \cdot ||E_n||}$$

and

$$||F_n|| \leq ||E_n||(1 + ||W_n||) + ||W_n|| \cdot ||J_n||.$$

References

[1] D. V. Anosov, *Geodesic Flows and Closed Riemannian Manifolds with Negative Curvature*, Proc. Steklov Inst. Math. **90** (1967).

[2] W.-J. Beyn, *On the Numerical Approximation of Phase Portraits Near Stationary Points*, SIAM J. Numer. Anal. **24** (1987), 1095–1113.

[3] R. Bowen, *ω-limit sets for Axiom A Diffeomorphisms*, J. Diff. Eq. **18** (1975), 333–339.

[4] B. A. Coomes, H. Kocak and K. J. Palmer, *A Shadowing Theorem for Ordinary Differential Equations*, preprint (1993).

[5] S. N. Chow, X. B. Lin and K. J. Palmer, *A Shadowing Lemma for Maps in Infinite Dimensions*, Differential Equations: Proceedings of the EQUADIFF Conference (C. M. Daffermos, G. Ladas and G. Papanicolaou, eds.), Marcel Dekker, New York, 1989, pp. 127–136.

[6] ———, *A Shadowing Lemma with Applications to Semilinear Parabolic Equations*, SIAM J. Math. Anal. **20** (1989), 547–557.

[7] S. N. Chow and K. J. Palmer, *On the Numerical Computation of Orbits of Dynamical Systems: the One-Dimensional Case*, Dynamics and Diff. Eqn. **3** (1991), 361-380.

[8] ———, *On the Numerical Computation of Orbits of Dynamical Systems: the Higher Dimensional Case*, J. of Complexity **8** (1992), 398–423.

[9] S. N. Chow and E. S. Van Vleck, *A Shadowing Lemma for Random Diffeomorphisms*, Random & Computational Dynamics **1(2)** (1992), 197–218.

[10] ———, *A Shadowing Lemma Approach to Global Error Analysis for Initial Value ODEs*, SIAM J. Sci. Comp. **15** (1994).

[11] W. A. Coppel, *Dichotomies in Stability Theory, Lecture Notes in Mathematics 629*, Springer-Verlag, New York, 1978.

[12] H. E. Cook, D. de Fontaine, and J. E. Hilliard, *A Model for Diffusion on Cubic Lattices and its Application to the Early Stages of Ordering*, Acta Metallurgica **17** (1969), 765–773.

[13] F. R. de Hoog and R. M. M. Mattheij, *On Dichotomy and Conditioning in BVP*, SIAM J. Numer. Anal. **24** (1987), 89–105.

[14] J. E. Franke and J. F. Selgrade, *Hyperbolicity and Chain Recurrence*, J. Diff. Eqn. **26** (1977), 27-36.

[15] S. Hammel, J. A. Yorke and C. Grebogi, *Do Numerical Orbits of Chaotic Dynamical Processes Represent True Orbits?*, J. Complexity **3** (1987), 136–145.

[16] ———, *Numerical Orbits of Chaotic Processes Represent True Orbits*, Bull. Am. Math. Soc. **19** (1988), 465–470.

[17] K. Kaneko, *Spatiotemporal Chaos in One- and Two-Dimensional Coupled Map Lattices*, Physica D **37** (1989), 60–82.

[18] R. M. M. Mattheij and G. W. M. Staarink, *An Efficient Algorithm for Solving General Linear Two-Point BVP*, SIAM J. Sci. Stat. Comput. **5** (1984), 745–763.

[19] K. J. Palmer, *A Perturbation Theorem for Exponential Dichotomies*, J. Royal Society of Edinburgh **106A** (1987), 25–37.

[20] J. M. Sanz-Serna and S. Larsson, *Shadows, Chaos and Saddles*, Appld. Numer. Math. (to appear).

[21] T. Sauer and J. A. Yorke, *Rigorous Verification of Trajectories for the Computer Simulation of Dynamical Systems*, Nonlinearity **4** (1991), 961–979.

[22] G. W. Stewart, *Perturbation Bounds for the QR factorization of a Matrix*, SIAM J. Numer. Anal. **14** (1977), 509–518.

[23] J. H. Wilkinson, *The Algebraic Eigenvalue Problem*, Oxford University Press, 1965.

School of Mathematics, Georgia Institute of Technology, Atlanta, Georgia 80332

E-mail address: chow@math.gatech.edu

Department of Mathematical and Computer Sciences, Colorado School of Mines, Golden, Colorado 80401

E-mail address: erikvv@lyapunov.mines.colorado.edu

Contemporary Mathematics
Volume **172**, 1994

Periodic Shadowing

Brian A. Coomes, Hüseyin Koçak, and Kenneth J. Palmer

ABSTRACT: We prove a general theorem for establishing the existence of a periodic orbit of an autonomous system of ordinary differential equations near a numerically computed pseudo orbit whose final point is sufficiently close to its initial point. The conditions of the theorem are given in terms of quantities which can be computed directly from the pseudo orbit and the vector field. As an application, the existence, as well as the stability types, of one asymptotically stable and one unstable periodic orbit of the Lorenz equations for certain parameter values is rigorously established.

1. Introduction

The search for periodic orbits at large in specific differential equations is a notoriously difficult dynamical problem. To be sure, there has been considerable success in this endeavor in certain special situations. In general, numerical computations remain one of the most common methods of hunting for periodic orbits. In such a pursuit, however, it is always difficult to infer rigorously the existence of a periodic orbit from the numerical computations. We present here a new method for establishing the existence of a true periodic orbit of an autonomous system of ordinary differential equations near a computed apparent periodic orbit.

Our results in this paper have been inspired by our earlier work [2, 4] on *shadowing* orbits of ordinary differential equations where we established, in a rather general setting, shadowing of a pseudo orbit, an approximate solution, by an associated true orbit with albeit somewhat different initial data. Here we begin with a pseudo periodic orbit, that is, a pseudo orbit where the last computed point is sufficiently close to the first one, and show, under computable conditions, the existence of a true periodic orbit near the pseudo periodic orbit.

As an application of our Periodic Shadowing Theorem, we show the existence of periodic orbits of the Lorenz equations for certain parameter values. With some

---

1991 *Mathematics Subject Classification.* Primary 34C25, 65L70; Secondary 65L07.

The second and third authors were supported in part by NSF and the second author was supported in part by DOE (Office of Scientific Computing).

This paper is in final form and no version of it will be submitted for publication elsewhere.

additional numerical computations we also establish that these periodic orbits are hyperbolic and determine their stability types. We emphasize that our shadowing method presented in this paper works well for establishing the existence of both stable and unstable periodic orbits. The main concern in the unstable case is to generate a sufficiently accurate pseudo periodic orbit. This problem is overcome by using a Newton-like method as described in, for example, [5, 7].

Finally, we should mention that numerical computation of periodic orbits has been the subject of numerous investigations. A small sample of semi-empirical studies is contained in the references [5, 8, 11, 15, 16] and of rigorous results in [1, 6, 9, 13, 14].

2. Statement of the Periodic Shadowing Theorem

In this section we first recast the notions of a pseudo orbit and shadowing in the context of periodic orbits. Then we present the statement of our main theorem which guarantees the existence of a true periodic orbit of a system of ordinary differential equations near an appropriate pseudo orbit.

We consider the autonomous system

$$\dot{\mathbf{x}} = f(\mathbf{x}), \tag{1}$$

where $f : \mathbb{R}^n \to \mathbb{R}^n$ is a C^2 vector field, with its associated flow φ^t. Throughout this paper, unless otherwise stated, we use the Euclidean norm for vectors and the relevant operator norm for matrices and linear operators.

DEFINITION. *For a given positive number δ, a sequence of points $\{\mathbf{y}_k\}_{k=0}^N$, with $f(\mathbf{y}_k) \neq \mathbf{0}$ for all k, is said to be a δ pseudo periodic orbit of Eq. (1) if there is an associated sequence $\{h_k\}_{k=0}^N$ of positive times such that*

$$\|\mathbf{y}_{k+1} - \varphi^{h_k}(\mathbf{y}_k)\| \leq \delta \quad \text{for } k = 0, \ldots, N-1, \tag{2}$$

and

$$\|\mathbf{y}_0 - \varphi^{h_N}(\mathbf{y}_N)\| \leq \delta. \tag{3}$$

Next, we introduce the notion of shadowing a pseudo periodic orbit by a true periodic orbit.

DEFINITION. *For a given positive number ε, a δ pseudo periodic orbit $\{\mathbf{y}_k\}_{k=0}^N$ with associated times $\{h_k\}_{k=0}^N$ is said to be ε-shadowed by a true periodic orbit if there are points $\{\mathbf{x}_k\}_{k=0}^N$ on the true periodic orbit and positive times $\{t_k\}_{k=0}^N$ with $\varphi^{t_k}(\mathbf{x}_k) = \mathbf{x}_{k+1}$ for $k = 0, \ldots, N-1$, and $\mathbf{x}_0 = \varphi^{t_N}(\mathbf{x}_N)$ such that*

$$\|\mathbf{x}_k - \mathbf{y}_k\| \leq \varepsilon \quad \text{and} \quad |t_k - h_k| \leq \varepsilon \quad \text{for } k = 0, \ldots, N.$$

To state our theorem we need to introduce certain sequences of matrices and also some constants.

Let $\{\mathbf{y}_k\}_{k=0}^N$ be a δ pseudo periodic orbit of Eq. (1) with associated times $\{h_k\}_{k=0}^N$. Also suppose that we have a sequence $\{Y_k\}_{k=0}^N$ of $n \times n$ matrices such that

$$\|Y_k - D\varphi^{h_k}(\mathbf{y}_k)\| \leq \delta \qquad \text{for } k = 0, \ldots, N. \tag{4}$$

For $k = 0, \ldots, N$, we let S_k be an $n \times (n-1)$ matrix chosen so that its columns form an "almost orthonormal" basis for the subspace orthogonal to $f(\mathbf{y}_k)$, that is,

$$\|S_k^* f(\mathbf{y}_k)\| \le \delta_1, \qquad \|S_k^* S_k - I\| \le \delta_1 \tag{5}$$

for a positive number δ_1, where $*$ denotes transpose. Then, we choose $(n-1)\times(n-1)$ matrices A_k satisfying

$$\begin{aligned} &\|A_k - S_{k+1}^* Y_k S_k\| \le \delta_1 \qquad \text{for } k = 0, \ldots, N-1, \\ &\|A_N - S_0^* Y_N S_N\| \le \delta_1. \end{aligned} \tag{6}$$

Notice that we introduced the quantity δ_1 in the inequalities in Eqs. (5) and (6) to account for possible round-off errors in the necessary matrix computations.

As the last piece of our notational collection, we define various constants. Let U be a convex set containing $\{\mathbf{y}_k\}_{k=0}^N$ in its interior. For such a U, we define

$$M_0 = \sup_{\mathbf{x}\in U} \|f(\mathbf{x})\|, \qquad M_1 = \sup_{\mathbf{x}\in U} \|Df(\mathbf{x})\|, \qquad M_2 = \sup_{\mathbf{x}\in U} \|D^2 f(\mathbf{x})\|.$$

Then we define

$$h_{\min} = \inf_{0\le k\le N} h_k, \qquad h_{\max} = \sup_{0\le k\le N} h_k.$$

Next, we choose a positive number $\varepsilon_0 \le h_{\min}$ such that for $k = 0, \ldots, N$ and $\|\mathbf{x} - \mathbf{y}_k\| \le \varepsilon_0$ the solution $\varphi^t(\mathbf{x})$ is defined and remains in U for $0 \le t \le h_k + \varepsilon_0$. Finally, we define

$$\Delta = \inf_{0\le k\le N} \|f(\mathbf{y}_k)\|, \quad \overline{M}_0 = \sup_{0\le k\le N} \|f(\mathbf{y}_k)\|, \quad \overline{M}_1 = \sup_{0\le k\le N} \|Df(\mathbf{y}_k)\|,$$

and

$$\Theta = \sup_{0\le k\le N} \|Y_k\|.$$

Now, we can state our main theorem:

PERIODIC SHADOWING THEOREM. *Let $\{\mathbf{y}_k\}_{k=0}^N$ be a δ pseudo periodic orbit of the autonomous system $\dot{\mathbf{x}} = f(\mathbf{x})$ such that the matrix*

$$L = I - A_N \cdots A_0$$

is invertible. Let

$$C_1 = \max_{k=0}^{N} \Big(\|A_{k-1} \ldots A_0\| \, \|L^{-1}\| \, \big(1 + \sum_{m=1}^{N} \|A_N \ldots A_m\|\big) + \sum_{m=1}^{k} \|A_{k-1} \ldots A_m\| \Big),$$

$$C = \max\Big\{ \Delta^{-1}(\Theta C_1(1+\delta_1) + 1),\ C_1\sqrt{1+\delta_1} \Big\},$$

$$\delta_{\mathcal{K}} = C\Big((M_1 + \sqrt{1+\delta_1})\delta + 3\delta_1(\sqrt{1+\delta_1} + \Delta^{-1})/(1 - \delta_1(1+\Delta^{-2})) \Big),$$

and

$$\begin{aligned} \overline{M} = {} & (\overline{M}_0 + M_1\nu\delta)(\overline{M}_1 + M_2\nu\delta) + 2(\overline{M}_1 + M_2\nu\delta)\sqrt{1+\delta_1}\, e^{M_1(h_{\max}+\varepsilon_0)} \\ & + M_2(h_{\max}+\varepsilon_0)(1+\delta_1)e^{2M_1(h_{\max}+\varepsilon_0)}, \end{aligned}$$

where

$$\nu = 2C(e^{M_1(h_{\max}+\varepsilon_0)}\sqrt{1+\delta_1} + M_0)(1-\delta_{\mathcal{K}})^{-1} + 1.$$

If these quantities together with δ, δ_1, Δ, and ε_0 satisfy the inequalities

(i) $(1+\Delta^{-2})\delta_1 < 1$,

(ii) $\delta_{\mathcal{K}} < 1$,

(iii) $2C(1-\delta_{\mathcal{K}})^{-1}\sqrt{1+\delta_1}\,\delta < \varepsilon_0$, *and*

(iv) $2\overline{M}C^2(1-\delta_{\mathcal{K}})^{-2}\delta < 1$,

then the pseudo periodic orbit $\{\mathbf{y}_k\}_{k=0}^N$ is ε-shadowed by a true periodic orbit $\{\mathbf{x}_k\}_{k=0}^N$ with

$$\varepsilon \le 2C(1-\delta_{\mathcal{K}})^{-1}\sqrt{1+\delta_1}\,\delta.$$

We shall provide a proof of this theorem in Section 4. First, however, we would like to illustrate its utility in a practical situation.

3. Periodic Orbits of the Lorenz Equations.

Here we apply the Periodic Shadowing Theorem to rigorously establish the existence of one asymptotically stable and one unstable periodic orbit of the Lorenz equations [10, 15]

$$\begin{aligned}\dot{x} &= \sigma(y-x)\\ \dot{y} &= \rho x - y - xz\\ \dot{z} &= xy - \beta z\end{aligned}$$

for certain values of the parameters σ, ρ, and β (see also [1, 5, 6, 14]). Moreover, with some additional computations as presented in Section 5, we rigorously establish the stability types of these periodic orbits.

We first present the results of our computations for the asymptotically stable periodic orbit. It appears numerically that the Lorenz equations have an asymptotically stable periodic orbit for the parameter values $\sigma = 10$, $\rho = 100.5$, $\beta = 8/3$. To generate a good pseudo periodic orbit we numerically computed the orbit for some time with a rough guess of initial conditions. Then using refined initial data we computed the pseudo periodic orbit as seen in Fig. 1. The important quantities and the necessary inequalities pertaining to this pseudo periodic orbit are tabulated in Fig. 2. These numbers manifest the fact that our Periodic Shadowing Theorem can be invoked to prove the existence of a true periodic orbit within $\varepsilon = 3.299220544489139846e-12$ of the pseudo periodic orbit.

We now briefly describe the details of the computation of the key quantities listed in Fig. 2. It is well-known that the Lorenz equations are dissipative. Using the Liapunov function

$$V(x,y,z) = \rho x^2 + \sigma y^2 + \sigma(z-2\rho)^2,$$

it is not difficult to establish that the set

$$U = \{(x,y,z) : \rho x^2 + \sigma y^2 + \sigma(z-2\rho)^2 \le \sigma\rho^2\beta^2/(\beta-1)\}$$

is forward invariant under the flow of the Lorenz equations for $\sigma \ge 1$, $\rho > 0$, and $\beta > 1$. The pseudo periodic orbits $\{\mathbf{y}_k\}_{k=0}^N$ of the Lorenz equations to which we apply the Periodic Shadowing Theorem will lie inside this forward invariant ellipsoid U. We take $\varepsilon_0 = 0.00001$ and check at each step that dist$(\mathbf{y}_k, \partial U) > \varepsilon_0$ so that

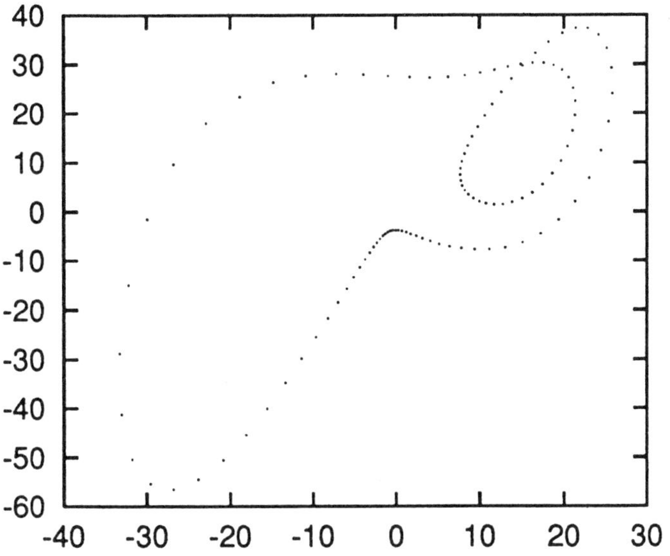

Fig 1. *A numerically computed asymptotically stable pseudo periodic orbit of the Lorenz equations for the parameter values* $\sigma = 10$, $\rho = 100.5$, $\beta = 8/3$, *and the initial data* $x(0) = 1.758904452774827471$, $y(0) = -4.480910873458781704$, *and* $z(0) = 80.99267161483650640$ *(projected onto the* (x, y)*-plane).*

if $\|\mathbf{x} - \mathbf{y}_k\| \leq \varepsilon_0$ then the solution $\varphi^t(\mathbf{x})$ lies in U for $t \geq 0$. This choice of ε_0 was guided by the fact that it should be smaller than any prospective $h_{\min}$ but not smaller than the desired shadowing distance ε.

We use a Taylor series method of order 61 with a constant step size to generate the pseudo periodic orbit; simultaneously, we use the same method on the linear variational equation to compute the matrices Y_k's. In order to have the final computed point on the pseudo periodic orbit sufficiently close to the initial point we need to estimate the period of the orbit and adjust the step size accordingly. With step size 9.765625×10^{-3} we computed $N + 1 = 112$ steps at which time $\mathbf{y}_{112}$ was close to $\mathbf{y}_0$. Then we approximated $\varphi^\tau(\mathbf{y}_{112})$ by its Taylor polynomial of 60th-order and used Newton's method to solve for τ in $< \varphi^\tau(\mathbf{y}_{112}) - \mathbf{y}_0, (1, 0, 0) >= 0$. Thus we were able to find an approximation to the period, and we divided this by 112 to obtain the constant step size $h_k = 9.7878465498488 32861 \times 10^{-3}$ to be used in the application of the Periodic Shadowing Theorem.

The choice of the method and the step size were, in part, motivated by the fact that N should not be too large in order to keep down the size of the estimated error in the matrix products in the stability calculations. By using a large-order method we can still keep δ small. The details of the implementation of the Taylor method for the Lorenz equations are given in [4].

The δ in Eqs. (2) and (4) is the sum of the local discretization error and the

Parameters: $\sigma = 10.0, \quad \beta = 8/3, \quad \rho = 100.5$

$$\begin{aligned}
\text{Taylor order} &= 61, \\
h_k &= 9.787846549848832861e-03, \\
\mathbf{y}_0 &= (1.758904452774827471, \\
&\quad -4.480910873458781704, \\
&\quad 80.99267161483650640), \\
N &= 111, \\
\text{Approx. period} &= 1.0962388136, \\
\mathbf{y}_{112} &= (1.758904452774827483, \\
&\quad -4.480910873458781727, \\
&\quad 80.99267161483650648), \\
\|\mathbf{y}_0 - \mathbf{y}_{112}\| &= 8.052703168109228659e-17, \\
\varepsilon_0 &= 0.000010, \\
M_0 &\le 36718.936567,
\end{aligned}$$

$$\begin{aligned}
M_1 &\le 308.429465, \\
M_2 &= \sqrt{2.0}, \\
\varepsilon_M &= 1.084202172485504434e-19, \\
\delta &= 1.378519452130205557e-15, \\
\delta_1 &= 1.048170747110693910e-14, \\
\Delta &\ge 1.743352294289366523e+02, \\
\overline{M}_0 &\le 1.513524418260484468e+03, \\
\overline{M}_1 &\le 7.226612380920388238e+01, \\
\Theta &\le 1.386875738727617454, \\
\|L^{-1}\| &\le 1.006455169892370874e+00, \\
C &\le 1.196653604554502446e+03, \\
\delta_{\mathcal{K}} &\le 5.482827840694540808e-10, \\
\overline{M} &\le 1.123498039473411361e+05,
\end{aligned}$$

Inequalities:

$$\begin{aligned}
&(i) && \delta_1(1+\Delta^{-2}) \le 1.048205234611193110e-14, \\
&(ii) && \delta_{\mathcal{K}} \le 5.482827840694540808e-10, \\
&(iii) && 2C(1-\delta_{\mathcal{K}})^{-1}\sqrt{1+\delta_1\delta} \le 3.299220544489139846e-12, \\
&(iv) && 2\overline{M}C^2(1-\delta_{\mathcal{K}})^{-2}\delta \le 4.435597402371528755e-04,
\end{aligned}$$

Shadowing distance $\varepsilon \le 3.299220544489139846e-12$,

Compiler: Think C v6.0.1; Operating System: Apple Mac OS v7.0.1; CPU/FPU: 68030/68882

Fig 2. *Important numbers associated with the pseudo periodic orbit of the Lorenz equations in Fig. 1.*

round-off error, as in [4]. In Eq. (3), however, since

$$\|\mathbf{y}_0 - \varphi^{h_N}(\mathbf{y}_N)\| \le \|\mathbf{y}_0 - \mathbf{y}_{N+1}\| + \|\mathbf{y}_{N+1} - \varphi^{h_N}(\mathbf{y}_N)\|,$$

we have to add to δ the quantity $\|\mathbf{y}_0 - \mathbf{y}_{N+1}\|$.

We choose S_0 so that $[f(\mathbf{y}_0)/\|f(\mathbf{y}_0)\| \mid S_0]$ is almost orthogonal in the sense that the inequalities in Eq. (5) are satisfied. Once S_0 is chosen, we apply the Gram-Schmidt procedure to get

$$[f(\mathbf{y}_{k+1}) \mid Y_k S_k] = [f(\mathbf{y}_{k+1})/\|f(\mathbf{y}_{k+1})\| \mid S_{k+1}]R_k$$

where $[f(\mathbf{y}_{k+1})/\|f(\mathbf{y}_{k+1})\| \mid S_{k+1}]$ is almost orthogonal and R_k is upper triangular with positive diagonal entries. Then we take A_k for $k = 0, \ldots, N-1$ to be the 2×2 upper triangular matrix in the lower right hand corner of R_k; we take A_N to be the computed product $S_0^* Y_N S_N$. We obtain δ_1 using the techniques described in [4].

To establish the invertibility of L and to determine the value of C_1 we need to rigorously estimate the norms of products of matrices. To make these estimates we must account for floating point errors. We obtain a rigorous upper bound for the floating point error in each product as follows: If $\widehat{A} = A + E_A$ and $\widehat{B} = B + E_B$ are approximations to the $n \times n$ matrices A and B and if $\mathrm{fl}\,(\widehat{A}\widehat{B})$ denotes the computed product $\widehat{A}\widehat{B}$ then

$$\mathrm{fl}\,(\widehat{A}\widehat{B}) = AB + E_{AB},$$

with the error term E_{AB} satisfying the bound (conf. [18])

$$(7) \qquad |E_{AB}| \leq |\widehat{A}|\,|E_B| + |E_A|\,|\widehat{B}| + |E_A|\,|E_B| + (135/128)n\varepsilon_M|\widehat{A}|\,|\widehat{B}|,$$

where the absolute value of a matrix is taken entrywise and ε_M is the machine epsilon which is assumed to obey $n\varepsilon_M < 0.1$, and the entries of $\mathrm{fl}\,(\widehat{A}\widehat{B})$ are computed in the fashion described in [17, p.18].

To minimize the estimated error in the product $A_N \cdots A_0 = I - L$ we proceed as follows: When $N + 1$ is a power of 2, we compute the product $A_N \cdots A_0$ by computing the pairs $B_0 = A_1A_0$, $B_1 = A_3A_2$, $B_3 = A_5A_4$, etc., then the pairs B_1B_0, B_3B_2, and so on (when $N + 1$ is not a power of 2 we use a slight variation of this). Then to compute a bound for $\|L^{-1}\|$ we use essentially the same method as that used with the matrix R in Section 5.

Now, it is a routine matter to compute the remaining quantities C, $\delta_{\mathcal{K}}$, and $\overline{M}$, and check the four inequalities in the hypothesis of the Periodic Shadowing Theorem. For our pseudo periodic orbit we find that it is shadowed by a true periodic orbit within $\varepsilon \leq 3.2992205444891398 46e - 12$. In Section 5, we show that this periodic orbit is hyperbolic and asymptotically stable.

We conclude this section with a brief summary of our computations on an unstable periodic orbit. It appears numerically that the Lorenz equations have an unstable periodic orbit [5] for the parameter values $\sigma = 10$, $\rho = 28$, $\beta = 8/3$. To generate a good pseudo periodic orbit we used a Newton-like method; see, for example [5, 7]. This pseudo periodic orbit is plotted in Fig. 3 and the associated important quantities are tabulated in Fig. 4. As is evident from this table, this pseudo periodic orbit is shadowed by a true periodic orbit within $\varepsilon \leq 1.799087099871078045e - 12$ of the pseudo periodic orbit.

The computation of the necessary quantities associated with the unstable pseudo periodic orbit is similar to that in the case of the asymptotically stable pseudo periodic orbit.

We should record in closing this section that during the computations outlined above, we have carefully accounted for roundoff errors by doing interval arithmetic. Since our programming environment supports rounding mode control we were able to rigorously establish the existence of these periodic orbits.

4. Proof of the Periodic Shadowing Theorem

We begin with a δ pseudo periodic orbit $\{\mathbf{y}_k\}_{k=0}^N$ of Eq. (1) and an associated sequence $\{Y_k\}_{k=0}^N$ of $n \times n$ matrices satisfying Eq. (4). We wish to show that $\{\mathbf{y}_k\}_{k=0}^N$ shadows a true periodic orbit containing $\{\mathbf{x}_k\}_{k=0}^N$, with $\mathbf{x}_k$ being contained in a hyperplane $\mathcal{H}_k$ through $\mathbf{y}_k$ and approximately normal to $f(\mathbf{y}_k)$. In fact, we will find a sequence of times $\{t_k\}_{k=0}^N$ and a sequence of points $\{\mathbf{x}_k\}_{k=0}^N$ with $\mathbf{x}_k \in \mathcal{H}_k$ and near $\mathbf{y}_k$ such that $\mathbf{x}_{k+1} = \phi^{t_k}(\mathbf{x}_k)$ for $k = 0, \ldots, N-1$ and $\mathbf{x}_0 = \varphi^{t_N}(\mathbf{x}_N)$.

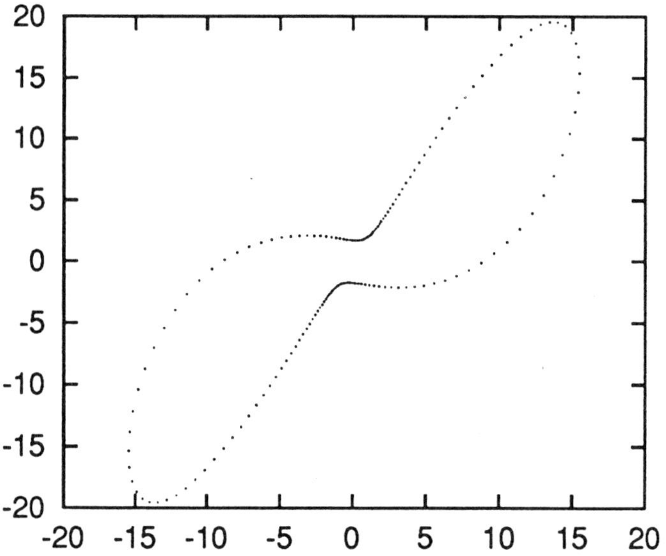

Fig 3. *A numerically computed unstable pseudo periodic orbit of the Lorenz equations (projected onto the (x, y)-plane).*

We first define $\mathcal{H}_k$ as the image of $\mathbb{R}^{n-1}$ via the map $\mathbf{z} \mapsto \mathbf{y}_k + S_k\mathbf{z}$. The problem of finding appropriate sequences of t_k's and $\mathbf{x}_k$'s becomes that of finding a sequence of times $\{t_k\}_{k=0}^N$ and a sequence of points $\{\mathbf{z}_k\}_{k=0}^N$ in $\mathbb{R}^{n-1}$ such that

$$\begin{aligned} \mathbf{y}_{k+1} + S_{k+1}\mathbf{z}_{k+1} &= \varphi^{t_k}(\mathbf{y}_k + S_k\mathbf{z}_k), \qquad k = 0, \ldots, N-1, \\ \mathbf{y}_0 + S_0\mathbf{z}_0 &= \varphi^{t_N}(\mathbf{y}_N + S_N\mathbf{z}_N). \end{aligned}$$

We next introduce the space $X = (\mathbb{R})^{N+1} \times (\mathbb{R}^{n-1})^{N+1}$ with norm

$$\|(\{s_k\}_{k=0}^N, \{\mathbf{w}_k\}_{k=0}^N)\| = \max\{\max_{0\le k\le N} |s_k|, \max_{0\le k\le N} \|\mathbf{w}_k\|\}$$

and the space $Y = (\mathbb{R}^n)^{N+1}$ with norm

$$\|\{\mathbf{g}_k\}_{k=0}^N\| = \max_{0\le k\le N} \|\mathbf{g}_k\|.$$

Now, we let B be the open set in X consisting of those $\mathbf{v} = (\{s_k\}_{k=0}^N, \{\mathbf{w}_k\}_{k=0}^N)$ with $|s_k - h_k| < \varepsilon_0$ and $\|\mathbf{w}_k\| < \varepsilon_0/\sqrt{1+\delta_1}$, and introduce the function $\mathcal{G}: B \to Y$ given by

$$(8) \qquad \begin{aligned} [\mathcal{G}(\mathbf{v})]_k &= \mathbf{y}_{k+1} + S_{k+1}\mathbf{w}_{k+1} - \varphi^{s_k}(\mathbf{y}_k + S_k\mathbf{w}_k), \qquad k = 0, \ldots, N-1, \\ [\mathcal{G}(\mathbf{v})]_N &= \mathbf{y}_0 + S_0\mathbf{w}_0 - \varphi^{s_N}(\mathbf{y}_N + S_N\mathbf{w}_N). \end{aligned}$$

Since $\|S_k\| \le \sqrt{1+\delta_1}$ it follows that $\mathcal{G}$ is well defined and that the Periodic Shadowing Theorem will be proved if we can find a solution $\bar{\mathbf{v}} = (\{t_k\}_{k=0}^N, \{\mathbf{z}_k\}_{k=0}^N)$ of the equation

$$(9) \qquad \mathcal{G}(\bar{\mathbf{v}}) = \mathbf{0}$$

Parameters: $\sigma = 10.0, \quad \beta = 8/3, \quad \rho = 28.0$

$$
\begin{aligned}
\text{Taylor order} &= 31, \\
h_k &= 9.741576316976092037e-03, \\
\mathbf{y}_0 &= (-1.278619065852397651e+01, \\
&\quad -1.936418793711800464e+01, \\
&\quad 2.400000000000000000e+01), \\
N &= 159, \\
\text{Approx. period} &= 1.558652210, \\
\mathbf{y}_{160} &= (-1.278619065852397643e+01, \\
&\quad -1.936418793711800458e+01, \\
&\quad 2.399999999999999983e+01), \\
\|\mathbf{y}_0 - \mathbf{y}_{160}\| &= 1.966311164336872854e-16, \\
\varepsilon_0 &= 0.000010, \\
M_0 &\le 5546.180770,
\end{aligned}
$$

$$
\begin{aligned}
M_1 &\le 87.040363, \\
M_2 &= \sqrt{2.0}, \\
\varepsilon_M &= 1.084202172485504434e-19, \\
\delta &= 2.979855631325919248e-16, \\
\delta_1 &= 1.526332344596874036e-15, \\
\Delta &\ge 4.167300646976115312e+01, \\
\overline{M}_0 &\le 2.105953303518521667e+02, \\
\overline{M}_1 &\le 3.191981600198188167e+01, \\
\Theta &\le 1.112237007307779190e+00, \\
\|L^{-1}\| &\le 1.003103891488919268e+00, \\
C &\le 3.018748762171755367e+03, \\
\delta_{\mathcal{K}} &\le 9.335067929495991118e-11, \\
\overline{M} &\le 6.871419725661555817e+03,
\end{aligned}
$$

Inequalities:

$$
\begin{aligned}
&(i) && \delta_1(1+\Delta^{-2}) \le 1.527211244548487840e-15, \\
&(ii) && \delta_{\mathcal{K}} \le 9.335067929495991118e-11, \\
&(iii) && 2C(1-\delta_{\mathcal{K}})^{-1}\sqrt{1+\delta_1}\delta \le 1.799087099871078045e-12, \\
&(iv) && 2\overline{M}C^2(1-\delta_{\mathcal{K}})^{-2}\delta \le 3.731862525830518931e-05,
\end{aligned}
$$

Shadowing distance $\varepsilon \le 1.799087099871078045e-12$,

Compiler: Think C v6.0.1; Operating System: Apple Mac OS v7.0.1; CPU/FPU: 68030/68882

Fig 4. *Important numbers associated with the pseudo periodic orbit of the Lorenz equations in Fig. 3.*

in the closed ball of radius $\varepsilon/\sqrt{1+\delta_1}$ with $\varepsilon \le 2C(1-\delta_{\mathcal{K}})^{-1}\sqrt{1+\delta_1}\delta$ about $\mathbf{v}_0 = (\{h_k\}_{k=0}^{N}, \mathbf{0})$. To solve Eq. (9) we use the following lemma:

LEMMA. *Let X and Y be finite dimensional real vector spaces of the same dimension, let B be an open subset of X, and let $\mathcal{G}: B \to Y$ be a C^2 function satisfying the following properties:*

(i) *The derivative $D\mathcal{G}(\mathbf{v}_0)$ at $\mathbf{v}_0 \in B$ has an inverse $\mathcal{K}$;*

(ii) *The closed ball about $\mathbf{v}_0$ with radius $\bar{\varepsilon}$, where*

$$\bar{\varepsilon} = 2\,\|\mathcal{K}\|\,\|\mathcal{G}(\mathbf{v}_0)\|,$$

is contained in B;

(iii) *The inequality*

$$2M\,\|\mathcal{K}\|^2\,\|\mathcal{G}(\mathbf{v}_0)\| < 1$$

holds, where

$$M = \sup\left\{\,\|D^2\mathcal{G}(\mathbf{v})\| : \|\mathbf{v}-\mathbf{v}_0\| \le \bar{\varepsilon}\,\right\}.$$

Then there is a unique solution $\bar{\mathbf{v}}$, satisfying $\|\bar{\mathbf{v}}-\mathbf{v}_0\| \le \bar{\varepsilon}$, of the equation

$$\mathcal{G}(\bar{\mathbf{v}}) = \mathbf{0}.$$

A proof of this lemma can be constructed, using the contraction mapping principle, in a way similar to the one in [3].

We now proceed to verify that the map $\mathcal{G}$ in Eq. (8), which is clearly C^2, does indeed satisfy the hypotheses of the lemma.

Verification of hypothesis (i) of Lemma. Notice that for $\mathbf{u} = (\{\tau_k\}_{k=0}^N, \{\xi_k\}_{k=0}^N)$ in X, the derivative of $\mathcal{G}$ at $\mathbf{v}_0$ is given by

$$\begin{aligned} [D\mathcal{G}(\mathbf{v}_0)\mathbf{u}]_k &= -\tau_k f(\varphi^{h_k}(\mathbf{y}_k)) + S_{k+1}\xi_{k+1} - D\varphi^{h_k}(\mathbf{y}_k)S_k\xi_k, \qquad k = 0, \dots, N-1, \\ [D\mathcal{G}(\mathbf{v}_0)\mathbf{u}]_N &= -\tau_N f(\varphi^{h_N}(\mathbf{y}_N)) + S_0\xi_0 - D\varphi^{h_N}(\mathbf{y}_N)S_N\xi_N. \end{aligned}$$

We will approximate $D\mathcal{G}(\mathbf{v}_0)$ by another operator T. In order to define T, we need to prove the invertibility of the operator

$$\begin{bmatrix} f(\mathbf{y}_k)^* \\ S_k^* \end{bmatrix},$$

which we will take to be a linear operator from $\mathbb{R}^n$ equipped with the Euclidean norm to $\mathbb{R} \times \mathbb{R}^{n-1}$ equipped with the norm $\|(s, \mathbf{w})\| = \max\{|s|, \|\mathbf{w}\|\}$.

Notice that

$$\begin{bmatrix} f(\mathbf{y}_k)^* \\ S_k^* \end{bmatrix} \left[f(\mathbf{y}_k)/\|f(\mathbf{y}_k)\|^2 \mid S_k \right] = I - H, \tag{10}$$

where

$$H = \begin{bmatrix} 0 & -f(\mathbf{y}_k)^* S_k \\ -S_k^* f(\mathbf{y}_k)/\|f(\mathbf{y}_k)\|^2 & I - S_k^* S_k \end{bmatrix}.$$

We regard H as an operator from $\mathbb{R} \times \mathbb{R}^{N-1}$ into itself with the norm as given above. Then using hypothesis (i) of the Periodic Shadowing Theorem we find that

$$\|H\| \le \delta_1(1 + \Delta^{-2}) < 1,$$

and thus, by [12], the operator $I - H$ is invertible with inverse having norm bounded by $(1 - \delta_1(1 + \Delta^{-2}))^{-1}$. It follows from Eq. (10) that $\begin{bmatrix} f(\mathbf{y}_k)^* \\ S_k^* \end{bmatrix}$ is invertible with inverse

$$\left[f(\mathbf{y}_k)/\|f(\mathbf{y}_k)\|^2 \mid S_k \right] (I - H)^{-1}$$

which has norm bounded by

$$(\|S_k\| + 1/\|f(\mathbf{y}_k)\|)(1 - \delta_1(1 + \Delta^{-2}))^{-1}.$$

Hence

$$\left\| \begin{bmatrix} f(\mathbf{y}_k)^* \\ S_k^* \end{bmatrix}^{-1} \right\| \le (\sqrt{1 + \delta_1} + \Delta^{-1})(1 - \delta_1(1 + \Delta^{-2}))^{-1}. \tag{11}$$

Now we define the operator $T: X \to Y$ for $\mathbf{u} \in X$ and $k = 0, \dots, N-1$ as

$$[T\mathbf{u}]_k = \begin{bmatrix} f(\mathbf{y}_{k+1})^* \\ S_{k+1}^* \end{bmatrix}^{-1} \begin{bmatrix} -\|f(\mathbf{y}_{k+1})\|^2 \tau_k - f(\mathbf{y}_{k+1})^* Y_k S_k \xi_k \\ \xi_{k+1} - A_k \xi_k \end{bmatrix}$$

and for $k = N$ as

$$[T\mathbf{u}]_N = \begin{bmatrix} f(\mathbf{y}_0)^* \\ S_0^* \end{bmatrix}^{-1} \begin{bmatrix} -\|f(\mathbf{y}_0)\|^2\tau_N - f(\mathbf{y}_0)^* Y_N S_N \xi_N \\ \xi_0 - A_N \xi_N \end{bmatrix}.$$

We obtain the inverse for $D\mathcal{G}(\mathbf{v}_0)$ from the inverse for T which is given, for $\mathbf{g} \in Y$, by

$$T^{-1}\mathbf{g} = (\{\tau_k\}_{k=0}^N, \{\xi_k\}_{k=0}^N),$$

where

$$\xi_0 = L^{-1}(S_0^* \mathbf{g}_N + \sum_{m=1}^{N} A_N \cdots A_m S_m^* \mathbf{g}_{m-1}),$$

$$\xi_k = A_{k-1} \cdots A_0 \xi_0 + \sum_{m=1}^{k} A_{k-1} \cdots A_m S_m^* \mathbf{g}_{m-1}, \qquad k = 1, \ldots, N,$$

and

$$\tau_k = -\frac{1}{\|f(\mathbf{y}_{k+1})\|^2} f(\mathbf{y}_{k+1})^* (Y_k S_k \xi_k + \mathbf{g}_k), \qquad k = 0, \ldots, N-1$$

$$\tau_N = -\frac{1}{\|f(\mathbf{y}_0)\|^2} f(\mathbf{y}_0)^* (Y_N S_N \xi_N + \mathbf{g}_N).$$

Notice that

(12) $$\|T^{-1}\| \leq C,$$

where the constant C is given in the statement of the Periodic Shadowing Theorem.

Now we use T^{-1} to construct the inverse $\mathcal{K}$ of $D\mathcal{G}(\mathbf{v}_0)$. If we can show

(13) $$\|T^{-1}(D\mathcal{G}(\mathbf{v}_0) - T)\| \leq \delta_{\mathcal{K}}$$

then it will follow from hypothesis (ii) of the Periodic Shadowing Theorem and [12] that the inverse $(I + T^{-1}(D\mathcal{G}(\mathbf{v}_0) - T))^{-1}$ exists and that

(14) $$\mathcal{K} = (I + T^{-1}(D\mathcal{G}(\mathbf{v}_0) - T))^{-1} T^{-1}$$

is the inverse of $D\mathcal{G}(\mathbf{v}_0)$, with

(15) $$\|\mathcal{K}\| \leq C/(1 - \delta_{\mathcal{K}}).$$

To establish the inequality in Eq. (13) it is convenient to introduce the operator $\overline{T}: X \to Y$ defined by

$$[\overline{T}\mathbf{u}]_k = -\tau_k f(\mathbf{y}_{k+1}) + S_{k+1}\xi_{k+1} - Y_k S_k \xi_k, \qquad k = 0, \ldots, N-1,$$

$$[\overline{T}\mathbf{u}]_N = -\tau_N f(\mathbf{y}_0) + S_0 \xi_0 - Y_N S_N \xi_N.$$

Now, by adding and subtracting $\overline{T}$, we can estimate the term on the left-hand side of Eq. (13) as follows:

(16) $$\|T^{-1}(D\mathcal{G}(\mathbf{v}_0) - T)\| \leq \|T^{-1}\|(\|D\mathcal{G}(\mathbf{v}_0) - \overline{T}\| + \|\overline{T} - T\|).$$

For $\mathbf{u}$ in X and $k = 0, \ldots, N-1$, we have

$$\begin{aligned}\|[(D\mathcal{G}(\mathbf{v}_0) - \overline{T})\mathbf{u}]_k\| &\le \big(\|f(\varphi^{h_k}(\mathbf{y}_k)) - f(\mathbf{y}_{k+1})\| + \|(D\varphi^{h_k}(\mathbf{y}_k) - Y_k)S_k\|\big)\|\mathbf{u}\| \\ &\le \big(M_1\|\varphi^{h_k}(\mathbf{y}_k) - \mathbf{y}_{k+1}\| + \sqrt{1+\delta_1}\|D\varphi^{h_k}(\mathbf{y}_k) - Y_k\|\big)\|\mathbf{u}\| \\ &\le (M_1 + \sqrt{1+\delta_1})\delta\|\mathbf{u}\|.\end{aligned}$$

When $k = N$ a similar inequality holds and thus

$$\|D\mathcal{G}(\mathbf{v}_0) - \overline{T}\| \le \big(M_1 + \sqrt{1+\delta_1}\big)\delta. \tag{17}$$

Notice also that for $k = 0, \ldots, N-1$, we have

$$[(\overline{T} - T)\mathbf{u}]_k = \begin{bmatrix} f(\mathbf{y}_{k+1})^* \\ S_{k+1}^* \end{bmatrix}^{-1} \begin{bmatrix} f(\mathbf{y}_{k+1})^* S_{k+1}\xi_{k+1} \\ (A_k - S_{k+1}^* Y_k S_k)\xi_k - S_{k+1}^* f(\mathbf{y}_{k+1})\tau_k - (I - S_{k+1}^* S_{k+1})\xi_{k+1} \end{bmatrix}$$

and hence, using the inequalities in Eqs. (5), (6), and (11), we get

$$\|[(\overline{T} - T)\mathbf{u}]_k\| \le 3\delta_1(\sqrt{1+\delta_1} + \Delta^{-1})(1 - \delta_1(1+\Delta^{-2}))^{-1}\|\mathbf{u}\|.$$

A similar inequality holds when $k = N$ and thus

$$\|\overline{T} - T\| \le 3\delta_1(\sqrt{1+\delta_1} + \Delta^{-1})(1 - \delta_1(1+\Delta^{-2}))^{-1}. \tag{18}$$

Combining the inequalities in Eqs. (12), (16), (17), and (18) we obtain the desired inequality in Eq. (13). Therefore, $\mathcal{K}$ defined in Eq. (14) is the inverse of $D\mathcal{G}(\mathbf{v}_0)$ and its norm satisfies the inequality in Eq. (15). Thus we have verified hypothesis (i) of the Lemma.

Verification of hypothesis (ii) of Lemma. Since

$$\bar{\varepsilon} = 2\|\mathcal{K}\|\|\mathcal{G}(\mathbf{v}_0)\| \le 2C\delta/(1 - \delta_{\mathcal{K}}), \tag{19}$$

it follows from hypothesis (iii) of the Periodic Shadowing Theorem that the closed ball of radius $\bar{\varepsilon}$ around $\mathbf{v}_0$ is contained in the open set B.

Verification of hypothesis (iii) of Lemma. If $\mathbf{v} = (\{s_k\}_{k=0}^N, \{\mathbf{w}_k\}_{k=0}^N)$, $\mathbf{u} = (\{\tau_k\}_{k=0}^N, \{\xi_k\}_{k=0}^N)$, and $\bar{\mathbf{u}} = (\{\sigma_k\}_{k=0}^N, \{\eta_k\}_{k=0}^N)$, one calculates, for $k = 0, \ldots, N$, that

$$\begin{aligned}[D^2\mathcal{G}(\mathbf{v})\mathbf{u}\bar{\mathbf{u}}]_k &= -\tau_k\sigma_k\, Df(\varphi^{s_k}(\mathbf{y}_k + S_k\mathbf{w}_k))\, f(\varphi^{s_k}(\mathbf{y}_k + S_k\mathbf{w}_k)) \\ &\quad - \tau_k\, Df(\varphi^{s_k}(\mathbf{y}_k + S_k\mathbf{w}_k))\, D\varphi^{s_k}(\mathbf{y}_k + S_k\mathbf{w}_k)\, S_k\eta_k \\ &\quad - \sigma_k\, Df(\varphi^{s_k}(\mathbf{y}_k + S_k\mathbf{w}_k))\, D\varphi^{s_k}(\mathbf{y}_k + S_k\mathbf{w}_k)\, S_k\xi_k \\ &\quad - D^2\varphi^{s_k}(\mathbf{y}_k + S_k\mathbf{w}_k)\,(S_k\xi_k)\,(S_k\eta_k).\end{aligned} \tag{20}$$

We use the following estimates, which follow easily from Gronwall's lemma and the variation of constants formula (see [2] for details)

$$\|D\varphi^t(\mathbf{x})\| \le e^{M_1 t}, \qquad \|D^2\varphi^t(\mathbf{x})\| \le M_2 t e^{2M_1 t} \tag{21}$$

when $0 \le t \le h_k + \varepsilon_0$ and $\|\mathbf{x} - \mathbf{y}_k\| \le \varepsilon_0$ for each $0 \le k \le N$ since $\varphi^t(\mathbf{x})$ remains in U for $0 \le t \le h_k + \varepsilon_0$.

Notice that if $\mathbf{v} = \left(\{s_k\}_{k=0}^N, \{\mathbf{w}_k\}_{k=0}^N\right)$ is in the ball in X of radius $\bar{\varepsilon}$ about $\mathbf{v}_0$, then for $k = 0, \ldots, N-1$

$$\begin{aligned}\|\varphi^{s_k}(\mathbf{y}_k + S_k\mathbf{w}_k) - \mathbf{y}_{k+1}\| &\le \|\varphi^{s_k}(\mathbf{y}_k + S_k\mathbf{w}_k) - \varphi^{s_k}(\mathbf{y}_k)\| \\ &\quad + \|\varphi^{s_k}(\mathbf{y}_k) - \varphi^{h_k}(\mathbf{y}_k)\| + \|\varphi^{h_k}(\mathbf{y}_k) - \mathbf{y}_{k+1}\| \\ &\le e^{M_1(h_k+\varepsilon_0)}\|S_k\mathbf{w}_k\| + M_0|s_k - h_k| + \delta \\ &\le (e^{M_1(h_{\max}+\varepsilon_0)}\sqrt{1+\delta_1} + M_0)\bar{\varepsilon} + \delta. \\ &\le (2C(e^{M_1(h_{\max}+\varepsilon_0)}\sqrt{1+\delta_1} + M_0)(1-\delta_{\mathcal{K}})^{-1} + 1)\delta.\end{aligned}$$

If we set $\nu = 2C(e^{M_1(h_{\max}+\varepsilon_0)}\sqrt{1+\delta_1} + M_0)(1-\delta_{\mathcal{K}})^{-1} + 1$, it follows that for $k = 0, \ldots, N-1$

$$\begin{aligned}\|f(\varphi^{s_k}(\mathbf{y}_k + S_k\mathbf{w}_k))\| &\le \|f(\mathbf{y}_{k+1})\| + \|f(\varphi^{s_k}(\mathbf{y}_k + S_k\mathbf{w}_k)) - f(\mathbf{y}_{k+1})\| \\ &\le \overline{M}_0 + M_1\nu\delta,\end{aligned} \tag{22}$$

and

$$\begin{aligned}\|Df(\varphi^{s_k}(\mathbf{y}_k + S_k\mathbf{w}_k))\| &\le \|Df(\mathbf{y}_{k+1})\| \\ &\quad + \|Df(\varphi^{s_k}(\mathbf{y}_k + S_k\mathbf{w}_k)) - Df(\mathbf{y}_{k+1})\| \\ &\le \overline{M}_1 + M_2\nu\delta.\end{aligned} \tag{23}$$

When $k = N$ we get the same estimates as above by simply replacing every occurrence of $\mathbf{y}_{k+1}$ by $\mathbf{y}_0$.

Thus from Eqs. (20), (21), (22), and (23), one sees that for $\|\mathbf{v} - \mathbf{v}_0\| \le \bar{\varepsilon}$

$$\begin{aligned}\|D^2\mathcal{G}(\mathbf{v})\| \le \overline{M} &= (\overline{M}_0 + M_1\nu\delta)(\overline{M}_1 + M_2\nu\delta) \\ &\quad + 2(\overline{M}_1 + M_2\nu\delta)\sqrt{1+\delta_1}\, e^{M_1(h_{\max}+\varepsilon_0)} \\ &\quad + M_2(h_{\max} + \varepsilon_0)(1+\delta_1)e^{2M_1(h_{\max}+\varepsilon_0)}.\end{aligned}$$

Then hypothesis (iv) of the Periodic Shadowing Theorem implies that hypothesis (iii) of the Lemma is satisfied.

At last, all hypotheses of the Lemma have been verified and the conclusion of the Periodic Shadowing Theorem follows by taking $\varepsilon = \bar{\varepsilon}\sqrt{1+\delta_1}$ and using Eq. (19).

5. Stability of the Periodic Orbits

In this section we rigorously prove that the periodic orbits of the Lorenz equations found in Section 3 are indeed hyperbolic; one is stable and the other unstable.

As is well-known, the periodic orbit $\varphi^t(\mathbf{x}_0)$ with period T is asymptotically stable if all but one of the eigenvalues of the multiplier matrix $D\varphi^T(\mathbf{x}_0)$ lie inside the unit circle. Since $T = t_0 + \cdots + t_N$, by the chain rule, we have $D\varphi^T(\mathbf{x}_0) = D\varphi^{t_N}(\mathbf{x}_N) \cdots D\varphi^{t_0}(\mathbf{x}_0)$. We now approximate $D\varphi^T(\mathbf{x}_0)$ by the product $Y_N \cdots Y_0$.

For this purpose, we use the inequalities in Eq. (21) to estimate for $k = 0, \ldots, N$ that

$$
\begin{aligned}
\|Y_k - D\varphi^{t_k}(\mathbf{x}_k)\| &\leq \|Y_k - D\varphi^{h_k}(\mathbf{y}_k)\| + \|D\varphi^{h_k}(\mathbf{y}_k) - D\varphi^{h_k}(\mathbf{x}_k)\| \\
&\quad + \|D\varphi^{h_k}(\mathbf{x}_k) - D\varphi^{t_k}(\mathbf{x}_k)\| \\
&\leq \delta + M_2 h_k e^{2M_1 h_k} \|\mathbf{y}_k - \mathbf{x}_k\| + M_1 e^{M_1(h_k+\varepsilon)} |h_k - t_k| \\
&\leq \delta + \left[M_2 h_{\max} e^{2M_1 h_{\max}} + M_1 e^{M_1(h_{\max}+\varepsilon)} \right] \varepsilon.
\end{aligned}
$$

(24)

For the pseudo periodic orbit of the Lorenz equations at the parameter values $\sigma = 10$, $\rho = 100.5$, $\beta = 8/3$ we compute that the product $Y_{111} \cdots Y_0$ is approximately the matrix

$$
\begin{bmatrix}
0.1148298293844557445 & 0.1407215703402642672 & 0.2711131402272012236 \\
-0.2071367041800333639 & -0.1999042129801062231 & -0.1501919431735588229 \\
0.5964535334427969472 & 0.6576303795399115119 & 0.9477065512018716121
\end{bmatrix}.
$$

In an attempt to diagonalize this matrix we consider the matrix

$$
R = \begin{bmatrix}
-0.2648354535051923767 & -0.2347679420252267349 & -0.7750378044745185866 \\
0.1646461037014009290 & -1.026627029777592155 & 0.9121741860885619132 \\
-0.9501335922398737178 & 0.7512581695968794036 & -0.1451916627281834783
\end{bmatrix}
$$

and its approximate computed inverse

$$
R_I = \begin{bmatrix}
-0.5362206239393113422 & -0.6163398302158273953 & -1.009824015609569126 \\
-0.8427820946186850489 & -0.6979375534432878082 & 0.1139691095204763533 \\
-0.8517410971951629847 & 0.4220207061438613757 & 0.3105408619402872189
\end{bmatrix}.
$$

Notice that

$$
|I - RR_I| \leq |I - fl(RR_I)| + |fl(RR_I) - RR_I|
$$

where, since $3\varepsilon_M < 0.1$, by [18] and Eq. (7),

$$
|fl(RR_I) - RR_I| \leq 3(135/128)\varepsilon_M |R| |R_I|
$$

and we obtain a rigorous upper bound for $|I - fl(RR_I)|$ using interval arithmetic. Therefore, $\|I - RR_I\| \leq \| |I - RR_I| \| \leq 1.312708516166413284 \times 10^{-18}$. Since this number is less than 1, the matrix RR_I is invertible, R has the inverse $R^{-1} = R_I(RR_I)^{-1}$, and
(25)
$$
\|R^{-1} - R_I\| \leq \|R_I\| (1 - \|I - RR_I\|)^{-1} \|I - RR_I\| \leq 2.060352391909747594 \times 10^{-18}.
$$

Using the same approach as that described in Section 3 for the computation of the product $A_N \cdots A_0$, we compute $\mathrm{fl}(R_I Y_{111} \cdots Y_0 R)$ and obtain the almost diagonal matrix

$$
V = \begin{bmatrix}
1.000000000000000004 & 5.095750210681870840e{-18} & 1.041003492175535117e{-18} \\
9.757819552369539906e{-19} & -0.1373655648671290939 & -2.718975760686304088e{-18} \\
1.123589204532829400e{-18} & 3.415660359802966117e{-19} & -2.267526649762505285e{-6}
\end{bmatrix}.
$$

Now the technique embodied in Eq. (7) together with the inequalities in Eqs. (24) and (25) enable us to estimate that

$$|V - R^{-1} D\varphi^T(\mathbf{x}_0) R| \\ \leq \begin{bmatrix} 1.077917416083018056e{-4} & 3.331731421683138055e{-4} & 1.269720799321621539e{-5} \\ 1.769175581848260459e{-4} & 8.163683313532197181e{-4} & 3.058830423944958537e{-5} \\ 8.596132816290476153e{-6} & 4.212223020875922222e{-6} & 8.568644116331711265e{-8} \end{bmatrix}.$$

By Gerschgorin's Theorem, the eigenvalues of $R^{-1}D\varphi^T(\mathbf{x}_0)R$, and hence of $D\varphi^T(\mathbf{x}_0)$, are contained in the union of the three disks with centers 1.000000000000000004, -0.1373655648671290939, $-2.267526649762505285 \times 10^{-6}$ and radii $4.536620917698379632 \times 10^{-4}$, $1.023874193777499045 \times 10^{-3}$, $1.289404227833118065 \times 10^{-5}$. These three disks are disjoint and thus there is exactly one real eigenvalue in each disk.

One of the eigenvalues of $D\varphi^T(\mathbf{x}_0)$ is 1. Also their product is

$$\det D\varphi^T(\mathbf{x}_0) = e^{\int_0^T \nabla\cdot f(\varphi^t(\mathbf{x}_0))dt} = e^{-(\sigma+1+\beta)T}.$$

Since $\sigma = 10$, $\beta = 8/3$ and $|T - 1.096238813583069280| < 3.695127009827836627 \times 10^{-10}$, we see that $3.11480077 \times 10^{-7} \leq \det D\varphi^T(\mathbf{x}_0) \leq 3.11480081 \times 10^{-7}$. Now the third eigenvalue is just $\det D\varphi^T(\mathbf{x}_0)$ divided by the second eigenvalue which lies in the interval $[-0.139, -0.136]$. So the third eigenvalue lies in the interval $[-0.00000230, -0.00000224]$. Consequently, our periodic orbit is hyperbolic and asymptotically stable.

For the periodic orbit in the case $\sigma = 10$, $\beta = 8/3$, and $\rho = 28$ we find that the two other (besides 1) eigenvalues lie in the intervals $[1.190002 \times 10^{-10}, 1.190008 \times 10^{-10}]$ and $[4.712938, 4.712956]$. Hence this periodic orbit is hyperbolic and unstable. The details of the computations are akin to the ones already given for the asymptotically stable periodic orbit.

References

1. E. Adams, *The reliability question for discretization of evolution equations*, in *Scientific computing with automatic result verification,* edited by E. Adams and U. Kulisch, Academic Press, San Diego (1993), 423–526.
2. B.A. Coomes, H. Koçak, and K.J. Palmer, *Shadowing orbits of ordinary differential equations*, to appear in J. Comp. Appl. Math.
3. B.A. Coomes, H. Koçak, and K.J. Palmer, *A shadowing theorem for ordinary differential equations*, preprint.
4. B.A. Coomes, H. Koçak, and K.J. Palmer, *Rigorous computational shadowing of orbits of ordinary differential equations*, preprint.
5. J.H. Curry, *An algorithm for finding closed orbits*, in *Global Theory of Dynamical Systems,* edited by Z. Nitecki and C. Robinson, Lecture Notes in Math. **819** (1979), 111–120.
6. S. De Gregorio, *The study of periodic orbits of dynamical systems. The use of a computer*, J. Stat. Phys. **38** (1985), 947–972.
7. P. Deuflhard, *Computation of periodic solutions of nonlinear ODEs*, BIT **24** (1984), 456–466.

8. V. Franceschini, *Numerical methods for studying periodic and quasiperiodic orbits in dissipative differential equations*, in *Applications of Mathematics in Technology*, Teubner, Stuttgart (1984), 212–226.
9. J.E. Franke and J.F. Selgrade, *A computer method for verification of asymptotically stable periodic orbits*, SIAM J. Math. Anal. **10** (1979), 614–628.
10. E.N. Lorenz, *Deterministic nonperiodic flow*, J. Atmos. Sci. **20** (1963), 130–141.
11. E. Reithmeier, *Periodic Solutions of Nonlinear Dynamical Systems,* Lecture Notes in Math. **1483**, Springer-Verlag, Berlin, 1991.
12. W. Rudin, *Functional Analysis*, McGraw-Hill, Inc., New York, 1991.
13. I.B. Schwartz, *Estimating regions of existence of unstable periodic orbits using computer-based techniques*, SIAM J. Numer. Anal. **20** (1983), 106–120.
14. Ja. G. Sinai and E.B. Vul, *Discovery of closed orbits of dynamical systems with the use of computers*, J. Stat. Phys. **23** (1980), 27–47.
15. C. Sparrow, *The Lorenz Equations: Bifurcations, Chaos, and Strange Attractors*, Springer-Verlag, New York, 1982.
16. M. Urabe, *Nonlinear Autonomous Oscillations,* Academic Press, New York, 1967.
17. J.H. Wilkinson, *Rounding Errors in Algebraic Processes,* Prentice-Hall, Englewood Cliffs, New Jersey, 1963.
18. J.H. Wilkinson, *The Algebraic Eigenvalue Problem,* Clarendon Press, Oxford, 1965.

Department of Mathematics and Computer Science,
University of Miami,
Coral Gables, Florida 33124
E-mail address: coomes@mthvax.cs.miami.edu, hk@math.miami.edu, kjp@paris.cs.miami.edu

Contemporary Mathematics
Volume **172**, 1994

On well–posed problems for connecting orbits in dynamical systems

W.-J. BEYN

ABSTRACT. We develop formulations of well–posed problems for orbits which connect steady states to periodic orbits or periodic orbits to each other in a dynamical system. It turns out that the property of asymptotic phase on the periodic side plays a crucial role for the resulting boundary value problem on the real line. Our approach is closely related to a paper by Hale and Lin [HaLi 86] where Liapunov–Schmidt type methods and associated bifurcation functions have been developed for periodic–to–periodic connections in functional differential equations. In our formulation we avoid any non– autonomous transformation of the independent variable and we keep the periodic orbits as part of the problem. The boundary value problems thus obtained, are directly amenable to numerical approximation schemes on finite intervals.

1. Introduction

In this introduction we consider the general case of two compact invariant manifolds which are connected by an orbit of a given parametrized dynamical system. Our aim here is to establish a relation between the dimensions of the unstable manifolds of the invariant manifolds and the number of parameters for which we expect such a connecting orbit to occur generically. We will also outline how the formulation of a well–posed problem should look like for connecting orbits of this general type.

While these considerations will be mainly nonrigorous the main body of the paper is designed to provide the analytical details for the special case in which the connected manifolds are stationary points or periodic orbits. We are particularly interested in the formulation of well–posed boundary value problems which can be tackled numerically.

1991 *Mathematics Subject Classification.* Primary 58F22, 65L10; Secondary 58F18.
This papers is in final form and no version of it will be submitted for publication elsewhere.

Consider a parametrized dynamical system

$$\dot{x} = f(x, \lambda), \quad x(t) \in \mathbb{R}^m, \ \lambda \in \mathbb{R}^p \tag{1.1}$$

where $f : \mathbb{R}^m \times \mathbb{R}^p \to \mathbb{R}^m$ is assumed to be sufficiently smooth. In many instances it will be convenient to work with the variables $z = (x, \lambda) \in \mathbb{R}^{m+p}$ and to rewrite (1.1) as

$$\dot{z} = g(z), \quad g(x, \lambda) = (f(x, \lambda), 0). \tag{1.2}$$

Any compact invariant set $M \subset \mathbb{R}^{m+p}$ of this system is trivially foliated

$$M = \bigcup_{\lambda \in \Lambda} (M(\lambda) \times \{\lambda\}) \tag{1.3}$$

where $\Lambda \subset \mathbb{R}^p$ is compact and the $M(\lambda)$ are compact invariant sets of the system (1.1).

Let $M_+, M_- \subset \mathbb{R}^{m+p}$ be two such compact invariant sets and let $z(t) = (x(t), \lambda(t))$, $t \in \mathbb{R}$ be a solution of (1.2). Then the orbit $\gamma = \{(x(t), \ \lambda(t)) : t \in \mathbb{R}\}$ will be called a **connecting orbit from M_- to M_+ if**

$$\text{dist } (z(t), M_\pm) \to 0 \ \text{ as } \ t \to \pm\infty. \tag{1.4}$$

In particular, the α– and ω–limit sets satisfy

$$\alpha(\gamma) \subset M_-, \quad \omega(\gamma) \subset M_+. \tag{1.5}$$

In case $\alpha(\gamma) = \omega(\gamma)$ we call γ a **homoclinic orbit** and a **heteroclinic orbit** otherwise.

Of course we may rephrase (1.4) as

$$\text{dist } (x(t), \ M_\pm(\lambda(0))) \to 0 \ \text{ as } \ t \to \pm\infty$$

where $M_\pm(\lambda)$ are the sets in the decomposition of $M_\pm$ corresponding to (1.3).

Now let us assume in addition, that $M_\pm$ are smooth invariant manifolds of dimension $m_{\pm c} + p$ where $m_{\pm c}$ is the dimension of the manifolds $M_\pm(\lambda)$ in the decomposition (1.3). We want to determine the number p of parameters for which we expect the connecting orbits above to be isolated and stable phenomena in the system (1.1). In this case we also expect the connecting orbits to occur in a generic sense.

Let $M_\pm(\lambda)$, $\lambda \in \Lambda$, have stable and unstable manifolds $M_{\pm s}(\lambda)$, $M_{\pm u}(\lambda)$ which are of dimension $m_{\pm c} + m_{\pm s}$ and $m_{\pm c} + m_{\pm u}$ respectively and let these be independent of $\lambda \in \Lambda$ (e.g. let $M_\pm(\lambda)$ have a hyperbolic structure, see [Irw 80], [HPS 77]). Then we have $m = m_{\pm c} + m_{\pm s} + m_{\pm u}$ and

$$\gamma \subset M_{-u} \cap M_{+s} \ \text{ where } \ M_{-u} = \bigcup_{\lambda} (M_{-u}(\lambda) \times \{\lambda\}), \ M_{+s} = \bigcup_{\lambda} (M_{+s}(\lambda) \times \{\lambda\}).$$

M_{-u} and M_{+s} are in fact center–unstable resp. center-stable manifolds for the system (1.2).

We expect γ to be an isolated connecting orbit if

$$(1.6) \qquad T_{z(t)}M_{-u} \cap T_{z(t)}M_{+s} = T_{z(t)}\gamma = \text{span } \{\dot{z}(t)\} \quad \text{for all} \quad t \in \mathbb{R}$$

holds for the tangent spaces at $z(t)$. Moreover, the connecting orbit should persist in p–parameter systems if the intersection is transversal, i.e.

$$(1.7) \qquad T_{z(t)}M_{-u} + T_{z(t)}M_{+s} = \mathbb{R}^{m+p}.$$

Counting dimensions we obtain from (1.6) and (1.7)

$$p + m_{-u} + m_{-c} + m_{+s} + m_{+c} + p - 1 = m + p$$

and thus

$$(1.8) \qquad p = m_{+u} - m_{-u} - m_{-c} + 1.$$

We notice that this relation also makes sense in the framework of partial differential evolution equations provided the center and center–unstable manifolds are finite dimensional. Further, it can easily be seen, that any of the three statements (1.6) to (1.8) is implied by the other two.

In the steady case we have $m_{-c} = m_{+c} = 0$ and (1.8) is the relation discussed in [Bey 90a]. In cases where periodics are involved the right hand side of (1.8) turns out to be the negative of the index as defined in [HaLi 86]. In the general case we define the index of the connecting orbit as

$$\text{ind } (\gamma) = m_{-u} + m_{-c} - 1 - m_{+u}$$

and this turns out to be the Fredholm index of some suitable linearization around γ (see Proposition 2.8).

If ind $(\gamma)+p = 0$, i.e. (1.8) holds, then we set up a well–posed problem for γ. In the case ind $(\gamma)+p < 0$ we should add another $-$ind $(\gamma) - p$ parameters in order to have a well–posed problem while in case ind $(\gamma)+p > 0$ there is a ($p+$ind (γ))–dimensional manifold of connecting orbits and we may add $p+$ind (γ) conditions for its parametrization.

In the following let us assume that $z(t) = (x(t), \lambda(t))$ is an orbit connecting M_- to M_+ and let (1.6)–(1.8) hold. In view of the general theory of asymptotic stability with rate constants ([Fen 74], [Fen 77]) we look for solutions y_- and y_+ of

$$(1.9) \qquad \dot{y}_\pm = f(y_\pm, \lambda)$$

which lie on $M_{\pm}(\lambda)$ and which have the same **asymptotic phase** as $x(t)$, i.e. for some $\epsilon > 0$

$$||x(t) - y_{\pm}(t)|| = O(e^{-\epsilon|t|}) \quad \text{as} \quad t \to \pm\infty. \tag{1.10}$$

Altogether, (1.2) and (1.9) comprises a set of $3m + p$ differential equations for which we need an appropriate number of boundary conditions. We use a scalar condition which fixes the phase of the triple (x, y_-, y_+)

$$\Psi(x, y_-, y_+, \lambda) = 0 \tag{1.11}$$

and some equation which specifies that $(y_{\pm}(0), \lambda)$ is in $M_{\pm}$

$$N_{\pm}(y_{\pm}(0), \lambda) = 0. \tag{1.12}$$

The last are $m - m_{-c} + m - m_{+c}$ equations. In (1.10) we require that $x(0)$ lies in the fiber which is in asymptotic phase with $y_{\pm}(0)$ and which has dimension m_{+s} and m_{-u} respectively. Therefore we get a total of

$$m_{-s} + m_{-c} + m_{+u} + m_{+c} + 1 + 2m - m_{-c} - m_{+c} = 3m + m_{+u} - m_{-u} - m_{-c} + 1$$

boundary conditions which is precisely $3m + p$ by (1.8).

In general it can be very difficult to set up the equation (1.12) because it requires the knowledge of the manifolds $M_{\pm}(\lambda)$. In this paper we are mainly interested in the case where y_- is a steady state and $y_+ = y$ is a periodic orbit. For appropriate $\epsilon > 0$ define the spaces

$$\begin{aligned} Z_0(\epsilon) = \{(x, y) : \; & x \in C(\mathbb{R}, \mathbb{R}^m) \text{ is bounded}, \; y \in C(\mathbb{R}, \mathbb{R}^m) \text{ is 1-periodic and} \\ & ||x(t) - y(t)|| \leq C\, e^{-\epsilon t} \text{ for } t \geq 0\}, \\ Z_1(\epsilon) = \{(x, y) : \; & (x, y) \in Z_0, \; (\dot{x}, \dot{y}) \in Z_0\} \end{aligned}$$

with norms

$$\begin{aligned} ||(x, y)||_{Z_0} &= \sup_{t\in\mathbb{R}} ||x(t)|| + \sup_{t\in\mathbb{R}} ||y(t)|| + \sup_{t\geq 0} e^{\epsilon t}\, ||x(t) - y(t)||, \\ ||(x, y)||_{Z_1} &= ||(x, y)||_{Z_0} + ||(\dot{x}, \dot{y})||_{Z_0}. \end{aligned}$$

We include the period T as an unknown in the connecting orbit problem and write it in the following form

$$0 = F(x, y, T, \lambda) = \begin{pmatrix} \dot{x} - Tf(x, \lambda) \\ \dot{y} - Tf(y, \lambda) \\ \Psi(x, y, T, \lambda) \end{pmatrix} = 0 \tag{1.13}$$

where F maps $Z_1(\epsilon) \times \mathbb{R} \times \mathbb{R}^p$ into $Z_0(\epsilon) \times \mathbb{R}$.

In the case of two periodic orbits y_- and y_+ we have to determine both periods T_- and T_+ and we split up x into x_- on $\mathbb{R}_-$ and x_+ on $\mathbb{R}_+$. In some appropriate spaces we then consider the equation

$$(1.14) \quad 0 = F(x_-, y_-, x_+, y_+, T_-, T_+, \lambda) = \begin{bmatrix} \dot{x}_- - T_- f(x_-, \lambda) \\ \dot{y}_- - T_- f(y_-, \lambda) \\ \dot{x}_+ - T_+ f(x_+, \lambda) \\ \dot{y}_+ - T_+ f(y_+, \lambda) \\ x_+(0) - x_-(0) \\ \Psi(x_-, y_-, x_+, y_+, T_-, T_+, \lambda) \end{bmatrix}.$$

For both cases we will show in Section 3 that regular solutions of these operator equations (i.e. solutions at which the linearization is homeomorphic) are directly related to the transversal intersection of the manifolds M_{-u} and M_{+s} above. In Section 2 we provide the necessary preparations about linear systems, in particular a result on a special type of exponential trichotomies (compare [HaLi 86], [ChLi 90]).

Finally, we treat an example of a point–to–periodic connection from the Lorenz system ([Spa 82]). It shows that it is rather straightforward to set up and solve a boundary value problem which approximates (1.13) on a finite interval. However, a detailed analysis of the errors involved in this approximation will be deferred to a subsequent paper.

2. Preliminaries on linear systems

We consider linear differential operators

$$Lx = \dot{x} - A(t)x$$

where $A \in C(J, \mathbb{R}^{m,m})$, $x \in C^1(J, \mathbb{R}^m)$ and $J \subset \mathbb{R}$ is some interval. By $S(t,s)$, $t, s \in J$ we denote the solution operator of L, i.e. $x(t) = S(t,s)\xi$ is the solution of $Lx = 0$, $x(s) = \xi$.

In [HaLi 86] the notion of an exponential dichotomy [Cop 78] was generalized to an exponential trichotomy where a certain center part was allowed to grow or decay at an exponential rate close to an intermediate value (see also [ChLi 90]). For the linearizations about point–to–periodic connections this center part will in fact be bounded in both time directions as in an ordinary dichotomy ([Cop 78], p. 10). This motivates the following definition in which the center part is required to have an exact exponential behaviour.

DEFINITION 2.1.

L has an **ordinary exponential trichotomy** on J with exponents $\alpha < \nu < \beta$ if there exists a constant $K > 0$ and projectors $P_\kappa(t)$, $t \in J$, $\kappa = s, c, u$ of rank m_κ such that $P_s + P_c + P_u = I$ in J and such that the following conditions hold for all $t \geq s$ in J

$$(2.1) \qquad S(t,s)P_\kappa(s) = P_\kappa(t)S(t,s), \;\; \kappa = s, c, u$$

$$\begin{aligned} &||S(t,s)P_s(s)|| \le Ke^{\alpha(t-s)}, \; ||S(s,t)P_u(t)|| \le Ke^{-\beta(t-s)} \\ &||S(t,s)P_c(s)|| \le Ke^{\nu(t-s)}, \; ||S(s,t)P_c(t)|| \le Ke^{-\nu(t-s)}. \end{aligned} \tag{2.2}$$

As in [HaLi 86] we speak of a **shifted exponential dichotomy** if the center part is trivial ($P_c = 0$) and of an **exponential dichotomy** if in addition $\alpha < 0 < \beta$.

An easy calculation shows that (2.1) holds if and only if P_κ satisfies Liapunov's equation

$$P_\kappa = AP_\kappa - P_\kappa A \text{ in } J. \tag{2.3}$$

Further, if $P_\kappa(0) = X(0)\Phi(0)^T$ holds for some matrices $X(0), \Phi(0) \in \mathbb{R}^{m,m_\kappa}$ then we also have

$$P_\kappa(t) = X(t)\Phi(t)^T \text{ where } X(t) = S(t,0)X(0), \Phi(t) = S(0,t)^T\Phi(0). \tag{2.4}$$

Finally, we notice that if L has an ordinary exponential trichotomy, then so has the adjoint $L^* = \frac{d}{dt} + A(t)^T$ with solution operator $S^*(t,s) = S(s,t)^T$, exponents $-\beta < -\nu < -\alpha$ and projectors $P_s^*(t) = P_u(t)^T$, $P_c^*(t) = P_c(t)^T, P_u^*(t) = P_s(t)^T$.

Shifting the indices in a trichotomy is described in the following lemma which is straightforward to verify.

LEMMA 2.2.

For $\gamma \in \mathbb{R}$ let $L_\gamma x = Lx + \gamma x$ with solution operator $S_\gamma(t,s)$. Then the following holds

(i) $e^{\gamma t}L_\gamma x = L(e^{\gamma t}x)$ for $t \in J, x \in C^1(J,\mathbb{R}^m)$
(ii) $S_\gamma(t,s) = e^{\gamma(s-t)}S(t,s)$ for $t,s \in J$
(iii) If L has an ordinary exponential trichotomy on J with exponents $\alpha < \nu < \beta$ then L_γ has an ordinary exponential trichotomy on J with exponents $\alpha - \gamma < \nu - \gamma < \beta - \gamma$ and the same projectors.

Another important tool is the roughness theorem for exponential dichotomies ([Cop78], p. 34) to which we add some estimates on the projectors which are useful later on.

PROPOSITION 2.3.

Let L have a shifted exponential dichotomy on $J = [\tau, \infty)$ with exponents $\alpha < \beta$, with constant K and with projectors $P_s(t), P_u(t), t \in J$. Assume that $B \in C(J, \mathbb{R}^{m,m})$ satisfies

$$\frac{8K^2 b_\infty}{\beta - \alpha} < 1 \text{ where } b_\infty = \sup_{t \ge \tau} \parallel B(t) \parallel . \tag{2.5}$$

Then the perturbed operator $\widetilde{L}x = Lx - B(t)x$ has a shifted exponential dichotomy on J with exponents

$$\widetilde{\alpha} = \alpha + 2b_\infty K < \widetilde{\beta} = \beta - 2b_\infty K,$$

with constant $\widetilde{K} = \frac{5}{2}K^2$ and with projectors $\widetilde{P}_s(t), \widetilde{P}_u(t), t \in J$ which satisfy

$$\| \widetilde{P}_\kappa(t) - P_\kappa(t) \| \leq 2K\widetilde{K} \int_\tau^\infty e^{-(\widetilde{\beta}-\alpha)|t-s|} \| B(s) \| \, ds, \quad \kappa = s, u. \tag{2.6}$$

PROOF.

It is sufficient to prove this for an exponential dichotomy with $\alpha = -\beta$. In the general case we first shift by $\gamma = \frac{1}{2}(\alpha + \beta)$, apply Lemma 2.2 and the known results and then shift back by $\gamma = -\frac{1}{2}(\alpha + \beta)$.
In the case $\alpha = -\beta$ the result is proved in [Cop78], p. 34 with the exception of (2.6). In the proof there, $Y_1(t) = \widetilde{S}(t,\tau)\widetilde{P}_s(\tau)$ is constructed as the solution of the integral equation

$$Y_1(t) = (GY_1)(t) + S(t,\tau)P_s(\tau), t \geq \tau. \tag{2.7}$$

where

$$(GY_1)(t) = \int_\tau^t S(t,s)P_s(s)B(s)Y_1(s)ds - \int_t^\infty S(t,s)P_u(s)B(s)Y_1(s)ds.$$

Then $Y_1(\tau) = \widetilde{P}_s(\tau)$ turns out to be a projector satisfying

$$N(\widetilde{P}_s(\tau)) = N(P_s(\tau)) \tag{2.8}$$

We set $\widetilde{P}_s(t) = \widetilde{S}(t,\tau)\widetilde{P}_s(\tau)\widetilde{S}(\tau,t) = Y_1(t)\widetilde{S}(\tau,t)$, $\widetilde{P}_u(t) = I - \widetilde{P}_s(t)$ for $t \in J$ and find

$$\widetilde{P}_s(t) - P_s(t) = P_u(t)\widetilde{P}_s(t) - P_s(t)\widetilde{P}_u(t). \tag{2.9}$$

Using (2.7) and (2.1) we obtain

$$\begin{aligned} P_u(t)\widetilde{P}_s(t) = P_u(t)Y_1(t)\widetilde{S}(\tau,t) &= -\int_t^\infty S(t,s)P_u(s)B(s)Y_1(s)ds\widetilde{S}(\tau,t) \\ &= -\int_t^\infty S(t,s)P_u(s)B(s)\widetilde{S}(s,t)\widetilde{P}_s(t)ds \end{aligned}$$

and thus by the exponential dichotomies

$$\| P_u(t)\widetilde{P}_s(t) \| \leq K\widetilde{K} \int_t^\infty e^{(\widetilde{\beta}-\alpha)(s-t)} \| B(s) \| \, ds.$$

For the second term in (2.9) we use the formula

$$P_s(t)\tilde{P}_u(t) = \int_{\tau}^{t} S(t,s)P_s(s)B(s)\tilde{P}_u(s)\tilde{S}(s,t)ds \tag{2.10}$$

and again the dichotomies to get (2.6). For the proof of (2.10) we notice that both sides have the value 0 at $t = \tau$ due to (2.8) and both satisfy the same differential equation as may be easily seen with the help of (2.3).

∎

Similar to [Pal 84] we now replace the smallness assumption (2.5) by an asymptotic estimate on the perturbation

$$\| B(t) \| \leq Ce^{-\varepsilon t} \text{ for } t \geq 0 \text{ and some } \varepsilon > 0. \tag{2.11}$$

This will be the interesting case with our applications but we notice that it is easy to derive generalized statements under the assumption $B(t) \to 0$ as $t \to \infty$.

PROPOSITION 2.4.

Let L have a shifted exponential dichotomy on $J = [\tau, \infty)$ with exponents $\alpha < \beta$, with constant K and with projectors $P_s(t), P_u(t), t \in J$ and let $B \in C(J, \mathbb{R}^{m,m})$ satisfy (2.11) for some $0 < \varepsilon < \beta - \alpha$.
The the perturbed operator $\tilde{L}x = Lx - B(t)x$ has a shifted exponential dichtomy on J with exponents $\tilde{\alpha} < \tilde{\beta}$, which may be chosen arbitrarily close to $\alpha < \beta$, and with constants $\tilde{K}$ depending on $\tilde{\alpha}, \tilde{\beta}$. For the projectors $\tilde{P}_s(t), \tilde{P}_u(t)$, and the solution operator $\tilde{S}(t,s)$ the following estimates hold for all $t \geq s$ in J

$$\| \tilde{P}_\kappa(t) - P_\kappa(t) \| \leq Ce^{-\varepsilon t}, \kappa = s, u \tag{2.12}$$

$$\| \tilde{S}(t,s)\tilde{P}_s(s) - S(t,s)P_s(s) \| \leq C\left(e^{-\varepsilon t + \tilde{\alpha}(t-s)} + e^{-\varepsilon s + \alpha(t-s)}\right) \tag{2.13}$$

$$\| \tilde{S}(s,t)\tilde{P}_u(t) - S(s,t)P_u(t) \| \leq C(e^{-\varepsilon s - \tilde{\beta}(t-s)} + e^{-\varepsilon t - \beta(t-s)}). \tag{2.14}$$

PROOF.
We choose $\tau_1 > \tau$ such that $b_\infty = \sup_{t \geq \tau_1} \| B(t) \|$ satisfies (2.5) and $2b_\infty K < \text{Min}(\varepsilon, \beta - \alpha - \varepsilon)$. From Proposition 2.3 we obtain the shifted exponential dichotomy for $\tilde{L}$ on $J_1 = [\tau_1, \infty)$. Moreover, by (2.6) and (2.11)

$$\| \tilde{P}_\kappa(t) - P_\kappa(t) \| \leq C\left[\int_{\tau_1}^{t} e^{-(\tilde{\beta}-\alpha)(t-s)-\varepsilon s}ds + \int_{t}^{\infty} e^{-(\tilde{\beta}-\alpha)(s-t)-\varepsilon s}ds\right] \leq Ce^{-\varepsilon t}.$$

For the estimate (2.13) we use the identity

$$\widetilde{S}(t,s)\widetilde{P}_s(s) - S(t,s)P_s(s) = P_s(t)(\widetilde{S}(t,s) - S(t,s))\widetilde{P}_s(s) + (\widetilde{P}_s(t) - P_s(t))\widetilde{S}(t,s)\widetilde{P}_s(s) + P_s(t)S(t,s)(\widetilde{P}_s(s) - P_s(s)).$$

The last two terms can be estimated by the dichotomies and (2.12). The first term is zero at $t = s$ and satisfies a linear inhomogenous differential equation which upon integration gives

$$P_s(t)(\widetilde{S}(t,s) - S(t,s))\widetilde{P}_s(s) = \int_s^t S(t,\sigma)P_s(\sigma)B(\sigma)\widetilde{S}(\sigma,s)\widetilde{P}_s(s)d\sigma.$$

Again the dichotomies and (2.11) yield the desired result. The proof of (2.14) proceeds in an analogous way. Finally, we notice that the shifted exponential dichotomy as well as the estimates (2.12)–(2.14) easily carry over from J_1 to J. ∎

Unlike the case of exponential trichotomies ([HaLi 86], Lemma 4.3) the exact exponential behaviour of the center part in an ordinary exponential trichotomy is generally not preserved under perturbation. The following lemma describes a specialized situation where this is the case.

LEMMA 2.5.

Let L have an ordinary exponential trichotomy on $J = [0, \infty)$ with exponents $\alpha < \nu < \beta$ and projectors $P_\kappa(t), \kappa = s, c, u$. Further assume that $P_c(t)$ is of rank $m_c = 1$ and has the form $P_c(t) = z(t)\psi(t)^T$ where $Lz = 0$ and

$$C_1 e^{\nu t} \leq \| z(t) \| \leq C_2 e^{\nu t}, t \in J, \quad C_1, C_2 > 0 \tag{2.15}$$

Let $B \in C(J, \mathbb{R}^{m,m})$ satisfy (2.11) for some $\varepsilon < \mathrm{Min}(\nu - \alpha, \beta - \nu)$, and assume that $\widetilde{L}x = Lx - B(t)x = 0$ has a solution $\widetilde{z}$ which satisfies

$$\| z(t) - \widetilde{z}(t) \| \leq C e^{(\nu - \varepsilon)t}, t \in J. \tag{2.16}$$

Then $\widetilde{L}$ has an ordinary exponential trichotomy on J with exponents $\widetilde{\alpha} < \nu < \widetilde{\beta}$ where $\widetilde{\alpha}, \widetilde{\beta}$ may be chosen arbitrarily close to α, β. Moreover, the estimates (2.12) for $\kappa = s, c, u$ and (2.13), (2.14) hold and $\widetilde{P}_c(t)$ can be represented as $\widetilde{P}_c(t) = \widetilde{z}(t)\widetilde{\psi}(t)^T$ for some $\widetilde{\psi}$ with $\widetilde{L}^*\widetilde{\psi} = 0$ and

$$\| \psi(t) - \widetilde{\psi}(t) \| \leq C e^{(\nu - \varepsilon)t}, t \in J. \tag{2.17}$$

PROOF.

In a first step we apply Prop. 2.4 to the two shifted exponential dichotomies of L with exponents $\alpha < \nu$, $\nu < \beta$ and projectors $Q_s = P_s, Q_u = P_c + P_u$ and $R_s = P_s + P_c$, $R_u = P_u$ respectively. Let $\widetilde{Q}_\kappa$, $\widetilde{R}_\kappa$ be the corresponding projectors for $\widetilde{L}$. We define $\widetilde{P}_s(t) = \widetilde{Q}_s(t), \widetilde{P}_u(t) = \widetilde{R}_u(t)$ so that (2.13), (2.14) and (2.12) for $\kappa = s, u$ have been proved. Moreover, the range of $\widetilde{R}_s(0)$ is uniquely determined and we have $R(\widetilde{Q}_s(0)) \subset R(\widetilde{R}_s(0))$ with the codimension being $m_c = 1$.

Next we show $\widetilde{z}(0) \notin R(\widetilde{Q}_s(0))$. If this is false then

$$\begin{aligned}\| z(0) \| = \| S(0,t)P_c(t)z(t) \| &\leq Ke^{-\nu t}(\| z(t) - \widetilde{z}(t) \| + \| \widetilde{S}(t,0)\widetilde{z}(0) \|) \\ &\leq Ke^{-\nu t}(Ce^{(\nu-\varepsilon)t} + \widetilde{K}e^{\widetilde{\alpha}t} \| \widetilde{z}(0) \|) \to 0 \text{ as } t \to \infty\end{aligned}$$

and we arrive at a contradiction.

Thus we have the decomposition

$$\mathbb{R}^m = R(\widetilde{Q}_s(0)) \oplus \text{span}\{\widetilde{z}(0)\} \oplus R(\widetilde{R}_u(0))$$

and we let $\widetilde{P}_c(0)$ be the projector onto span $\{\widetilde{z}(0)\}$. If we write $\widetilde{P}_c(0) = \widetilde{z}(0)\widetilde{\psi}(0)^T$ where $\widetilde{\psi}(0)^T\widetilde{z}(0) = 1$ and let $\widetilde{\psi}(t) = \widetilde{S}(0,t)^T\widetilde{\psi}(0)$ then $\widetilde{P}_c(t) + \widetilde{P}_u(t) + \widetilde{P}_s(t) = I$ holds for $\widetilde{P}_c(t) = \widetilde{z}(t)\widetilde{\psi}(t)^T$ (see (2.4)). Therefore the estimate (2.12) is also valid for $\kappa = c$.

Finally, we notice that (e. g. in Euclidean norm)

$$\| P_c(t) \| = \| z(t)\psi(t)^T \| = \| z(t) \| \| \psi(t) \|$$

so that the estimates in (2.2) are equivalent to

$$\| z(t)e^{-\nu t} \|, \| \psi(s)e^{\nu s} \| \leq C \text{ for } t, s \in J. \tag{2.18}$$

From (2.15) and (2.16) we then find $\| e^{-\nu t}\widetilde{z}(t) \| \geq c > 0$ and hence by (2.12), (2.16)

$$\begin{aligned}C \| \psi(t) - \widetilde{\psi}(t) \| &\leq \| e^{-\nu t}\widetilde{z}(t)(\psi(t) - \widetilde{\psi}(t))^T \| \\ &= \| e^{-\nu t}(P_c(t) - \widetilde{P}_c(t)) + e^{-\nu t}(\widetilde{z}(t) - z(t))\psi(t)^T \| \\ &\leq Ce^{(-\nu-\varepsilon)t}. \quad \blacksquare\end{aligned}$$

We apply Lemma 2.5 to the periodic case. Assume that $A(t)$, $t \in \mathbb{R}$ is 1–periodic and that $Lx = \dot{x} - A(t)x$ has the simple Floquet multiplier 1 and no further multipliers on the unit circle. By classical Floquet theory we have a fundamental matrix of the form

$$(Z_s(t)\ z(t)\ Z_u(t)) \ \exp\left(t\begin{pmatrix} B_s & 0 & 0 \\ 0 & 0 & 0 \\ 0 & 0 & B_u \end{pmatrix}\right)$$

where Z_s, z, Z_u are 1–periodic and where the Floquet exponents, given by the spectra of B_s, B_u, satisfy for some $\alpha < \beta$

$$\text{Re } \sigma(B_s) < \alpha < 0 < \beta < \text{Re } \sigma(B_u). \tag{2.19}$$

Setting $(\Psi_s(t)\ \psi(t)\ \Psi_u(t)) = (Z_s\ (t)\ z(t)\ Z_u(t))^{T-1}$ we find for the solution operator

$$(2.20)\qquad S(t,s) = Z_s(t)e^{(t-s)B_s}\Psi_s(s)^T + z(t)\psi(s)^T + Z_u(t)e^{(t-s)B_u}\Psi_u(s)^T.$$

Therefore L has an ordinary exponential trichotomy on any interval $J \subset \mathbb{R}$ with exponents $\alpha < \nu = 0 < \beta$ and with projectors

$$P_\kappa(t) = Z_\kappa(t)\Psi_\kappa(t)^T,\ \kappa = s, u,\ P_c(t) = z(t)\psi(t)^T.$$

We notice that the projectors P_s, P_u are always real operators but the matrices $Z_\kappa, \Psi_\kappa,\ \kappa = s, u$ may be complex in general.

Lemma 2.5 then yields the following

COROLLARY 2.6.

Let $L\ x = \dot{x} - A(t)x$ be as above and assume that $\widetilde{L}x = \dot{x} - \widetilde{A}(t)x,\ \widetilde{A} \in C([0,\infty), \mathbb{R}^{m,m})$ satisfies

$$(2.21)\qquad ||A(t) - \widetilde{A}(t)|| \le Ce^{-\epsilon t} \text{ for some } 0 < \epsilon < \text{Min}\ (-\alpha, \beta)$$

and that there exists a solution $\widetilde{z}$ of $\widetilde{L}\widetilde{z} = 0$ such that

$$(2.22)\qquad ||z(t) - \widetilde{z}(t)|| \le Ce^{-\epsilon t},\ t \ge 0.$$

Then $\widetilde{L}$ has an ordinary exponential trichotomy on $[0,\infty)$ with exponents $\widetilde{\alpha} < 0 < \widetilde{\beta}$ arbitrarily close to α and β. The projectors $\widetilde{P}_s, \widetilde{P}_c, \widetilde{P}_u$ satisfy the estimates (2.12) — (2.14) and we have $\widetilde{P}_c(t) = \widetilde{z}(t)\widetilde{\psi}(t)^T$ where $\widetilde{L}^*\widetilde{\psi} = 0$ and

$$(2.23)\qquad ||\psi(t) - \widetilde{\psi}(t)|| \le Ce^{-\epsilon t} \text{ for } t \ge 0.$$

In the situation of this corollary we consider the differential operator $\Gamma : Z_1^+(\epsilon) \to Z_0^+(\epsilon)$ defined by

$$\Gamma(x,y) := (\widetilde{L}x, Ly)$$

with spaces and norms given by

$$Z_0^+(\epsilon) = \{(x,y) : x \in C([0,\infty), \mathbb{R}^m) \text{ bounded},\ y \in C([0,\infty), \mathbb{R}^m) \text{ 1-periodic}$$
$$||x(t) - y(t)|| \le Ce^{-\epsilon t} \text{ for } t \ge 0\},$$
$$||(x,y)||_{Z_0^+} := \sup_{t\ge0} ||x(t)|| + \sup_{t\ge0} ||y(t)|| + \sup_{t\ge0} e^{\epsilon t}||x(t) - y(t)||,$$
$$Z_1^+(\epsilon) = \{(x,y) \in Z_0^+(\epsilon) : (\dot{x}, \dot{y}) \in Z_1^+(\epsilon)\}$$
$$||(x,y)||_{Z_1^+} := ||(x,y)||_{Z_0^+} + ||(\dot{x}, \dot{y})||_{Z_0^+}.$$

PROPOSITION 2.7.

Under the assumptions of Corollary 2.6 the operator

$$\Gamma : Z_1^+(\epsilon) \to Z_0^+(\epsilon),\ \Gamma(x,y) = (\widetilde{L}x, Ly)$$

is Fredholm of index m_{+s}, which is the number of stable Floquets multipliers of L. Moreover, for the range and nullspace we have

$$R(\Gamma) = \{(r,\rho) \in Z_0^+(\epsilon) : \int_0^1 \psi(t)^T \rho(t) dt = 0\} \tag{2.24}$$

$$N(\Gamma) = \text{span}\ \{(\widetilde{z}, z),\ (\widetilde{S}(t,0)\xi, 0)\ \text{ where }\ \xi \in R(\widetilde{P}_s(0))\}. \tag{2.25}$$

REMARK.

In case $(r,\rho) \in R(\Gamma)$ we give a representation of a special solution $(\widehat{x}, \widehat{y})$ of $\Gamma(x,y) = (r,\rho)$ which will be used later on.

PROOF.

It is enough to show (2.24), (2.25) since this implies codim $R(\Gamma) = 1$ and dim $N(\Gamma) = 1 + m_{+s}$.

For $(r,\rho) = (\widetilde{L}x, Ly) \in R(\Gamma)$ we obtain in the standard way

$$\int_0^1 \psi(t)^T \rho(t) dt = \int_0^1 \psi(t)^T Ly(t) dt = -\int_0^1 (L^*\psi(t))^T y(t) dt + [\psi(t)^T y(t)]_0^1 = 0.$$

Now suppose that $(r,\rho) \in Z_0^+(\epsilon)$ and $\int_0^1 \psi(t)^T \rho(t) dt = 0$ hold. Then there exists a 1–periodic solution $\widehat{y}$ of $Ly = \rho$. For example, we may define $\widehat{y}$ by

$$P_s(t)\widehat{y}(t) = S(t,0)\xi + \int_0^t S(t,s) P_s(s)\rho(s) ds \tag{2.26}$$

$$P_c(t)\widehat{y}(t) = z(t) \int_0^t \psi(s)^T \rho(s) ds \tag{2.27}$$

$$P_u(t)\widehat{y}(t) = -\int_t^\infty S(t,s) P_u(s)\rho(s) ds \tag{2.28}$$

where ξ is the unique solution in $R(P_s(0))$ of the linear system

$$(2.29)\qquad (I - S(1,0))\xi = \int_0^1 S(1,s)P_s(s)\rho(s)ds = P_s(0)\int_0^1 S(1,s)\rho(s)ds.$$

Equation (2.29) guarantees that $P_s(t)\widehat{y}(t)$ is 1–periodic. For $P_c(t)\widehat{y}(t)$ the periodicity follows from our assumption while for $P_u(t)\widehat{y}(t)$ it is a consequence of the relation

$$S(t+1, s+1) = S(t,s).$$

For the solution $\widehat{x}$ of $\widetilde{L}x = r$ we use an analogous set of formulae

$$(2.30)\qquad \widetilde{P}_s(t)\widehat{x}(t) = \int_0^t \widetilde{S}(t,s)\widetilde{P}_s(s)r(s)ds$$

$$(2.31)\qquad \widetilde{P}_c(t)\widehat{x}(t) = \omega\widetilde{z}(t) + \widetilde{z}(t)\int_0^t \widetilde{\psi}(s)^T r(s)ds$$

$$(2.32)\qquad \widetilde{P}_u(t)\widehat{x}(t) = -\int_t^\infty \widetilde{S}(t,s)\widetilde{P}_u(s)r(s)ds,$$

where ω will be determined so that

$$\widehat{x}(t) - \widehat{y}(t) = O(e^{-\epsilon t}).$$

For the center part we find

$$P_c(t)\widehat{y}(t) - \widetilde{P}_c(t)\widehat{x}(t) = (z(t) - \widetilde{z}(t))\int_0^t \psi(s)^T\rho(s)ds$$
$$+ \widetilde{z}(t)\left[\int_0^t (\psi(s)^T\rho(s) - \widetilde{\psi}(s)^T r(s))ds - \omega\right].$$

The first term behaves like $O(e^{-\epsilon t})$ by (2.22) and our assumption and so does the second if we set

$$(2.33)\qquad \omega = \int_0^\infty (\psi(s)^T\rho(s) - \widetilde{\psi}(s)^T r(s))ds.$$

Notice that (2.33) and $r(t) - \rho(t) = O(e^{-\epsilon t})$ yield

$$\psi(s)^T\rho(s) - \widetilde{\psi}(s)^T r(s) = (\psi(s) - \widetilde{\psi}(s))^T\rho(s) + \widetilde{\psi}(s)^T(\rho(s) - r(s)) = O(e^{-\epsilon s})$$

and therefore

$$|\omega - \int_0^t (\psi(s)^T\rho(s) - \widetilde{\psi}(s)^T r(s))ds| \le C \int_t^\infty e^{-\epsilon s}ds = Ce^{-\epsilon t}.$$

Thus we have shown

$$P_c(t)\widehat{y}(t) - \widetilde{P}_c(t)\widehat{x}(t) = O(e^{-\epsilon t}).$$

In particular $\widetilde{P}_c\widehat{x}$ is bounded and by the exponential trichotomy this is also true for $\widetilde{P}_s\widehat{x}$ and $\widetilde{P}_u\widehat{x}$. Hence $\widehat{x}$ is a bounded solution of $\widetilde{L}x = r$.

Furthermore, by using (2.13) we obtain

$$||P_s(t)\widehat{y}(t) - \widetilde{P}_s(t)\widehat{x}(t)|| \le ||S(t,0)\xi|| +$$

$$\int_0^t ||S(t,s)P_s(s)||\ ||\rho(s) - r(s)|| + ||S(t,s)P_s(s) - \widetilde{S}(t,s)\widetilde{P}_s(s)||\ ||r(s)||ds$$

$$\le C\{e^{\alpha t} + \int_0^t (e^{-\epsilon s + \alpha(t-s)} + e^{-\epsilon t + \widetilde{\alpha}(t-s)})ds\} \le Ce^{-\epsilon t}.$$

In a similar way we use (2.14) to show

$$||P_u(t)\widehat{y}(t) - \widetilde{P}_u(t)\widehat{x}(t)|| \le Ce^{-\epsilon t}.$$

For the proof of (2.25) we first notice that

$$(\widetilde{z}, z), (\widetilde{S}(t,0)\xi, 0) \in N(\Gamma) \quad \text{for} \quad \xi \in R(\widetilde{P}_s(0))$$

is obvious. Suppose that $(x, y) \in N(\Gamma)$, then $Ly = 0$ and $y = cz$ for some $c \in \mathbb{R}$ follows. But the trichotomy of $\widetilde{L}$ implies

$$x(t) = \widetilde{S}(t,0)\xi + \beta\widetilde{z} \quad \text{for some} \quad \xi \in R(\widetilde{P}_s(0)), \beta \in \mathbb{R}$$

and

$$(c - \beta)z(t) = y(t) - x(t) + \beta(\widetilde{z}(t) - z(t)) + \widetilde{S}(t,0)\xi = O(e^{-\epsilon t}).$$

Hence $c = \beta$ and we have the form

$$(x, y) = c(\widetilde{z}, z) + (\widetilde{S}(t,0)\xi, 0).$$

■

Lemma 2.7 will be used for the linerization of initial value problems.

The corresponding result for the boundary value case which uses the spaces $Z_1(\epsilon), Z_0(\epsilon)$ from the introduction is given in the following

PROPOSITION 2.8.

Let the differential operators $Lx = \dot{x} - A(t)x$ and $\widetilde{L}x = \dot{x} - \widetilde{A}(t)x,\ t \in \mathbb{R}$ satisfy the assumptions of Corollary 2.6. Further assume that $\widetilde{z}(t)$ is bounded for $t \leq 0$ and

$$\widetilde{A}(t) \to A_- \quad \text{as} \quad t \to -\infty \tag{2.34}$$

where A_- is hyperbolic with stable subspace of dimension m_{-s} and unstable subspace of dimension $m_{-u} = m - m_{-s}$.

Then the operator

$$\Gamma : Z_1(\epsilon) \to Z_0(\epsilon),\ \Gamma(x,y) = (\widetilde{L}x, Ly)$$

is Fredholm of index $m_{-u} - m_{+u} - 1$ where m_{+u} is the number of unstable Floquet multipliers for L.

The operator $\widetilde{L}$ has an exponential dichotomy on $\mathbb{R}_-$ with projectors $Q_s(t)$, $Q_u(t)$ and

$$N(\Gamma) = \{(\widetilde{S}(t,0)\xi, 0) + c(\widetilde{z}(t), z(t)) : c \in \mathbb{R},\ \xi \in R(\widetilde{P}_s(0)) \cap R(Q_u(0))\}. \tag{2.35}$$

Moreover, with $\Gamma^*(x,y) = (\widetilde{L}^*x, L^*y)$ we have that $(r,\rho) \in Z_0(\epsilon)$ is in $R(\Gamma)$ if and only if the following two conditions are satisfied

$$\int_0^1 \varphi(t)^T \rho(t) dt = 0 \ \text{ for all } \ \varphi \in N(L^*) \tag{2.36}$$

(2.37)

$$\int_{-\infty}^0 \widetilde{\varphi}(t)^T r(t) dt + \int_0^\infty (\widetilde{\varphi}(t)^T r(t) - \varphi(t)^T \rho(t)) dt = 0 \ \text{ for all } \ (\widetilde{\varphi}, \varphi) \in N(\Gamma^*).$$

PROOF.
The exponential dichotomy on $\mathbb{R}_-$ with projectors Q_s, Q_u of rank m_{-s} and m_{-u} follows from (2.34) and the roughness theorem. Since $\widetilde{z}$ is a bounded solution of $\widetilde{L}x = 0$ on $\mathbb{R}$ the representation (2.35) is easily obtained as in the previous proposition.

Now suppose that $(r,\rho) = \Gamma(x,y) \in R(\Gamma)$. Then (2.36) is clear and we consider some $(\widetilde{\varphi}, \varphi) \in N(\Gamma^*)$. It follows that

$$\begin{aligned}\widetilde{\varphi}(0) &\in R(\widetilde{P}_s^*(0) + \widetilde{P}_c^*(0)) \cap R(Q_u^*(0)) \\ &= R(\widetilde{P}_u(0)^T + \widetilde{P}_c(0)^T) \cap R(Q_s(0)^T) \\ &= (R(\widetilde{P}_u(0)^T) \oplus \ \text{span}\ \{\widetilde{\psi}(0)\}) \cap R(Q_s(0)^T).\end{aligned}$$

From $\varphi \in N(L^*)$ we obtain $\varphi = c\psi$ for some $c \in \mathbb{R}$ and hence by (2.23)

$$(\widetilde{\varphi} - c\widetilde{\psi})(t) = (\widetilde{\varphi} - \varphi)(t) + c(\psi - \widetilde{\psi})(t) = O(e^{-\epsilon t}).$$

Integration by parts then yields

$$\int_{-\infty}^{0} \widetilde{\varphi}(t)^T r(t)dt = \widetilde{\varphi}(0)^T x(0) = (\widetilde{\varphi}(0) - c\widetilde{\psi}(0))^T x(0) + c\widetilde{\psi}(0)^T x(0)$$

$$= -\int_{0}^{\infty} (\widetilde{\varphi}(t) - c\widetilde{\psi}(t))^T r(t)dt + c\widetilde{\psi}(0)^T x(0).$$

By Proposition 2.7 we may assume (see (2.26)–(2.32))

$$y = \widehat{y},\ x(t) = \widehat{x}(t) + \widetilde{S}(t,0)\xi \ \text{ for some } \ \xi \in R(\widetilde{P}_s(0)) \ \text{ and } \ t \geq 0.$$

From (2.31) and (2.33) we then find

$$\widetilde{\psi}(0)^T x(0) = \omega\widetilde{\psi}(0)^T \widetilde{z}(0) = \omega$$

and hence the assertion (2.37)

$$\int_{-\infty}^{0} \widetilde{\varphi}(t) r(t)dt = -\int_{0}^{\infty} (\widetilde{\varphi}(t) - c\widetilde{\Psi}(t))^T r(t)dt + c\int_{0}^{\infty} (\Psi(s)^T \rho(s) - \widetilde{\Psi}(s)^T r(s))ds$$

$$= -\int_{0}^{\infty} (\widetilde{\varphi}(t)^T r(t) - \varphi(t)^T \rho(t))dt.$$

For the converse statement let us assume that $(r, \rho) \in Z_0^+(\epsilon)$ satisfies (2.36), (2.37). We seek a solution (x, y) of $\Gamma(x, y) = (r, \rho)$ in the form

$$y = \widehat{y},\ x(t) = \begin{cases} \widehat{x}(t) + \widetilde{S}(t,0)Q_u(0)\xi, & t < 0 \\ \widehat{x}(t) + \widetilde{S}(t,0)\widetilde{P}_s(0)\xi, & t \geq 0 \end{cases}$$

where $\widehat{y}$ is given by (2.26)–(2.29) and $\widehat{x}(t)$ is defined for $t \geq 0$ through (2.30)–(2.33). For $t < 0$ we set

$$\widehat{x}(t) = \int_{-\infty}^{t} \widetilde{S}(t,s)Q_s(s)r(s)ds - \int_{t}^{0} \widetilde{S}(t,s)Q_u(s)r(s)ds$$

and ξ will be determined so that x is continuous at 0, i.e.

$$(Q_u(0) - \widetilde{P}_s(0))\xi = \widehat{x}(0+) - \widehat{x}(0-). \tag{2.38}$$

This equation has a solution if for any η satisfying $\eta^T(Q_u(0) - \widetilde{P}_s(0)) = 0$ we can show $\eta^T(\widehat{x}(0+) - \widehat{x}(0-)) = 0$.
Let $c = \eta^T \widetilde{z}(0)$ and define

$$\widehat{\psi}(t) = \begin{cases} \widetilde{S}^*(t,0)Q_u^*(0)\eta & \text{for } t < 0 \\ \widetilde{S}^*(t,0)\widetilde{P}_s^*(0)\eta + c\widetilde{\psi}(t) & \text{for } t \geq 0 \end{cases}.$$

$\widehat{\psi}$ is in fact continuous at 0 since

$$\begin{aligned}\widehat{\psi}(0+)^T - \widehat{\Psi}(0-)^T &= \eta^T(\widetilde{P}_u(0) - Q_s(0)) + c\widetilde{\psi}(0)^T \\ &= \eta^T(I - \widetilde{P}_s(0) - \widetilde{P}_c(0) - (I - Q_u(0) + \widetilde{z}(0))\widetilde{\psi}(0)^T) = 0.\end{aligned}$$

Therefore, $\widehat{\psi}$ is a bounded solution of $\widetilde{L}^*\widehat{\psi} = 0$ and $(\widehat{\psi}, c\psi) \in N(\Gamma^*)$. An application of (2.37) yields

$$\begin{aligned}0 &= \int_{-\infty}^{0} \widehat{\psi}(t)^T r(t)dt + \int_{0}^{\infty} \widetilde{\psi}(t)^T r(t) - c\psi(t)^T \rho(t)dt \\ &= \eta^T \Bigg[\int_0^\infty \widetilde{z}(0)\widetilde{\psi}(t)^T r(t) + \widetilde{P}_u(0)\widetilde{S}(0,t)r(t) - \widetilde{z}(0)\psi(t)^T\rho(t)dt + \\ &\quad \int_{-\infty}^{0} Q_s(0)\widetilde{S}(0,t)r(t)dt\Bigg] \\ &= \eta^T(\widehat{x}(0-) - \widehat{x}(0+)).\end{aligned}$$

Let us finally compute the Fredholm index of Γ. From the ordinary exponential trichotomy and (2.24) we get

$$\dim\ N(\Gamma) = \dim\ (R(\widetilde{P}_s(0) + \widetilde{P}_c(0)) \cap R(Q_u(0))).$$

Since$\widetilde{z}(0) \in R(\widetilde{P}_c(0)) \cap R(Q_u(0))$ by our assumption we find

$$\dim\ N(\Gamma) = \dim\ (V \cap W) + 1 \quad \text{where} \quad V = R(\widetilde{P}_s(0)),\ W = R(Q_u(0)).$$

Similarly

$$\dim\ N(\Gamma^*) = \dim\ (R(\widetilde{P}_u(0)^T + \widetilde{P}_c(0)^T) \cap R(Q_s(0)^T)) = \dim\ (V^\perp \cap W^\perp).$$

The formulae (2.36), (2.37) show that $R(\Gamma)$ is the null space of a space of linear functionals which has dimension $\dim N(\Gamma^*) + 1$.
Then we conclude as in [Pal 84]

$$\begin{aligned}\mathrm{ind}(\Gamma) &= \dim(V \cap W) - \dim(V^\perp \cap W^\perp) \\ &= \dim(V \cap W) - \dim((V + W)^\perp) \\ &= \dim(V \cap W) - m + \dim(V + W) \\ &= \dim V + \dim W - m = m_{+s} + m_{-u} - m = m_{-u} - m_{+u} - 1.\end{aligned}$$

■

3. Characterizations of well–posedness

Throughout this section we assume that we have a $\overline{T}$–periodic hyperbolic orbit

$$\overline{\gamma} = \{\widehat{y}(t) : t \in \mathbb{R}\}$$

of the system (1.1) at some $\lambda = \overline{\lambda}$. Let m_{+s} and m_{+u} denote the number of its stable and unstable Floquet multipliers respectively.

In our first step we repeat the construction of the local stable manifold and its foliation induced by the asymptotic phase by using the results of section 2 and the implicit function theorem. This approach is similar to [Irw 80, Ch. 6, II] where the time $\overline{T}$–map is employed but different from the more standard approach via the Poincaré map (cf. [Har 64, Ch. IX], (Hal 69, Ch. VI]) and we take some care in relating these two approaches.

The function $\overline{y}(t) = \widehat{y}(t\,\overline{T})$ is a 1–periodic solution of

$$\dot{x} = \overline{T}\ f(x, \overline{\lambda}) \tag{3.1}$$

and the linear operator

$$L = \frac{d}{dt} - A(t),\ A(t) = \overline{T}\ \frac{\partial f}{\partial x}\ (\overline{y}(t), \overline{\lambda})$$

has an ordinary exponential trichotomy with projectors $P_\kappa(t)$, $\kappa = s, c, u$ and solution operator $S(t,s)$.

Let C_1^k $(k \geq 0)$ denote the 1-periodic C^k–functions from $\mathbb{R}$ to $\mathbb{R}^m$.
We can continue $(\overline{y}(\cdot), \overline{T})$ to $(y(\cdot, \lambda),\ T(\lambda)) \in C_1^1 \times \mathbb{R}$, λ in some neighbourhood $U(\overline{\lambda})$, by an application of the implicit function theorem to the equation

$$F_1(y, T, \lambda) = \begin{pmatrix} \dot{y} - T\ f(y, \lambda) \\ \chi(y) \end{pmatrix} = 0. \tag{3.2}$$

Here F_1 maps $C_1^1 \times \mathbb{R}^{p+1}$ into $C_1^0 \times \mathbb{R}$ and $\chi : C_1^1 \to \mathbb{R}$ is a C^1–phase condition satisfying

$$\chi(\overline{y}) = 0,\ \chi'(\overline{y})\ \dot{\overline{y}} \neq 0. \tag{3.3}$$

Let $\varphi^t(\cdot, \lambda)$ denote the t–flow of the scaled system

$$\dot{x} = T(\lambda) f(x, \lambda) \tag{3.4}$$

and let $\Phi^t(x, \lambda) = (\varphi^t(x, \lambda), \lambda)$ denote the t–flow in $\mathbb{R}^{m+p}$ obtained by adding $\dot{\lambda} = 0$ (see (1.2)).

For a suitable neighbourhood $\overline{V}$ of $\overline{\gamma}$ the local stable set of $\overline{\gamma}$

$$\begin{aligned} M_{+s}(\overline{V}, \overline{\gamma}) = \{x \in \overline{V} : \varphi^t(x, \overline{\lambda}) \in \overline{V} \\ \text{for } t \geq 0 \text{ and dist } (\varphi^t(x, \overline{\lambda}), \overline{\gamma}) \to 0 \text{ as } t \to \infty\} \end{aligned}$$

is known to be an $(m_{+s}+1)$–dimensional smooth manifold (cf. [Hal 69, Ch. VI]). Then the local stable manifolds of the periodic orbits $\gamma(\lambda) = \{y(t,\lambda) : t \in \mathbb{R}\}$, $\lambda \in U(\overline{\lambda})$ can be put together to form an $(m_{+s}+1+p)$–dimensional manifold

$$M_{+s}(V) = \{(x,\lambda) \in V = \overline{V} \times U(\overline{\lambda}) : \varphi^t(x,\lambda) \in \overline{V} \text{ for } t \geq 0 \text{ and } \operatorname{dist}(\varphi^t(x,\lambda),\gamma(\lambda)) \to \infty \text{ as } t \to \infty\}$$

after possibly adjusting $\overline{V}$ and $U(\overline{\lambda})$.

Usually, the differentiable structure on $M_{+s}(V)$ is defined via the Poincaré map P with respect to a transversal section

$$\Sigma = \{\overline{y}(0) + \eta : \eta \in U(0) \subset R(P_s(0) + P_u(0))\}.$$

We have a representation

$$M_{+s}(V) \cap (\Sigma \times U(\overline{\lambda})) = \{(\overline{y}(0) + \xi + h(\xi,\lambda),\lambda) : \xi \in U(0) \subset R(P_s(0)),\ \lambda \in U(\overline{\lambda})\}$$

where h is a smooth mapping into $R(P_u(0))$ which satisfies $h(0,\overline{\lambda}) = 0$.

The charts on $M_{+s}(V)$ are then given by the inverses of the local parametrizations

$$\pi(t,\xi,\lambda) = \Phi^t(\overline{y}(0) + \xi + h(\xi,\lambda),\lambda) \tag{3.5}$$

where $\xi \in U(0) \subset R(P_s(0))$, $\lambda \in U(\overline{\lambda})$ and t is in some open interval $J \subset \mathbb{R}_+$ of length less than 1.

THEOREM 3.1.

Let $\overline{\gamma}$ be a hyperbolic periodic orbit as above and let $\epsilon <$ Min $(-\alpha,\beta)$ where α,β are bounds on the real parts of the Floquet exponents as in (2.19).

Then, for $\xi \in R(P_s(0))$ and $\lambda - \overline{\lambda}$ sufficiently small the operator equation

$$F_+(x,y,T,\xi,\lambda) = \begin{pmatrix} \dot{x} - T\ f(x,\lambda) \\ \dot{y} - T\ f(y,\lambda) \\ P_s(0)(x(0) - \overline{y}(0)) - \xi \\ \chi(y) \end{pmatrix} = 0 \tag{3.6}$$

has a unique solution $(x_+(\cdot,\xi,\lambda),\ y(\cdot,\lambda),\ T(\lambda))$ in some neighbourhood of $(\overline{y}(\cdot),\overline{y}(\cdot),\overline{T})$ in $Z_1^+(\epsilon) \times \mathbb{R}$ satisfying

$$x_+(t,0,\overline{\lambda}) = \overline{y}(t).$$

The local inverses of the mappings

$$\beta(t,\xi,\lambda) = (x_+(t,\xi,\lambda),\lambda),\ t \in U(0),\ \xi \in U(0) \subset R(P_s(0)),\ \lambda \in U(\overline{\lambda}) \tag{3.7}$$

are admissible charts of the local stable manifold $M_{+s}(V)$.

PROOF.

Clearly, $F_+(\overline{y},\overline{y},\overline{T},0,\overline{\lambda}) = 0$ and

$$F_+ : Z_1^+(\epsilon) \times \mathbb{R} \times R(P_s(0)) \times \mathbb{R}^p \to Z_0^+(\epsilon) \times R(P_s(0)) \times \mathbb{R}$$

is a smooth operator with

$$
(3.8) \qquad K := \frac{\partial F_+}{\partial(x,y,T)}(\overline{y},\overline{y},\overline{T},0,\overline{\lambda}) = \begin{bmatrix} \Gamma & & \begin{matrix} -f(\overline{y},\overline{\lambda}) \\ -f(\overline{y},\overline{\lambda}) \end{matrix} \\ P_s(0)E_0 & 0 & \\ & & 0 \\ 0 & \chi'(\overline{y}) & \end{bmatrix},
$$

where $\Gamma(x,y) = (L\,x, L\,y)$ and $E_0 x = x(0)$. By Proposition 2.7 Γ has Fredholm index m_{+s} and hence by the bordering lemma [Bey 90b] K has Fredholm index

$$
\text{ind}\,(K) = \text{ind}(\Gamma) + 1 - (m_{+s}+1) = 0.
$$

From (2.24) we find

$$
\begin{pmatrix} f(\overline{y},\overline{\lambda}) \\ f(\overline{y},\overline{\lambda}) \end{pmatrix} = \begin{pmatrix} \dot{\overline{y}} \\ \dot{\overline{y}} \end{pmatrix} \notin R(\Gamma)
$$

and moreover from (2.25) and (3.3)

$$
\begin{pmatrix} P_s(0)E_0 & 0 \\ 0 & \chi'(\overline{y}) \end{pmatrix} : N(\Gamma) \to R(P_s(0)) \times \mathbb{R}
$$

is nonsingular. Hence K is a linear homeomorphism and the implicit function theorem applies to (3.6). Since (3.2) has locally unique solutions the solutions of (3.6) are of the form

$$
(x_+(\cdot,\xi,\lambda),\ y(\cdot,\lambda),\ T(\lambda)) \ \text{ with } \ \xi \in U(0) \subset R(P_s(0)),\ \lambda \in U(\overline{\lambda}).
$$

By the construction of x_+ and by making $U(0), U(\overline{\lambda})$ sufficiently small we may assume that $\beta(t,\xi,\lambda) = (x_+(t,\xi,\lambda),\lambda) \in M_{+s}(V)$ holds for all $t \geq 0$.

Moreover, by implicit differentiation we obtain

$$
(3.9) \qquad \frac{\partial x_+}{\partial \xi}(t,0,\overline{\lambda}) = S(t,0)P_s(0)
$$

and hence

$$
(3.10) \qquad \frac{\partial \beta}{\partial(t,\xi,\lambda)}(0,0,\overline{\lambda}) = \begin{pmatrix} \dot{\overline{y}}(0) & I_{|R(P_s(0))} & \frac{\partial \beta}{\partial \lambda}(0,0,\overline{\lambda}) \\ 0 & 0 & I_p \end{pmatrix}
$$

has rull rank $m_{+s} + 1 + p$ and β is a local immersion.

Using the rank theorem (e.g. [Die 60, Ch. X]) we may write

$$
\beta = \sigma_2 \circ E \circ \sigma_1
$$

where σ_1 is a diffeomorphism from a neighbourhood $\widetilde{U} \subset \mathbb{R} \times R(P_s(0)) \times \mathbb{R}^p$ of $(0, 0, \overline{\lambda})$ onto the open unit ball in $\mathbb{R}^{m_{+s}+1+p}$, σ_2 is a diffeomorphism from the open unit ball in $\mathbb{R}^{m+p}$ onto $\beta(\widetilde{U})$ and

$$E(x_1, \dots, x_{m_{+s}+1+p}) = (x_1, \dots, x_{m_{+s}+1+p}, 0, \dots, 0) \in \mathbb{R}^{m+p}.$$

Therefore, we have a smooth representation of $\beta^{-1}\pi$ on the common domain of existence as

$$\beta^{-1}\pi(t, \xi, \lambda) = \sigma_2^{-1} \circ E^T \circ \sigma_1^{-1} \circ \pi(t, \xi, \lambda).$$

Furthermore, let $\tau(x, \lambda)$ be the unique return time in $-J$ for points of the form (3.5), then we find the smooth representation

$$\pi^{-1}\beta(t, \xi, \lambda) = (\tau \circ \beta(t, \xi, \lambda),\ P_s(0)(\varphi^{\tau\circ\beta(t,\xi,\lambda)}(\beta(t, \xi, \lambda)) - \overline{y}(0)), \lambda).$$

Thus β^{-1} is an admissible chart of $M_{+s}(V)$ near $(\overline{y}(0), \overline{\lambda})$. ∎

REMARK.

We used the implicit function theorem to construct the foliation of the stable manifold by asymptotic phase. This has some computational advantages. E.g. we can compute derivatives of

$$g(\xi) = x_+(0, \xi, \overline{\lambda}),\ g(0) = \overline{y}(0)$$

which parametrizes the fiber which is in asymptotic phase with $\overline{y}(0)$. Taking implicit derivatives in (3.6) and using (2.26)–(2.31) one finds

$$g'(0) = I_{|R(P_s(0))},\ P_s(0)g''(0) = 0 \ \text{ and}$$

$$\begin{aligned} P_\kappa(0)g''(0)(\xi_1, \xi_2) &= -\overline{T} \int_0^\infty P_\kappa(0)S(0, t)f_{xx}(\overline{y}(t), \overline{\lambda})(S(t, 0)\xi_1, S(t, 0)\xi_2)\, dt \\ &= -\int_0^\infty \widehat{P}_\kappa(0)\widehat{S}(0, t)f_{xx}(\widehat{y}(t), \overline{\lambda})(\widehat{S}(t, 0)\xi_1, \widehat{S}(t, 0)\xi_2)\, dt \end{aligned}$$

for $\kappa = c, u$ and $\xi_1, \xi_2 \in R(P_s(0))$.

Here $\widehat{S}$ and $\widehat{P}_\kappa$ denote the solution operator and projectors for the unscaled linearized problem

$$\dot{y} = \frac{\partial f}{\partial x}\,(\widehat{y}(t), \overline{\lambda})y \tag{3.11}$$

We now consider the analogue of Theorem 3.1 for the global stable manifold.

Let $\widehat{x}(t),\ t \geq 0$ be a solution of (1.1) at $\lambda = \overline{\lambda}$ such that

$$\text{dist}\ (\widehat{x}(t), \overline{\gamma}) \to 0 \ \text{ as } \ t \to \infty.$$

We set $\overline{x}(t) = \widehat{x}(t\overline{T})$ and find for some $\overline{t}$ sufficiently large

$$(\overline{x}(\overline{t}), \overline{\lambda}) \in M_{+s}(V).$$

Taking V sufficiently small we can cover $M_{+s}(V)$ by finitely many charts of the type (3.7), hence for some t_0

$$(3.12) \qquad ||\overline{x}(t) - \overline{y}(t+t_0)|| + ||\dot{\overline{x}}(t) - \dot{\overline{y}}(t+t_0)|| = O(e^{-\epsilon t})$$

where $\epsilon <$ Min $(-\alpha, \beta)$ as in (2.19), (2.21).

Without loss of generality we will assume $t_0 = 0$. All the assumptions of Proposition 2.7 are then satisfied with the settings

$$(3.13) \quad z = \dot{\overline{y}},\ L = \frac{\mathrm{d}}{\mathrm{dt}} - \overline{T}\frac{\partial f}{\partial x}(\overline{y}, \overline{\lambda}),\ \tilde{z} = \dot{\overline{x}},\ \tilde{L} = \frac{\mathrm{d}}{\mathrm{dt}} - \overline{T}\frac{\partial f}{\partial x}(\overline{x}, \overline{\lambda}).$$

In the following theorem we make use of the solution operator $\widetilde{S}$ and the projectors $\widetilde{P}_\kappa$, $\kappa = s, c, u$ associated with $\widetilde{L}$.

THEOREM 3.2.

Under the assumptions above $(\overline{x}, \overline{y}, \overline{T}, 0, \overline{\lambda}) \in Z_1^+(\epsilon) \times \mathbb{R} \times R(\widetilde{P}_s(0)) \times \mathbb{R}^p$ is a solution of the operator equation

$$(3.14) \qquad F_+(x, y, T, \xi, \lambda) = \begin{pmatrix} \dot{x} - Tf(x, \lambda) \\ \dot{y} - Tf(y, \lambda) \\ \widetilde{P}_s(0)(x(0) - \overline{x}(0)) - \xi \\ \chi(y) \end{pmatrix} = 0.$$

This equation has a unique solution

$$(x_+(\cdot, \xi, \lambda),\ y(\cdot, \lambda),\ T(\lambda)) \in Z_1^+(\epsilon) \times \mathbb{R}$$

in some neighbourhood of $(\overline{x}, \overline{y}, \overline{T})$ which depends smoothly on (ξ, λ) in some neighbourhood of $(0, \overline{\lambda})$.

The local inverse of

$$(3.15) \quad \beta(t, \xi, \lambda) = (x_+(t, \xi, \lambda), \lambda),\ t \in U(0),\ \xi \in U(0) \subset R(\widetilde{P}_s(0)),\ \lambda \in U(\overline{\lambda})$$

is an admissible chart of the global stable manifold

$$(3.16) \quad M_{+s} = \{(x, \lambda) \in \mathbb{R}^m \times U(\overline{\lambda}) :\ \text{dist}\ (\varphi^t(x, \lambda),\ \gamma(\lambda)) \to 0\ \text{ as }\ t \to \infty\}.$$

PROOF.
We are brief here because the implicit function theorem applies as in the proof of Theorem 3.1 with $(\overline{y}, \overline{y})$ replaced by $(\overline{x}, \overline{y})$ and P_s by $\widetilde{P}_s$.
The analogue of (3.10) is

$$\frac{\partial \beta}{\partial(t, \xi, \lambda)}(0, 0, \overline{\lambda}) = \begin{pmatrix} \dot{\overline{x}}(0) & I_{|R(\widetilde{P}_s(0))} & \frac{\partial \beta}{\partial \lambda}(0, 0, \overline{\lambda}) \\ 0 & 0 & I_p \end{pmatrix}$$

so that β is again a local immersion. The same is then true for $\widetilde{\beta} = \Phi^{\overline{t}} \circ \beta$ where we take $\overline{t}$ so large that the image of $\widetilde{\beta}$ is in $M_{+s}(V)$. As in the proof Theorem 3.1 $\widetilde{\beta}$ is an admissible chart of the local stable manifold. Therefore, $\beta^{-1} = \widetilde{\beta}^{-1} \circ \Phi^{\overline{t}}$ is admissible for the global stable manifold, because the differentiable structure on M_{+s} is defined by the charts of the local stable manifold transferred backwards in time by the flow. ∎

Now let $(\widehat{x}(t), \overline{\lambda})$, $t \in \mathbb{R}$ be an orbit connecting a hyperbolic steady state $y_-(\overline{\lambda})$ with stability indices m_{-s}, m_{-u} to the hyperbolic periodic orbit $\overline{\gamma}$.

Again, $y_-(\overline{\lambda})$ is contained in a smooth manifold of steady states

$$\{y_-(\lambda) : \lambda \in U(\overline{\lambda})\}$$

which have the same stability indices and we can form the global unstable manifold

(3.17) $$M_{-u} = \{(x, \lambda) \in \mathbb{R}^m \times U(\overline{\lambda}) : \varphi^t(x, \lambda) \to y_-(\lambda) \text{ as } t \to -\infty\}.$$

We set $\overline{x}(t) = \widehat{x}(t\overline{T})$, $t \in \mathbb{R}$ and

$$\widetilde{L} = \frac{\mathrm{d}}{\mathrm{dt}} - \frac{\partial f}{\partial x}(\overline{x}, \overline{\lambda})$$

so that $\widetilde{L}$ has an exponential dichotomy on $\mathbb{R}_-$ with projectors Q_s, Q_u and an ordinary exponential trichotomy on $\mathbb{R}_+$ with projectors P_s, P_c, P_u (see Proposition 2.8).

Similar to Theorem 3.2 we may parametrize M_{-u} near $(\overline{x}(0), \overline{\lambda})$ by

$$\beta(t, \eta, \lambda) = (x_-(0, \eta, \lambda), \lambda),\ \eta \in U(0) \subset R(Q_u(0)),\ \lambda \in U(\overline{\lambda}).$$

Here $x_-(\cdot, \eta, \lambda)$ is the solution of

(3.18) $$F_-(x, \eta, \lambda) = \begin{pmatrix} \dot{x} - T(\lambda) f(x, \lambda) \\ Q_u(0)(x(0) - \overline{x}(0)) - \eta \end{pmatrix} = 0$$

obtained from the implicit function theorem in a neighbourhood of $x = \overline{x}_{|\mathbb{R}_-}$, $\eta = 0$, $\lambda = \overline{\lambda}$ in the space of bounded C^1–functions (cf. [Bey 90b]).

Using $x_-(\cdot, \eta, \lambda)$ and $x_+(\cdot, \xi, \lambda)$ from Theorem 3.2 we define the $(m \times p)$ matrices

(3.19) $$E_\pm(t) = \frac{\partial x_\pm}{\partial \lambda}(t, 0, \overline{\lambda}),\ t \in \mathbb{R}_\pm.$$

These satisfy the variational equations

(3.20) $$\dot{E}_\pm - \overline{T}\frac{\partial f}{\partial x}(\overline{x}, \overline{\lambda})E_\pm = f(\overline{x}, \overline{\lambda})T'(\overline{\lambda}) + \overline{T}\frac{\partial f}{\partial \lambda}(\overline{x}, \overline{\lambda}) \text{ in } \mathbb{R}_\pm$$

subject to

$$Q_u(0)E_-(0) = 0,\ \widetilde{P}_s(0)E_+(0) = 0$$

and

$$E_+(t) = \frac{\partial y}{\partial \lambda}(t, \overline{\lambda}) + O(e^{-\epsilon t}) \text{ as } t \to \infty.$$

Here $T'(\overline{\lambda})$ and $Y(\cdot) = \frac{\partial y}{\partial \lambda}(\cdot, \overline{\lambda})$, can be obtained from (3.2) by implicit differentiation

$$LY = f(\overline{y}, \overline{\lambda})T'(\overline{\lambda}) + \overline{T}\,\frac{\partial f}{\partial \lambda}\,(\overline{y}, \overline{\lambda}),\ \chi'(\overline{y})Y = 0. \tag{3.21}$$

The tangent spaces at $\overline{z}(0) = (\overline{x}(0), \overline{\lambda})$ can now be written as

$$T_{\overline{z}(0)}M_{-u} = \{(\eta + E_-(0)\lambda, \lambda) : \eta \in R(Q_u(0), \lambda \in \mathbb{R}^p\} \tag{3.22a}$$

$$T_{\overline{z}(0)}M_{+s} = \{(\xi + E_+(0)\lambda + c\,\dot{\overline{x}}(0), \lambda) : \xi \in R(\widetilde{P}_s(0)), \lambda \in \mathbb{R}^p,\ c \in \mathbb{R}\}. \tag{3.22b}$$

With these preparations we can state the main result.

THEOREM 3.3.

As above let $\overline{\gamma} = \{\overline{y}(t) = \widehat{y}(t\overline{T}) : t \in \mathbb{R}\}$ be a $\overline{T}$-periodic hyperbolic orbit at $\lambda = \overline{\lambda}$ and let $\overline{z}(t) = (\overline{x}(t), \overline{\lambda}) = (\widehat{x}(t\overline{T}), \overline{\lambda})$ be an orbit connecting a hyperbolic steady state $y_-(\overline{\lambda})$ to $\overline{\gamma}$. Further, let $\Psi \in C^1(Z_1(\epsilon) \times \mathbb{R}^{p+1}, \mathbb{R})$ be given such that

$$\Psi(\overline{x}, \overline{y}, \overline{T}, \overline{\lambda}) = 0,\ \frac{\partial \Psi}{\partial x}\,(\overline{x}, \overline{y}, \overline{T}, \overline{\lambda})\,\dot{\overline{x}} + \frac{\partial \Psi}{\partial y}\,(\overline{x}, \overline{y}, \overline{T}, \overline{\lambda})\,\dot{\overline{y}} \neq 0. \tag{3.23}$$

Then the following conditions are equivalent.

(i) The manifolds M_{-u} and M_{+s} intersect transversely in the strong sense that for all $t \in \mathbb{R}$

$$T_{\overline{z}(t)}M_{-u} + T_{\overline{z}(t)}M_{+s} = \mathbb{R}^{m+p},\ T_{\overline{z}(t)}M_{-u} \cap T_{\overline{z}(t)}M_{+s} = \text{span}\ \{\dot{\overline{z}}(t)\}. \tag{3.24}$$

(ii) The linear mapping

$$B(\eta, \xi, \lambda) = \xi - \eta + (E_+(0) - E_-(0))\lambda \tag{3.25}$$

is a bijection from $R(Q_u(0)) \times R(\widetilde{P}_s(0)) \times \mathbb{R}^p$ into $\mathbb{R}^m$.

(iii) $(\overline{x}, \overline{y}, \overline{T}, \overline{\lambda}) \in Z_1(\epsilon) \times \mathbb{R}^{p+1}$ is a regular solution of the system (1.13), i.e. $F'(\overline{x}, \overline{y}, \overline{T}, \overline{\lambda}) : Z_1(\epsilon) \times \mathbb{R}^p \to Z_0(\epsilon) \times \mathbb{R}$ is a linear homeomorphism.

PROOF. (i) $\Rightarrow$ (ii)

From (3.24) we obtain the relation (see (1.8))

$$p = m_{+u} - m_{-u} + 1 \tag{3.26}$$

and this holds iff B is given by a quadratic matrix. Suppose $B(\eta, \xi, \lambda) = 0$ for some $\eta \in R(Q_u(0)),\ \xi \in R(\widetilde{P}_s(0)),\ \lambda \in \mathbb{R}^p$. Then

$$(\xi + E_+(0)\lambda, \lambda) = (\eta + E_-(0)\lambda, \lambda) \in T_{\overline{z}(0)}M_{-u} \cap T_{\overline{z}(0)}M_{+s}$$

holds and by (3.24) $\xi = \eta = c\,\dot{\overline{x}}(0)$ for some $c \in \mathbb{R}$. But $\dot{\overline{x}}(0) \notin R(\widetilde{P}_s(0))$ so that $c = 0$ and $\xi = \eta = 0$ follows.

(ii) $\Rightarrow$ (i)

The linearized flow $\frac{\partial \Phi^t}{\partial z}$ $(\overline{z}(t))$ maps the tangent spaces at $\overline{z}(0)$ onto those at $\overline{z}(t)$, hence it is sufficient to prove (3.24) at $t = 0$. Moreover, (3.26) follows from (ii) so that we need only prove the second relation in (3.24).

Suppose $(x, \lambda) \in T_{\overline{z}(0)} M_{-u} \cap T_{\overline{z}(0)} M_{+s}$ holds, then from (3.22) we obtain

$$x = \eta + E_-(0)\lambda = \xi + E_+(0)\lambda + c\dot{\overline{x}}(0)$$

for suitable ξ, η, λ. With $\widetilde{\eta} = \eta - c\,\dot{\overline{x}}(0) \in R(Q_u(0))$ we find $B(\widetilde{\eta}, \xi, \lambda) = 0$ and hence $\xi = 0,\ \lambda = 0,\ x = \eta = c\,\dot{\overline{x}}(0)$.

(ii) $\Rightarrow$ (iii)

By our assumptions $(\overline{x}, \overline{y}, \overline{T}, \overline{\lambda})$ is a solution of (1.13). For the derivative at this point (which we abbreviate as $(\cdot)$) we find

$$F'(\cdot) = \begin{bmatrix} \multicolumn{2}{c}{\multirow{2}{*}{Γ}} & -f(\overline{x}, \overline{\lambda}) & -\overline{T}\,\frac{\partial f}{\partial \lambda}\,(\overline{x}, \overline{\lambda}) \\ & & -f(\overline{y}, \overline{\lambda}) & -\overline{T}\,\frac{\partial f}{\partial \lambda}\,(\overline{y}, \overline{\lambda}) \\ \frac{\partial \Psi}{\partial x}\,(\cdot) & \frac{\partial \Psi}{\partial y}\,(\cdot) & \frac{\partial \Psi}{\partial T}\,(\cdot) & \frac{\partial \Psi}{\partial \lambda}\,(\cdot) \end{bmatrix}$$

where $\Gamma(x, y) = (\widetilde{L}x, Ly)$. By Proposition 2.8 Γ has Fredholm index $m_{-u} - m_{+u} - 1$ and from (3.26) we obtain with the help of the bordering lemma ([Bey 90a]) that $F'(\cdot)$ has Fredholm index 0.

Suppose $F'(\cdot)(x, y, T, \lambda) = 0$ for some $(x, y) \in Z_1(\epsilon),\ T \in \mathbb{R},\ \lambda \in \mathbb{R}^p$. By (3.3) we can choose $c \in \mathbb{R}$ such that $\widetilde{y} = y - c\,\dot{\overline{y}}$ satisfies $\chi'(\overline{y})\widetilde{y} = 0$.

Using this and the equation

$$L\widetilde{y} - Tf(\overline{y}, \overline{\lambda}) = \overline{T}\,\frac{\partial f}{\partial \lambda}\,(\overline{y}, \overline{\lambda})\lambda$$

we find that $(\widetilde{y}, T)$ and $(\frac{\partial y}{\partial \lambda}\,(\cdot, \overline{\lambda})\lambda,\ T'(\overline{\lambda})\lambda)$ satisfy the same system (see (3.21)) and hence

$$y = c\,\dot{\overline{y}} + \frac{\partial y}{\partial \lambda}\,(\cdot, \overline{\lambda})\lambda,\ T = T'(\overline{\lambda})\lambda. \tag{3.27}$$

We define $\xi = \widetilde{P}_s(0)x(0)$ and $\widetilde{x}(t) = x(t) - \widetilde{S}(t, 0)\xi - c\,\dot{\overline{x}}(t)$ for $t \geq 0$ so that

$$\widetilde{x}(t) - \widetilde{y}(t) = x(t) - y(t) - \widetilde{S}(t, 0)\xi = O(e^{-\epsilon t})$$

and $\widetilde{P}_s(0)\widetilde{x}(0) = 0$.

Therefore, $(\widetilde{x}, \widetilde{y}) \in Z_1^+(\epsilon)$ and we have shown, with the operator F_+ from (3.14), that

$$\frac{\partial F_+}{\partial(x, y, T)}\,(\cdot)(\widetilde{x}, \widetilde{y}, \widetilde{T}) = -\,\frac{\partial F_+}{\partial \lambda}\,(\cdot)\,\lambda$$

where $(\cdot)$ denotes evaluation at $(\overline{x}, \overline{y}, \overline{T}, 0, \overline{\lambda})$. Thus we obtain $\widetilde{x}(t) = E_+(t)\lambda$ and

$$x(0) = \xi + c\,\dot{\overline{x}}(0) + E_+(0)\lambda. \tag{3.28}$$

In a similar way we set $\eta = Q_u(0)x(0)$, $v(t) = x(t) - \widetilde{S}(t,0)\eta$ $(t \le 0)$ and find $v(t) = E_-(t)\lambda$ by using equation (3.18) for $x_-(\cdot,\eta,\lambda)$. Now we combine $x(0) = \eta + E_-(0)\lambda$ and (3.28) to get

$$B(\eta - c\,\dot{\overline{x}}(0), \xi, \lambda) = 0.$$

From assumption (ii) and (3.27) we conclude $\lambda = 0,\ \xi = 0,\ \eta = c\,\dot{\overline{x}}(0),\ T = 0,\ x = c\,\dot{\overline{x}}$ and $y = c\,\dot{\overline{y}}$. An application of (3.23) finally yields $c = 0$ and $x = 0,\ y = 0$.

(iii) $\Rightarrow$ (ii)

We know that $F'(\cdot)$ has Fredholm index 0, hence by Proposition 2.8 and the bordering lemma we have

$$0 = \text{ ind } (\Gamma) + p = m_{-u} - m_{+u} - 1 + p$$

i.e. (3.26) holds.

Now assume $B(\eta,\xi,\lambda) = 0$ and define

$$T = T'(\overline{\lambda}), \lambda,\ y(t) = \frac{\partial y}{\partial \lambda}\,(t,\overline{\lambda})\lambda + c\,\dot{\overline{y}}(t)$$

$$x(t) = \begin{cases} E_+(t)\lambda + \widetilde{S}(t,0)\xi + c\,\dot{\overline{x}}(t) & \text{for } t \ge 0 \\ E_-(t)\lambda + \widetilde{S}(t,0)\eta + c\,\dot{\overline{x}}(t) & \text{for } t < 0 \end{cases}$$

where c will be determined later on.

Using $B(\eta,\xi,\lambda) = 0$ we see that x is continuous at 0 and in fact $(x,y) \in Z_1(\epsilon)$. Moreover, the equation (3.20) yields

$$\Gamma\begin{pmatrix} x \\ y \end{pmatrix} - T\begin{pmatrix} f(\overline{x},\overline{\lambda}) \\ f(\overline{y},\overline{\lambda}) \end{pmatrix} - \overline{T}\begin{pmatrix} \frac{\partial f}{\partial \lambda}\,(\overline{x},\overline{\lambda})\lambda \\ \frac{\partial f}{\partial \lambda}\,(\overline{y},\overline{\lambda})\lambda \end{pmatrix} = 0.$$

Finally, by (3.23) we can determine c so that

$$0 = \frac{\partial \Psi}{\partial x}\,(\cdot)x + \frac{\partial \Psi}{\partial y}\,(\cdot)y + \frac{\partial \Psi}{\partial T}\,(\cdot)T + \frac{\partial \Psi}{\partial \lambda}\,(\cdot)\lambda.$$

Then (x,y,T,λ) is in the null space of $F'(\cdot)$ and we conclude $x = 0,\ y = 0,\ T = 0,\ \lambda = 0$. In particular, $\xi + c\,\dot{\overline{x}}\,(0) = 0 = \eta + c\,\dot{\overline{x}}(0)$ and $\xi = 0,\ c = 0$ because $\dot{\overline{x}}(0) \notin R(\widetilde{P}_s(0))$. ■

REMARK.

We can easily rewrite the nonsingularity of the matrix B in an equivalent form using the adjoint operator Γ^* from Proposition 2.8. If B is nonsingular then necessarily

(3.29) $$R(Q_u(0)) \cap R(\widetilde{P}_s(0)) = \{0\}.$$

Therefore

$$\dim N(\Gamma^*) = -\text{ind}\,(\Gamma) = m_{+u} - m_{-u} + 1 = p$$

and we can find a basis of the form

$$\begin{pmatrix} \widetilde{\psi} \\ \psi \end{pmatrix}, \begin{pmatrix} \varphi_2 \\ 0 \end{pmatrix}, \dots, \begin{pmatrix} \varphi_p \\ 0 \end{pmatrix}$$

where $\widetilde{L}^* \varphi_i = 0$ and $\varphi_i(t) = O(e^{-\epsilon t})$ for $i = 2, \dots, p$. Setting $\varphi_1 = \widetilde{\psi}$ we find from the proof of Proposition 2.8 that the columns of $\Phi(0) := (\varphi_1(0), \dots, \varphi_p(0))$ form a basis of
$(R(\widetilde{P}_s(0)) + R(Q_u(0)))^\perp$. Then the nonsingularity of B reduces to the nonsingularity of the $(p \times p)$–matrix

$$\Lambda = \Phi(0)^T (E_+(0) - E_-(0)). \tag{3.30}$$

This is the derivative with respect to λ of some generalized Melnikov–type function

$$\Phi(0)^T (x_+(0, 0, \lambda) - x_-(0, 0, \lambda)).$$

Of course, for the way back from the nonsingularity of Λ to that of B we need to assume conditions (3.29) and (3.26).

It is now quite straightforward to develop an anlogue of Theorem 3.3 for periodic–to–periodic connections. However, there are a few differences due to the fact that there is no simple scaling of the time axis which is suitable for both periodic orbits. This is the reason for the delicate nonautonomous transformations in [Ha Li 86]. Of course, the split formulation (1.14) is also some kind of nonautonomous transformation. But its treatment is more convenient, it is closer to numerical approximation schemes and it is easily covered by the theory developed so far.

Let $(\widehat{x}(t), \overline{\lambda})$, $t \in \mathbb{R}$ be an orbit connecting a $\overline{T}_-$–periodic hyperbolic orbit $\{\widehat{y}_-(t) : t \in \mathbb{R}\}$ to a $\overline{T}_+$–periodic one $\{\widehat{y}_+(t) : t \in \mathbb{R}\}$. We introduce the scaled functions

$$\overline{y}_\pm(t) = \widehat{y}_\pm(t\overline{T}_\pm), \ \overline{x}_\pm(t) = \widehat{x}(t\overline{T}_\pm) \ \text{ for } \ t \in \mathbb{R}_\pm$$

and the differential operators

$$L_\pm = \frac{\mathrm{d}}{\mathrm{dt}} - \overline{T}_\pm \frac{\partial f}{\partial x}(\overline{y}_\pm, \overline{\lambda}), \ \widetilde{L}_\pm = \frac{\mathrm{d}}{\mathrm{dt}} - \overline{T}_\pm \frac{\partial f}{\partial x}(\overline{x}_\pm, \overline{\lambda})$$

with solution operators $S_\pm$ and $\widetilde{S}_\pm$ respectively. L_- and $\widetilde{L}_-$ have ordinary exponential trichotomies on $\mathbb{R}_-$ and we denote the corresponding projectors by Q_κ and
$\widetilde{Q}_\kappa$, $\kappa = s, c, u$.

The unstable manifold M_{-u} is of the form

$$M_{-u} = \{(x, \lambda) \in \mathbb{R}^m \times U(\overline{\lambda}) : \ \text{dist}\,(\varphi^t(x, \lambda), \ \gamma_-(\lambda)) \to 0 \ \text{ as } \ t \to -\infty\}$$

where $\gamma_-(\lambda) = \{y_-(t,\lambda) : t \in \mathbb{R}\}$ are $T_-(\lambda)$–periodic orbits and $y_-(\cdot,\overline{\lambda}) = \overline{y}_-$. M_{-u} can be parametrized by $(x_-(t,\eta,\lambda),\lambda)$ is a way analogous to Theorem 3.2.

We will also assume that the phase conditions $\chi_\pm$ (compare (3.2)) for $y_\pm(\cdot,\lambda)$ have been chosen in such a way that

$$\overline{x}(t) - \overline{y}_\pm(t) = O(e^{-\epsilon|t|}) \quad \text{as} \quad t \to \pm\infty$$

holds. Finally, the spaces $Z_1^-(\epsilon)$, $Z_0^-(\epsilon)$ are defined analogously to $Z_1^+(\epsilon)$, $Z_0^+(\epsilon)$ and ϵ is chosen such that $[-\epsilon,\epsilon]$ contains no real parts of the Floquet exponents of $\gamma_\pm(\overline{\lambda})$.

THEOREM 3.4.

Let $\overline{z}(t) = (\overline{x}(t),\overline{\lambda})$ be a periodic–to–periodic connecting orbit as above and assume that $\Psi \in C^1(Z_1^+(\epsilon) \times Z_1^-(\epsilon) \times \mathbb{R}^{p+2},\ \mathbb{R})$ satisfies

$$\text{(3.31)} \qquad \begin{aligned} &\Psi(\overline{u}) = 0 \quad \text{for} \quad \overline{u} = (\overline{x}_-,\overline{y}_-,\overline{x}_+,\overline{y}_+,\overline{T}_-,\overline{T}_+,\overline{\lambda}) \\ &\frac{\partial\Psi}{\partial x_-}(\overline{u})\,\dot{\overline{x}}_- + \frac{\partial\Psi}{\partial y_-}(\overline{u})\,\dot{\overline{y}}_- + \frac{\partial\Psi}{\partial x_+}(\overline{u})\,\dot{\overline{x}}_+ + \frac{\partial\Psi}{\partial y_+}(\overline{u})\,\dot{\overline{y}}_+ \neq 0. \end{aligned}$$

Then the following conditions are equivalent.

(i) The manifolds M_{-u} and M_{+s} intersect transversely along $\overline{z}(t)$ in the strong sense (3.24)

(ii) The linear mapping

$$B(c,\eta,\xi,\lambda) = c\,\dot{\overline{x}}\,(0) + \xi - \eta + (E_+(0) - E_-(0))\lambda$$

is a bijection from $\mathbb{R} \times R(\widetilde{Q}_u(0)) \times R(\widetilde{P}_s(0)) \times \mathbb{R}^p$ into $\mathbb{R}^m$

(iii) $\overline{u}$ is a regular solution in $Z_1^+(\epsilon) \times Z_1^-(\epsilon) \times \mathbb{R}^{p+2}$ of the operator equation (1.14).

PROOF.

Since the proof is quite similar to that of Theorem 3.3 we only sketch the main differences.

(3.22a) becomes

$$T_{\overline{z}(0)}\, M_{-u} = \{(\eta + E_-(0)\lambda + c\,\dot{\overline{x}}\,(0),\lambda) : \eta \in R(\widetilde{Q}_u(0)),\ \lambda \in \mathbb{R}^p,\ c \in \mathbb{R}\}$$

and the relation (3.26) changes to (cf. (1.8))

$$\text{(3.32)} \qquad p = m_{+u} - m_{-u}.$$

From this the equivalence of (i) and (ii) follows immediately.

Let us assume (ii) and calculate

$$F'(\overline{u}) = \begin{bmatrix} \Gamma_- & 0 & \begin{matrix} -f(\overline{x}_-,\overline{\lambda}) \\ -f(\overline{y}_-,\overline{\lambda}) \end{matrix} & \begin{matrix} 0 \\ 0 \end{matrix} & \begin{matrix} -\overline{T}_- \frac{\partial f}{\partial \lambda}(\overline{x}_-,\overline{\lambda}) \\ -\overline{T}_- \frac{\partial f}{\partial \lambda}(\overline{y}_-,\overline{\lambda}) \end{matrix} \\ 0 & \Gamma_+ & \begin{matrix} 0 \\ 0 \end{matrix} & \begin{matrix} -f(\overline{x}_+,\overline{\lambda}) \\ -f(\overline{y}_+,\overline{\lambda}) \end{matrix} & \begin{matrix} -\overline{T}_+ \frac{\partial f}{\partial \lambda}(\overline{x}_+,\overline{\lambda}) \\ -\overline{T}_+ \frac{\partial f}{\partial \lambda}(\overline{y}_+,\overline{\lambda}) \end{matrix} \\ -E_0 \quad 0 & E_0 \quad 0 & 0 & 0 & 0 \\ \frac{\partial \Psi}{\partial(x_-,y_-)}(\overline{u}) & \frac{\partial \Psi}{\partial(x_+,y_+)}(\overline{u}) & \frac{\partial \Psi}{\partial T_-}(\overline{u}) & \frac{\partial \Psi}{\partial T_+}(\overline{u}) & \frac{\partial \Psi}{\partial \lambda}(\overline{u}) \end{bmatrix}$$

where $\Gamma_\pm(x,y) = (\widetilde{L}_\pm\ x, L_\pm\ y)$. The Fredholm index of $F'(\overline{u})$ is zero due to (3.32) and Theorem 3.2.

Let $u = (x_-, y_-, x_+, y_+, T_-, T_+, \lambda)$ be in the null space of $F'(\overline{u})$. Then we choose $c_\pm$ such that

$$\chi'_\pm(\overline{y}_\pm)(y_\pm - c_\pm\ \dot{\overline{y}}_\pm) = 0$$

and obtain as in (3.27)

$$y_\sigma = c\ \dot{\overline{y}}_\sigma + \frac{\partial y_\sigma}{\partial \lambda}\ (\cdot,\overline{\lambda})\lambda,\ T_\sigma = T'_\sigma(\overline{\lambda})\lambda \ \text{ for } \ \sigma = +,-\ .$$

The analogue of (3.28) is

$$x_\sigma(0) = \xi_\sigma + c_\sigma\ \dot{\overline{x}}\ (0) + E_\sigma(0)\lambda,\ \sigma = +,-\ ,$$

where $\xi_+ = \widetilde{P}_s(0)x_+(0)$ and $\xi_- = \widetilde{Q}_u(0)x_-(0)$.

From the equality $x_+(0) = x_-(0)$ we find

$$B(c_+ - c_-, \xi_-, \xi_+, \lambda) = 0$$

and hence $c_+ = c_-, \xi_- = 0,\ \xi_+ = 0,\ \lambda = 0$. Finally, using the last row in $F'(\overline{u})$ and assumption (3.31) we end up with $c_+ = c_- = 0$ and $x_\sigma = 0,\ y_\sigma = 0,\ T_\sigma = 0\ (\sigma = +,-)$.

For the converse statement we notice that (3.32) is a consequence of ind $(F'(\overline{u})) = 0$ and Theorem 3.2.

Assuming $B(c,\eta,\xi,\lambda) = 0$ we define $T_\pm = T'_\pm(\overline{\lambda})\lambda$ and

$$\begin{aligned}
x_+(t) &= E_+(t)\lambda + \widetilde{S}_+(t,0)\xi + (c+\widetilde{c})\ \dot{\overline{x}}\ (t),\ t \geq 0, \\
y_+(t) &= \frac{\partial y_+}{\partial \lambda}\ (t,\overline{\lambda})\lambda + (c+\widetilde{c})\ \dot{\overline{y}}_+\ (t), \\
x_-(t) &= E_-(t)\lambda + \widetilde{S}_-(t,0)\eta + \widetilde{c}\ \dot{\overline{x}}\ (t),\ t \leq 0, \\
y_-(t) &= \frac{\partial y_-}{\partial \lambda}\ (t,\overline{\lambda})\lambda + \widetilde{c}\ \dot{\overline{y}}_-\ (t),
\end{aligned}$$

where $\widetilde{c}$ is determined in such a way that the last row of $F'(\overline{u})$ applied to $u = (x_-, y_-, x_+, y_+, T_-, T_+, \lambda)$ vanishes.

Then $u \in N(F'(\overline{u}))$ and the assertion follows from $u = 0$ and $\dot{\overline{x}}\,(0) \notin R(\widetilde{P}_s(0)$, $\dot{\overline{x}}\,(0) \notin R(\widetilde{Q}_u(0))$. ■

This theorem holds also in the case $p = 0$ which e.g. occurs for a homoclinic periodic connection (see (3.32)). Then there are no parameters in the system (1.1) and all the statements involving λ trivialize in an obvious way.

Suppose in the homoclinic case that we can find a section Σ with Poincaré map P such that successive intersections of the orbit with Σ are obtained by an application of P. Then it can be shown that the transversality conditions of Theorem 3.4 hold if and only if the points of intersection with Σ are transversal homoclinic points of the map P (see e.g. [Pal 88] for an analysis of transversal homoclinic points). By this shooting type approach we have reduced the computation of homoclinic periodic connections to that of homoclinic points of maps. However, in the general homoclinic case it is not clear whether such a reduction is possible and we expect that one still has to tackle infinite boundary value problems of the form (1.14) in a numerical calculation.

4. A numerical example

In the well–posed formulations of section 3 the boundary conditions for the connecting orbits are hidden in the function spaces used.

For a numerical approximation we replace $(-\infty, \infty)$ by some large interval $J = (T_-, T_+)$ and we have to introduce finite boundary conditions at T_- and T_+.

In the case of stationary connecting orbits it is well–known how to set up these boundary conditions and how to estimate the errors involved, see [Bey 90a, Bey 90b, DoFr 89, FrDo 91, Kuz 90, Sch 93a, Sch 93b]. To our knowledge there are no numerical approaches to the periodic case and we will treat here only the point–to–periodic connection, i.e. equation (1.13).

We consider the following system of differential equations

(4.1) $$\dot{x} = f(x, \lambda), \;\; t \in J = [T_-, T_+]$$

(4.2) $$\dot{y} = T\; f(y, \lambda), \;\; t \in [0, 1]$$

(4.3) $$\dot{T} = 0, \;\; \dot{\lambda} = 0.$$

Here we have $2m + p + 1$ variables $u = (x, y, T, \lambda)$ and we need the same number of boundary conditions. We assume these to be of the following form

(4.4) $$B_-(x(T_-), \lambda) = 0$$

(4.5) $$B_+(x(T_+), y(0), \lambda) = 0$$

(4.6) $$\psi_J(x, y, T, \lambda) = 0$$

(4.7) $$y(1) - y(0) = 0$$

where $B_- : \mathbb{R}^{m+p} \to \mathbb{R}^{m_{-s}}$ and $B_+ : \mathbb{R}^{2m+p} \to \mathbb{R}^{m_{+u}+1}$ define the asymptotic boundary conditions at T_- and T_+ and ψ_J acts as a scalar phase condition.
The number of boundary conditions is

$$m_{-s} + m_{+u} + 2 + m = 2m - m_{-u} + m_{+u} + 2$$

which coincides with $2m + p + 1$ unter the assumption (3.26) (see Theorem 3.3).

The condition (4.4) requires $x(T_-)$ to lie in some approximation to the unstable manifolds of the steady states $x_-(\lambda)$ and a good choice are projection boundary conditions (see [Bey 90a])

(4.8) $$B_-(x, \lambda) = V_s(\lambda)(x - x_-(\lambda))$$

where the rows of $V_s(\lambda) \in \mathbb{R}^{m_{-s},m}$ depend smoothly on λ and span the stable subspace of $\frac{\partial f}{\partial x}\,(x_-(\lambda), \lambda)^T$, i.e.

$$V_s(\lambda)\,\frac{\partial f}{\partial x}\,(x_-(\lambda), \lambda) = G_-(\lambda)V_s(\lambda)$$

for some $G_-(\lambda) \in \mathbb{R}^{m_{-s},m_{-s}}$ with Re $\sigma(G_-(\lambda)) < 0$.
When $x_-(\lambda)$ is known, the matrix $V_s(\lambda)$ can easily be computed by an eigenvalue solver combined with a normalization procedure ([Bey 90a], [ChKu 93]).

Similarly, we should choose B_+ in such a way that $B_+(\cdot, y(0, \lambda), \lambda) = 0$ is an approximation to the fiber of dimension m_{+s} which is in asymptotic phase with $y(0, \lambda)$. Here, as in section 3, we denote by $y(\cdot, \lambda)$ the 1–periodic solutions of (4.2) with $T = T(\lambda)$ fixed by some suitable phase condition. The analogoue of (4.8) then is

(4.9) $$B_+(x(T_+), y(0), \lambda) = V_u(0, \lambda)(x(T_+) - y(0))$$

where $V_u(t, \lambda) \in \mathbb{R}^{m_{+u}+1,m}$ solves the adjoint variational equation

(4.10) $$\dot{V} = -V\;T(\lambda)\,\frac{\partial f}{\partial x}\,(y(\cdot, \lambda), \lambda)$$

(4.11) $$V_u(1, \lambda) = G_+(\lambda)V_u(0, \lambda) \quad \text{for some} \quad G_+(\lambda) \in \mathbb{R}^{m_{+u}+1,m_{+u}+1}$$

with $|\mu| \leq 1$ for all eigenvalues μ of $G_+(\lambda)$.
Notice that the spectrum of $G_+(\lambda)$ includes the trivial Floquet multiplier 1 and that the rows of $V_u(t, \lambda)$ should span the same space as those of

$$V_F(t, \lambda) = \begin{pmatrix} \psi(t, \lambda)^T \\ e^{-t\;B_u(\lambda)}\;\Psi_u(t, \lambda)^T \end{pmatrix}$$

which is obtained from the Floquet decomposition (2.20).

Since $y(\cdot,\lambda)$ is itself a result of the computation it is unrealistic to assume that $V_u(\cdot,\lambda)$ is known a–priori. One way out of this dilemma is to attach the variational equation (4.10) to the system (4.1)–(4.3) and to add some boundary conditions for V derived from the invariance condition (4.11). However, this blows up to the dimension of the boundary value problem by $m\cdot(m_{+u}+1)$ and is thus very costly.

In the example below we will use a simpler device, where (4.9) is replaced by

$$B_+(x(T_+),y(0),\lambda)=\widehat{V}_u(x(T_+)-y(0)) \tag{4.12}$$

and $\widehat{V}_u\in\mathbb{R}^{m_{+u}+1,m}$ is an approximation to $V_u(0,\lambda)$ obtained by solving (4.10), (4.11) with a periodic orbit $y(\cdot,\widehat{\lambda})$ for an initial guess $\widehat{\lambda}$. Of course, some accuracy is lost in this approach.

As an example we treat the well–known Lorenz equations (see [Spa 82])

$$\dot{x}_1=\sigma(x_2-x_1),\ \dot{x}_2=\lambda\,x_1-x_2-x_1x_3,\ \dot{x}_3=x_1x_2-bx_3. \tag{4.13}$$

For $\sigma>b+1$ this system has a subcritical Hopf bifurcation from the nontrivial steady states

$$\xi_\pm(\lambda)=(\pm(b(\lambda-1))^{1/2},\ \pm(b(\lambda-1))^{1/2},\ \lambda-1),\ \lambda>1$$

at

$$\lambda=\lambda_H=\frac{\sigma(\sigma+b+3)}{\sigma-b-1}.$$

The periodic orbits $\gamma_\pm(\lambda)$ shrinking to $\xi_\pm(\lambda)$ at λ_H have two–dimensional stable manifolds ($m_{+u}=m_{+s}=1$) and there is apparently a specific value

$$\lambda_A=\lambda_A(\sigma,b)<\lambda_H$$

at which the one–dimensional unstable manifold of the origin meets these stable manifolds, see [Spa 82, pp. 32–47] and in particular [Kuz 91, Lecture 7] for a nice illustration. (3.26) yields $p=1$ and so we expect this point to periodic connection to be a stable one parameter phenomenon. For $\lambda<\lambda_A$ the unstable manifold is attracted towards $\xi_-(\lambda)$ or $\xi_+(\lambda)$ while for $\lambda>\lambda_A$ it becomes part of the strange attractor. However, it is not clear whether λ_A is the precise value at which the strange invariant set becomes attracting.

Figures 1 and 2 show the different fate of two trajectories started on the linearized unstable manifold of the origin at values

$$\lambda_0=24.05<\lambda_A<\lambda_1=24.06,\ \sigma=10,\ b=\frac{8}{3}.$$

The trajectory with initial values

$$\widehat{x}(0)=(0.00435,0.009,0)$$

and $T_+ = -T_- = 0.8667$ was taken as initial approximation for the connecting orbit. Then a periodic orbit was computed with values

$$\widehat{\lambda} = 24.06, \quad \widehat{T} = 0.677,$$
$$\widehat{y}(1) = (-6.116, \ -4.799, \ 23.07)$$

and a suitably normalized $\widehat{V}_u$ was found from (4.10), (4.11) to be

$$\widehat{V}_u = \begin{pmatrix} 0, & 0.769, & 1 \\ 1, & 2.916, & 0 \end{pmatrix}.$$

The phase condition (4.6) was simply

$$y_3(0) - (\lambda - 1) = 0. \tag{4.14}$$

With these initial data the boundary value problem (4.1)–(4.7), (4.12) was solved with tolerance 10^{-8} by the code D02RAF (NAG–library, Oxford). The third component of the solutions x and y is shown in Figure 3 and the λ–value is

$$\lambda_A = 24.05790.$$

A good initial approximation is crucial in this example because the periodic orbit is close to the Hopf point and the periodic boundary value problem allows for the trivial solution $T = 0$, y constant.

Finally, in Figure 4 we show two further point–to–periodic connections obtained by increasing the parameter b. The continuation with respect to this parameter turns out to be rather sensitive. This is a well–known phenomenon due to the fact that the rigid phase condition (4.14) is not well–suited for mesh adaptation. Good alternatives are integral phase conditions [FrDo 91] or the use of the Gauss–Newton method [Deu 84]. In view of the theoretical results of section 3 it is clearly acceptable to have phase conditions which involve both the connecting and the periodic orbit.

References

[Bey 90a] W.–J. Beyn, *The numerical computation of connecting orbits in dynamical systems*, IMA J. Numer. Anal. **9** (1990), 379–405.

[Bey 90b] ______, *Global bifurcations and their numerical computation in: Continuation and Bifurcations Numerical Techniques and Applications (D. Roose, A. Spence, B. De Dier Eds.)* (1990), Kluwer, Dordrecht, 169–181.

[CHM 80] S.N. Chow, J.K. Hale, J. Mallet–Paret, *An example of bifurcation to homoclinic orbits*, J. Differential Equations **37** (1980), 351–373.

[ChKu 93] A.R. Champneys, Yu. A. Kuznetsov, *Numerical detection and continuation of codimension–two homoclinic bifurcations*, Report AM–R 9308, (1993), Amsterdam.

[ChLi 90] S.–N. Chow, X.–B. Lin, *Bifurcation of a homoclinic orbit with a saddle–node equilibrium*, Differential and Integral Equations **3** (1990), 435–466.

[Cop 78] W.A. Coppel, *Dichotomies in stability theory*, Lecture Notes in Mathematics **629** (1978).

[Deu 84] P. Deuflhard, *Computation of periodic solutions of nonlinear ODEs*, BIT **24** (1984), 456–466.

[Die 60] J. Dieudonné, *Foundations of Modern Analysis* (1960), Academic Press, New York.

[DoFr 89] E.J. Doedel, M. Friedman, *Numerical computation of heteroclinic orbits*, J. Comp. Appl. Math. **26** (1989), 159–170.

[Fen 74] N. Fenichel, *Asymptotic stability with rate constants*, Indiana Univ. Math. J. **23** (1974), 1109–1137.

[Fen 77] ______, *Asymptotic stability with rate constants II*, Indiana Univ. Math. J. **26** (1977), 81–93.

[FrDo 91] M. Friedman, E.J. Doedel, *Numerical computation of invariant manifolds connecting fixed points*, SIAM J. Numer. Anal. **28** (1991), 789–808.

[Hal 69] J.K. Hale, *Ordinary Differential Equations* (1969), Wiley & Sons, New York.

[HaLi 86] J. Hale, X.-B. Lin, *Heteroclinic orbits for retarded functional differential equations*, J. Differential Equations **65** (1986), 175–202.

[Har 64] Ph. Hartman, *Ordinary Differential Equations* (1964), Wiley & Sons, New York.

[HPS 77] M.W. Hirsch, C.C. Pugh, M. Shub, *Invariant Manifolds* **583** (1977), Springer Lecture Notes in Mathematics.

[Irw 80] M.C. Irwin, *Smooth dynamical systems* (1980), Academic Press, New York.

[Kuz 90] Y. A. Kuznetsov, *Computation of invariant manifold bifurcations in: Continuation and Bifurcations: Numerical Techniques and Applications* (1990), (D. Roose, A. Spence, B. De Dier Eds.), Kluwer, Dordrecht.

[Kuz 91] ______, *Elements of Nonlinear Dynamics* (1991), Lectures delivered at the Dipartimento di Elettronica di Politecnico di Milano, Italy.

[Pal 84] K.J. Palmer, *Exponential dichotomies and transversal homoclinic points*, J. Diff. Eqn. **55** (1984), 225–256.

[Pal 88] ______, *Exponential dichotomies, the shadowing lemma and transversal homoclinic points in: Dynamics Reported,* **Volume I** (1988), (U. Kirchgraber, H.O. Walther Eds.), Teubner, 265–306.

[Sch 93a] S. Schecter, *Numerical computation of saddle–node homoclinic bifurcation points*, To appear in SIAM J. Numer. Anal.

[Sche 93b] ______, *Rate of convergence of numerical approximations to homoclinic bifurcation points*, Preprint, North Carolina State University (1993).

[Spa 82] C. Sparrow, *The Lorenz Equations: Bifurcations, Chaos and Strange Attractors* (1982), Springer, New York.

Fakultät für Mathmematik
Universität Bielefeld
Postfach 100131
D–33501 Bielefeld
Germany

E-mail address: beyn@mathematik.uni-bielefeld.de

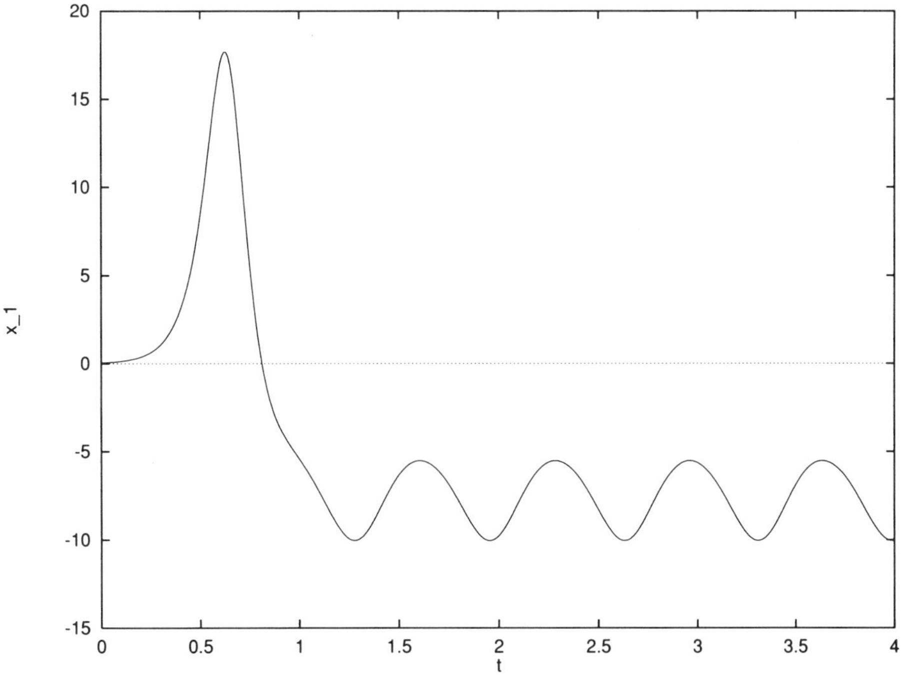

FIGURE 1A. Unstable manifold of the origin at $\lambda_0 = 24.05$, first component plotted over the time interval $[0, 4]$

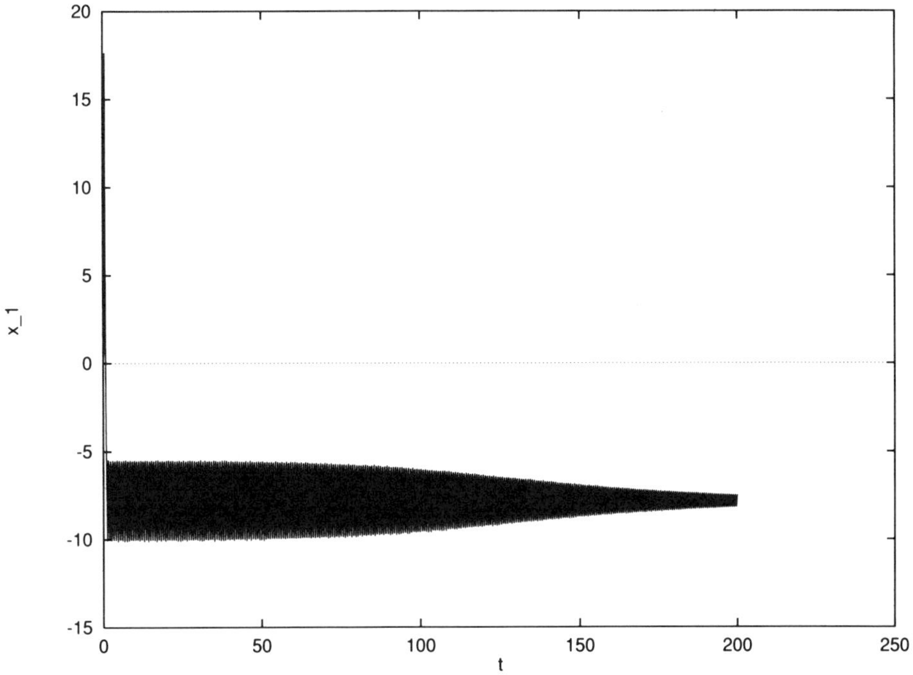

FIGURE 1B. The same as Figure 1a but with time interval $[0, 200]$

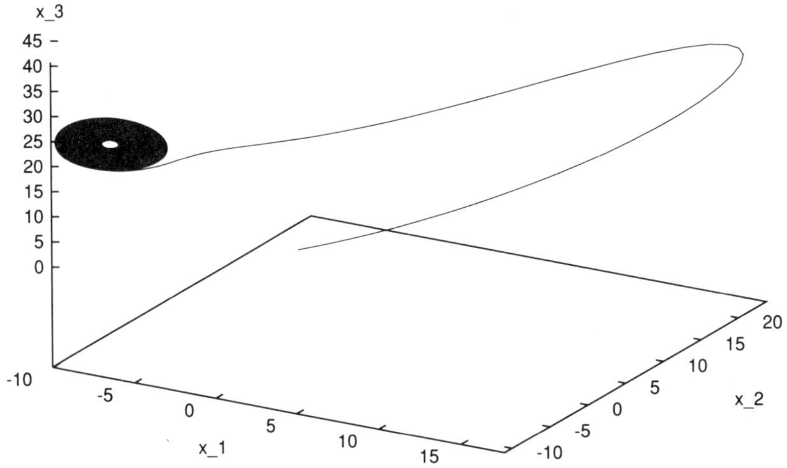

FIGURE 1C. The same as Figure 1b but plotted in phase space

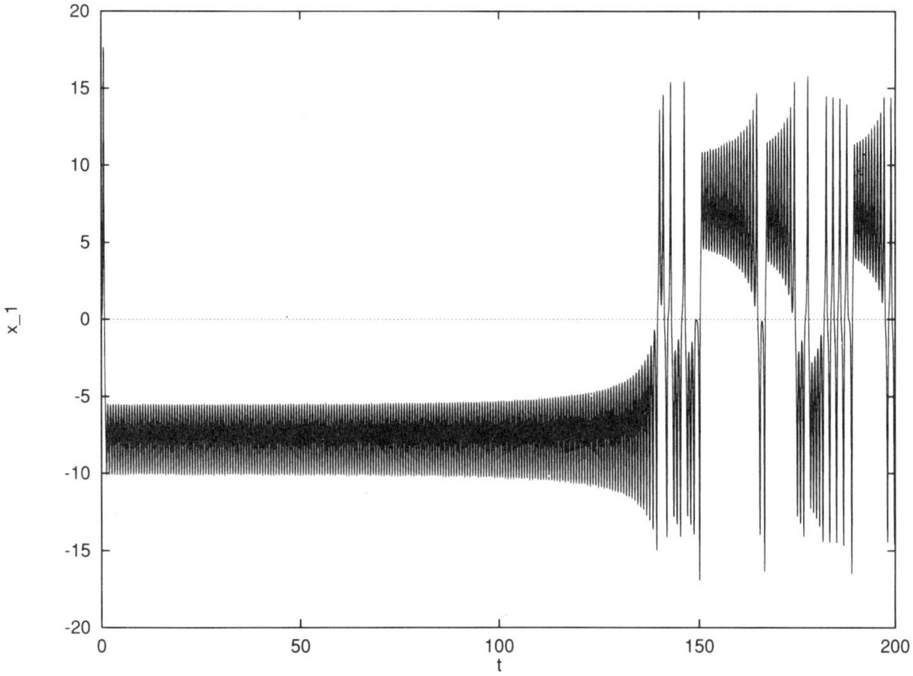

FIGURE 2A. Unstable manifold of the origin at $\lambda_1 = 24.06$, first component plotted over the time interval $[0, 200]$

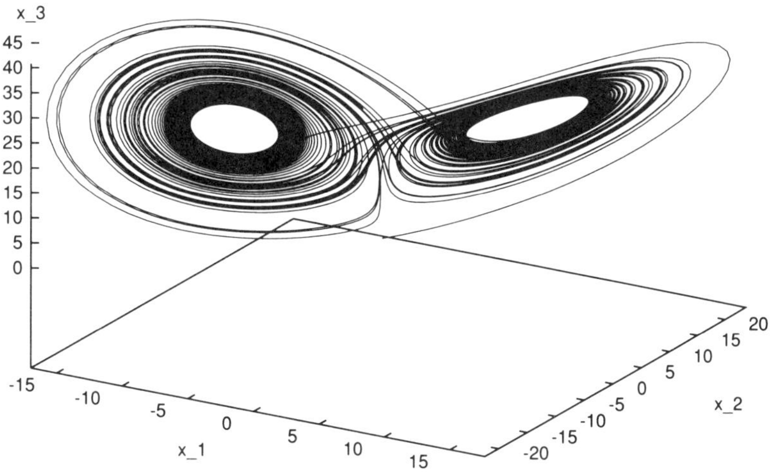

FIGURE 2B. The same as Figure 2a but plotted in phase space

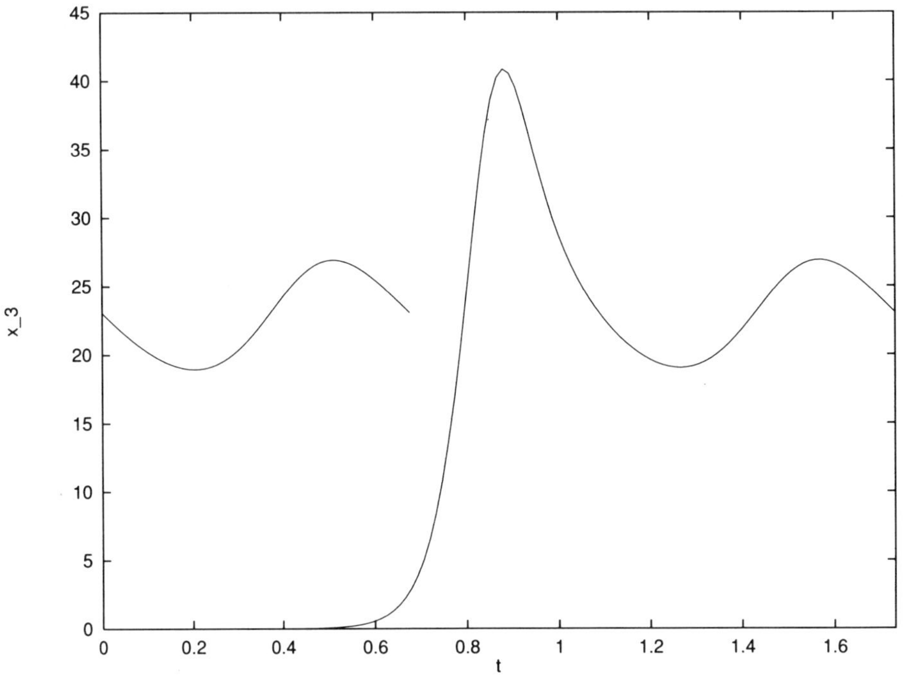

FIGURE 3. Third component of connecting and periodic orbit obtained numerically at $\lambda_A = 24.05790$

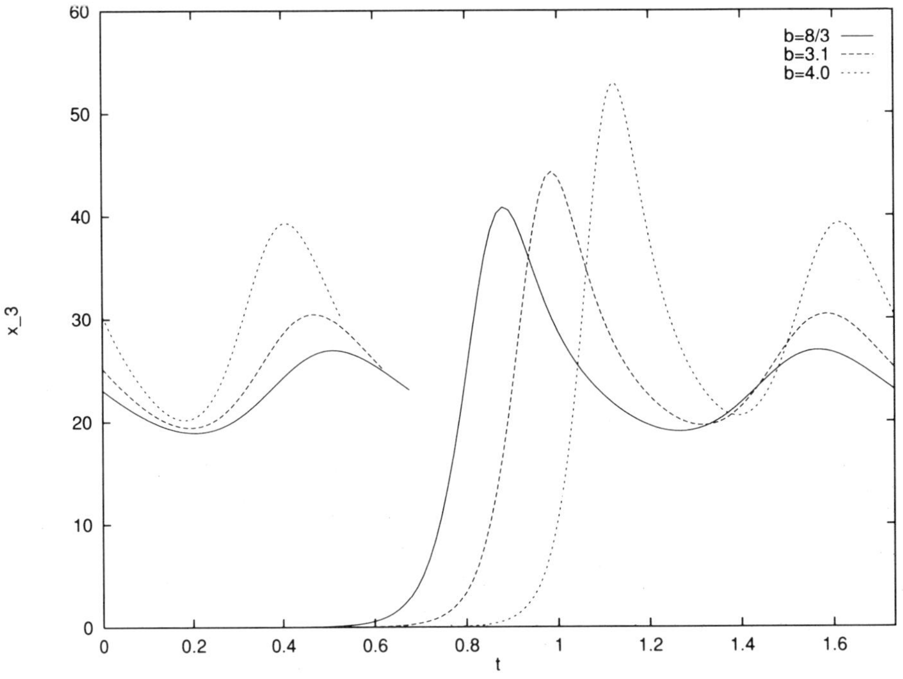

FIGURE 4. Third component of connecting and periodic orbits by continuation with respect to the parameter b.
$(b, \lambda_A) = (\frac{8}{3},\ 24.05790), (3.1, 26.16990), (4.0, 31.29453)$

Contemporary Mathematics
Volume **172**, 1994

Numerical computation of a branch of invariant circles starting at a Hopf bifurcation point

LAURENT DEBRAUX

ABSTRACT. In this paper we propose an original method, which uses a continuation technique, to compute a one-parameter family of smooth invariant circles starting at a Hopf bifurcation point. The method we describe can be used to compute stable invariant circles as well as unstable ones and employs a system of local coordinates removing any typical restrictions about the parameterization and works regardless of the dimension. We illustrate this technique with a map: the delayed logistic map, and with a periodically forced differential equation modelling ferroresonance in a transformer station.

1. Introduction

If we consider a discrete dynamical system generated by a one-parameter family of maps, we know that a generic phenomenon occurs when two simple conjugate complex eigenvalues cross the unit circle transversally, thereby spawning an invariant circle, that is the Hopf bifurcation for maps (see for instance [9] for the precise formulation). Similarly when we deal with the steady states of a one-parameter family of ordinary differential equations, a generic phenomenon is the well known Hopf bifurcation occuring when two complex conjugate eigenvalues cross transversally the imaginary axis and a branch of limit-cycles is generated. This branch of periodic solutions can itself further bifurcate into a branch of invariant T^2 tori when two Floquet multipliers cross the unit circle, that is the second Hopf bifurcation, sometimes called the "torus bifurcation". As a matter of fact, if we consider a family of Poincaré sections and maps associated with the periodic solutions this is again the invariant circle bifurcation.

There are now several methods to compute stable and unstable steady states and periodic states for differential equations and maps (see [7] for an example using

Mathematics Subject Classification. Primary 65L99, 65H99; Secondary 34C23, 34C40, 78A99.
The author wishes to thank E.D.F. (Electricité de France) and U.T.C. (Université de Technologie de Compiègne - France) for their supports.
This paper is in final form and no version of it will be submitted for publication elsewhere.

continuation techniques), but except for early works [5] and [12], only recently has attention been paid to the numerical computation of invariant circles. ([3], [4], [6] and [12] for instance). It should be noted that the difficulties with such a computation are one step beyond the former: for maps and steady states of flows, we have only one algebraic equation to solve; one step further, for periodic solutions of flows we have an ordinary differential equation to solve and naturally the solution is parameterized by time. But with invariant circles, there is no natural equation defining such a parameterization. Moreover we do not understand very well which kind of bifurcation can occur on a branch of invariant circles.

2. The algorithm

2.1. Formulation of the equations. We will consider now a one-parameter family of diffeomorphisms. Let Λ be an interval of $\mathbb{R}$ and let $(P_\lambda)_{\lambda \in \Lambda}$ be a family of diffeomorphisms, defined (at least locally) in $\mathbb{R}^n$:

$$(1) \qquad P_\lambda : \begin{cases} \Omega \to \Omega \\ x \mapsto P_\lambda(x) \end{cases} \text{ where } \Omega \text{ is an open subset of } \mathbb{R}^n,$$

and let's consider the discrete dynamical system generated by P_λ (sometimes named a "cascade"):

$$(2) \qquad x^{k+1} = P_\lambda(x^k) \text{ for any } x^0 \text{ belonging to } \Omega.$$

We suppose that for each λ belonging to a sub-interval Λ' of Λ, there exists an isolated closed curve globally invariant with the dynamics defined by P_λ. For obvious reason, we will use the terminology "torus" (for one dimensional torus $T^1 = \mathbb{R}/\mathbb{Z}$) as well as the terminology "circle".

An interesting case occurs when the diffeomorphisms are Poincaré mappings. Then, the T^1 tori are the traces of T^2 tori ($= \mathbb{R}^2/\mathbb{Z}^2$) in the related Poincaré section. When we deal with differential equations in $\mathbb{R}^n$, it may not be easy to find a Poincaré section but there is a well-known exception: differential equations with periodic driving.

In order to describe the proposed algorithm, we will restrict to the case where the invariant curves are at least C^2 but we will not make yet any assumption of stability or hyperbolicity.

Let T_λ be the invariant circle related to P_λ and let u_λ be a C^2 parameterization of it:

$$(3) \qquad u_\lambda : \begin{cases} [0, 1] \to \mathbb{R}^n \\ s \mapsto u_\lambda(s) \end{cases},$$

with the "periodicity" condition:

(4) $$u_\lambda(0) = u_\lambda(1),$$

and with :

(5) $$u'_\lambda(s) \neq 0_n \qquad \forall s \in [0, 1].$$

Even if we choose an "origin" on the invariant curve (a *phase condition*), it should be pointed out that such a parameterization is not unique because it is possible to compose it with any diffeomorphism of the circle leaving the "origin" fixed. This is one of the difficulties when computing invariant curves. As we want to use a

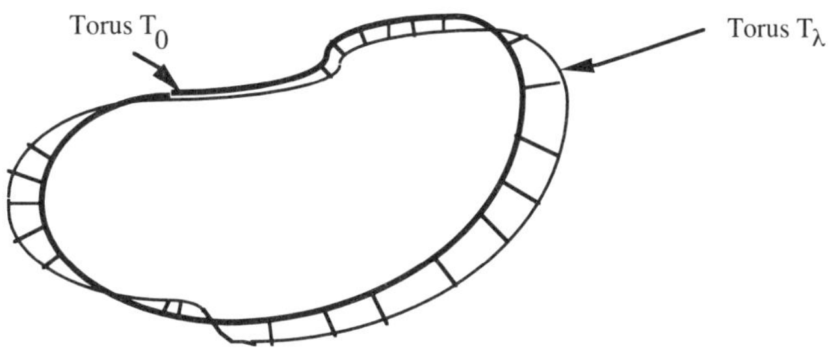

FIG. 1: *Principle of the parameterization*

continuation framework, we have to remove this indeterminacy in such a way that the family (u_λ) will be smooth with λ.

For some particular value $\lambda = \lambda_0$, we will simply use the notation T_0 instead of T_{λ_0} and u_0 instead of u_{λ_0}. Let the dynamics on the invariant torus be represented by the restriction R_λ of a smooth lift of P_λ restricted to T_λ, using the covering map defined by the periodic prolongation of u_λ:

(6) $$R_\lambda : \begin{cases} [0, 1] \to [0, 1] \\ s \mapsto R_\lambda(s) \end{cases} \quad \text{with} \quad P_\lambda(u_\lambda(s)) = u_\lambda(R_\lambda(s)).$$

Our method is based on the following lemma:

LEMMA: *Let T a C^k torus imbedded in $\mathbb{R}^n$ with $k \geqslant 2$ and let u be a C^k parameterization of this torus. Then $\exists \varepsilon > 0 \,/\, \forall M \in \mathbb{R}^n \,/ d(M,T) \leqslant \varepsilon$, thus $\exists !\, \sigma \in [0, 1)$ such that $u(\sigma)$ is the unique orthogonal projection of M on T in the ball $B(M, \varepsilon)$. Moreover this projection is C^{k-1}.*

This could be easily derived from results about moving orthonormal systems ([8]). This is indeed a very interesting and natural property. It means that we can actually use in some ways the parameterization of an "old" torus to compute a "new" one.

Let T_0 be the last torus "computed" and let $\varepsilon_0 > 0$ be related to this torus and the previous lemma. Let W_0 be the ball $B(0, \varepsilon_0)$ in $\mathbb{R}^n$. Let $T_0(\varepsilon_0)$ be an "ε_0 neighbourhood" of T_0:

(7) $$T_0(\varepsilon_0) = \{u_0(s)+x \mid s \in [0, 1) \text{ and } x \in W_0\}$$

Let T_λ be the torus we want to compute lying in $T_0(\varepsilon_0)$ for $|\lambda-\lambda_0|$ small enough. Let's write:

(8) $$u_\lambda(s) = u_0(s)+x(s)$$

with the "orthogonality condition"

(9) $$< x(s), u_0'(s) > = 0 \ \forall s \in [0, 1],$$

(we will use the abusive notation $x(.) \perp u_0(.)$ or even $x \perp u_0$ for this "orthogonality condition")

and with $\|x(s)\| \leqslant \varepsilon_0$.

So the situation is shown in FIG. 1.

Let the function $\mathcal{P}_\lambda$ be defined by:

(10) $$\mathcal{P}_\lambda : \begin{cases} [0, 1] \times W_0 \to \mathbb{R}^n \\ (s, x) \mapsto \mathcal{P}_\lambda(u_0(s)+x) \end{cases}$$

Let $\varepsilon_0' > 0$ and let V_0 be the ball $B(0, \varepsilon_0')$ in $\mathbb{R}^n$. Let's choose ε_0' such that $\varepsilon_0' \leqslant \varepsilon_0$ and $\mathcal{P}_\lambda([0, 1] \times W_0) \subset T_0(\varepsilon_0)$, this is again possible if we choose $|\lambda-\lambda_0|$ small enough because T_0 is invariant by P_0, and because P_λ depends continuously on λ.

We now define two new functions g_λ and h_λ:

(11) $$g_\lambda : \begin{cases} [0, 1] \times V_0 \to \mathbb{R} \\ (s, x) \mapsto g_\lambda(s, x) \end{cases} \quad \text{and} \quad h_\lambda : \begin{cases} [0, 1] \times V_0 \to \mathbb{R}^n \\ (s, x) \mapsto h_\lambda(s, x) \end{cases} \quad \text{by:}$$

(12) $$\mathcal{P}_\lambda(s, x) = u_0(g_\lambda(s, x)) + h_\lambda(s, x) \quad \text{with} \quad < \frac{du_0}{ds}(g_\lambda(s, x)), h_\lambda(s, x) > = 0.$$

Within the space $\mathcal{S}$ of the smooth mappings from $\mathbb{R}$ into $\mathbb{R}^n$ with period unity, let's define the sets X and Y:

$$X = \{f \in \mathcal{S} \mid \max(|f|) \in V_0\},$$

$$Y = \{f \in \mathcal{S} \mid \max(|f|) \in W_0\}.$$

Let's now define the function Φ_λ by :

(13) $$\Phi_\lambda : \begin{cases} X \to Y \\ x \mapsto \Phi_\lambda(x) \end{cases} \text{with } \Phi_\lambda(x)(s) = h_\lambda(s, x(s)) - x(g_\lambda(s, x(s))).$$

Then the torus is defined as the solution of:

(14) $$\Phi_\lambda(x) = 0_X \quad \text{and} \quad x \perp u_0$$

Hence we must have:

(15) $$\Phi_\lambda(x)(s) = 0\ , \forall\ s \in [0,1] \quad \text{with} \quad < x(s)\ , u_0'(s) > \ = 0$$

So the situation is as shown in FIG. 2.

But in fact this system is not well posed because the Φ_λ equation must be understood in the normal bundle. Hence we reformulate it using a projection in the normal space and writing the orthogonality condition in the tangent space.

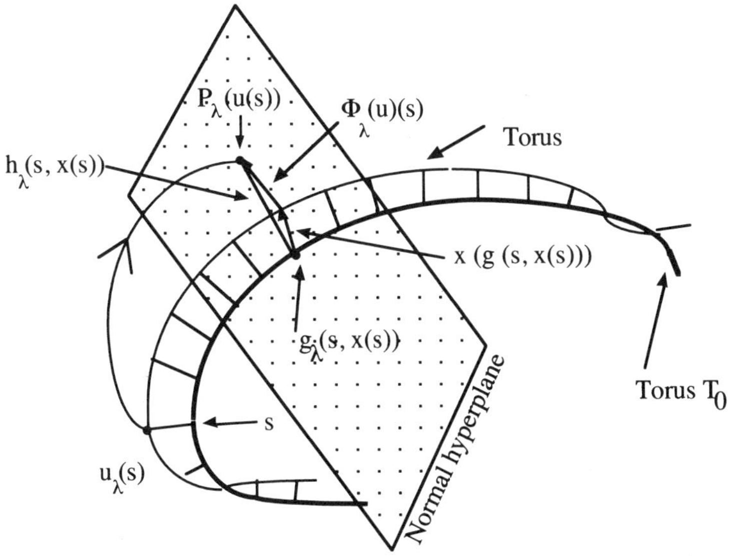

(The normal hyperplane is the one crossing T_0 at $u_0(g(s, x(s)))$

<u>FIG. 2: *Principle of the method*</u>

We now define the function Ψ_λ by :

(16a) $$\Psi_\lambda : \begin{cases} X \to Y \\ x \mapsto \Psi_\lambda(x) \end{cases} \quad \textit{with}$$

(16b) $$\forall\ s \in [0,1] \quad \Psi_\lambda(x)(s) =$$

$$\Phi_\lambda(x)(s) - \frac{< \Phi_\lambda(x)(s)\ ,\ u_0'(g_\lambda(s, x(s))) >}{< u_0'(g_\lambda(s, x(s)))\ ,\ u_0'(g_\lambda(s, x(s))) >}\ u_0'(g_\lambda(s, x(s))) + < x(s), u_0'(s) > u_0'(g_\lambda(s, x(s)))$$

So the torus equation is:

(17) $$\Psi_\lambda(x) = 0_Y,$$

hence we have:

(18) $$\Psi_\lambda(x)(s) = 0\ ,\ \forall\ s \in [0,1]\ .$$

Now it is possible to use a Newton method to solve the equation. We note that we used the torus T_0 in order to parameterize the torus T_λ. It could be simpler to use an approximation of T_λ in order to parameterize the torus T_λ itself.

2.2. Pseudo arclength continuation. Arclength continuation is indeed very simple to introduce; we only have to introduce one more equation and λ must now be treated as a free parameter:

(19) $$(u-u_0)\frac{u_0-u_{-1}}{\Delta l_0} + (\lambda-\lambda_0)\frac{\lambda_0-\lambda_{-1}}{\Delta l_0} - \Delta l = 0$$

where (u_0, λ_0) and (u_{-1}, λ_{-1}) are the last solutions computed and Δl_0 is the last step size and where Δl is the new one.

2.3. Starting procedure at a Hopf bifurcation point. Let (u_0, λ_0) denote a Hopf bifurcation point for the map P. Let $e^{\pm i\omega_0}$ be the pair of complex eigenvalues lying on the unit circle. Let $A+iB$ be a normalized eigenvector associated with $e^{i\omega_0}$. We can choose $A+iB$ such $<A, B> = 0$.

We then have:

(20) $$\begin{cases} u(t) = u_{\lambda_0}(t)+\varepsilon\varphi(t)+O(\varepsilon^2) \\ \omega(\varepsilon) = \omega_0+O(\varepsilon^2) \\ \lambda(\varepsilon) = \lambda_0+O(\varepsilon^2) \end{cases} \qquad \text{with } \varphi(t) = \cos(t)A - \sin(t)B$$

where ε locally parameterizes the branch of solutions, and where $\omega/2\pi$ is the rotation number. In fact the function $\varphi(t)$ is the nonzero normalized 2π-periodic solution of the linear homogenous equation:

(21) $$\varphi(t+\omega_0) = f_u(u_0, \lambda_0)\varphi(t).$$

This is an ellipse lying in the (u_{λ_0}, A, B) plane.

3. Practical discretization and implementation

We do not have room here to describe the complete discretization and its implementation; thus we will emphasise the main points without any details.

3.1. Periodic cubic splines with collocation. As we are going to use a Newton-like method, we have to employ at least a C^2 representation of the torus. We chose to use a periodic cubic spline. It should be noted we do not recommend the use of truncated Fourier series here. The torus is not necessarily C^{∞} and then it could be difficult to approximate it with such functions. There are reports about troubles with Fourier series in [3] and [12]. Furthermore for a given parameterization of the invariant curve, with a cubic spline, we can choose the mesh; but there is nothing equivalent if we use a truncated Fourier series because there is no mesh at all in this case. Moreover Fourier series would be more time consuming to compute.

We have chosen to use the nodes as collocation points; this is not the only possibility but it makes the computation simpler and we do not yet have any optimal way to choose the collocation points[1].

Unlike with other piecewise polynomials, with cubic splines we get a full Jacobian with respect to the nodal values. More precisely:

- let the nodes be $t_0, t_2, \dots t_N$ and
- let $\xi(.)$ be the cubic spline built with the nodal values $y_0, y_2 \dots y_N$,
- let $\tau \in \mathbb{R}$, with $t_j \leqslant \tau < t_{j+1}$, where $j \in \{0, 2, \dots N-1\}$, and let $i \in \{1, 2, \dots N\}$.

Then the scalar $\dfrac{d\xi(\tau)}{dy_i}$ is a priori non zero, unless $\tau = t_j$ with $j \neq i$. But as $|i-j|$ increases $\left\|\dfrac{d\xi(\tau)}{dy_i}\right\|$ decreases quickly, and we can neglect it for $|i-j| > b$, where $b \in \mathbb{N}$ has to be chosen[2,3,4].

So using this approximation it is interesting to give the structure of the Jacobian of the discretized equations we get from (18)+(19). One can look at the examples given in FIG 3

The last row corresponds to the pseudo arclength equation. The last column corresponds to the scalar λ. The thin bloc diagonal corresponds to the collocation points (it's thin because we use the nodes as collocation points). The "broken or more precisely periodic" twisted band corresponds to the "image" of the collocation points by the mapping g (it is thick because the image of a collocation point is not necessary another collocation point) The block bandwidth is $2b+1$. The aspect of this twisted band is related to the dynamics on the torus (and to the parameterization). It should be pointed out that we also have to compute numerically the dynamics on the discretized circle in order to get a first approximation for the computation of the function g_{λ} - in fact it's a bit more subtle: we have to interpolate the function defining the dynamics on

---

[1] The situation is rather different when we are using orthogonal collocation at gaussian points.

[2] Those facts are related to the inverse of the tridiagonal matrix occuring in the calculus of the cubic splines.

[3] We are using periodic cubic splines so the situation is a bit more subtle.

[4] A usual value we used was 5 for 40 nodes. But we also did computations too without using this approximation.

the approximation of the invariant circle T_0. We have to be careful with this projection in order to avoid the crossing of nodal locations. Moreover it's very interesting to know the dynamics on the torus: we can try then to compute the rotation number and look and see if it is like one of the first rationals in the Farey Tree (See [2]). See [3] for an interesting method of computing invariant circles with irrationnal rotation number.

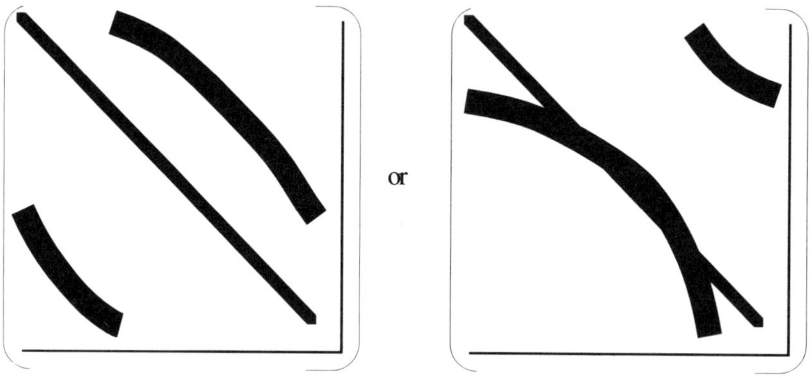

FIG 3 Possible structures of the Jacobian

3.2. Choices of a parameterization and choice of a mesh. It is necessary to point out that there are two distinct and different problems, but they are not independent of each other. Here we are merely raising questions and giving only partial answers.

3.2.1. *The Parameterization.* A natural parameterization for a curve is the arclength. But this is mainly a theoretical tool. Practically we only have access to the distances between nodes values or we can try to compute the arclength related to the discretization we use, but this does not seem satisfactory. Even if we can compute the arclength related to one solution with one kind of mesh, it is not easy to get a new parameterization.

In our implementation, we chose to base the parameterization on the distances between node values (the simplest way).

3.2.2. *The geometric mesh.*[5] Let's suppose that we have the parameterization (3) of the torus T_λ. In order to choose the mesh, a simple and general idea is to use the equidistribution of a positive continuous function on the invariant curve. Let's choose a function ζ:

[5] We use the word "geometric" to underline the fact that theoretically, we can choose a sequence of vertices (i.e. points) on the torus without referring to a parameterization. But practically, we have to deal with some parameterization and with its node values.

(22) $$\zeta : \begin{cases} [0, 1] \to \mathbb{R}^+ \\ s \mapsto \zeta(s) \end{cases} \quad \text{with } \zeta(0) = \zeta(1),$$

and let the function v be:

(23) $$v : \begin{cases} [0, 1] \to \mathbb{R}^+ \\ s \mapsto v(s) = \int_0^s \zeta(u)du \end{cases}$$

We can choose to define the nodes $t_0, \ldots, t_N$ with t_k as being the solution of

(24) $$v(t) = \frac{k}{N} \int_0^1 \zeta(u)du$$

If we choose:

(25) $$\zeta(.) = \|u'_\lambda(.)\|,$$

we have the simplest mesh: the equidistribution of the arclength.

From a geometrical point of view, it makes sense to use a linear combination of the arclength and curvature. But from a numerical point of view, it make sense to use simply:

(26) $$\zeta(.) = a\|u'_\lambda(.)\| + b\|u''_\lambda(s)\|, \text{ where } a, b > 0.$$

But we have to be very careful in order to avoid erroneous concentrations of points. In our implementation, we choose to use an equidistribution based upon the combination of the pseudo arclength and a limited curvature and furthermore requiring a minimal and a maximal step between two node values. This is done after every computation of a curve.

Those choices are purely geometric, and are not necessarily adapted for a general situation. It must be clear that the image of a "good geometric mesh" of the invariant curve is not always a good mesh![6] So it may be essential to consider the dynamics in the choice of the function to equidistribute: for instance, we can try to choose:

(27) $$\zeta(.) = \|u'_\lambda(.)\| + \|R'_\lambda(.)\, u'_\lambda(R_\lambda(.))\|.$$

In some other ways, it would be interesting to have an estimation of the error in the computation of the invariant curve and to equidistribute this quantity; in fact this is usually what is meant by "adaptive mesh". This will require further investigation. But

---

[6] Indeed this is the case if the rotation number is rational with a small denominator.

near the Hopf bifurcation point, we have essentially a rotation and the geometric mesh should be satisfactory as we will see in the numerical examples.

Lastly let's point out that we do not have access to a parameterization, and we can only deal with node values of an approximation of the invariant circle; thus it should be clear now that we have to choose mesh and parameterization together and this is not an easy task.

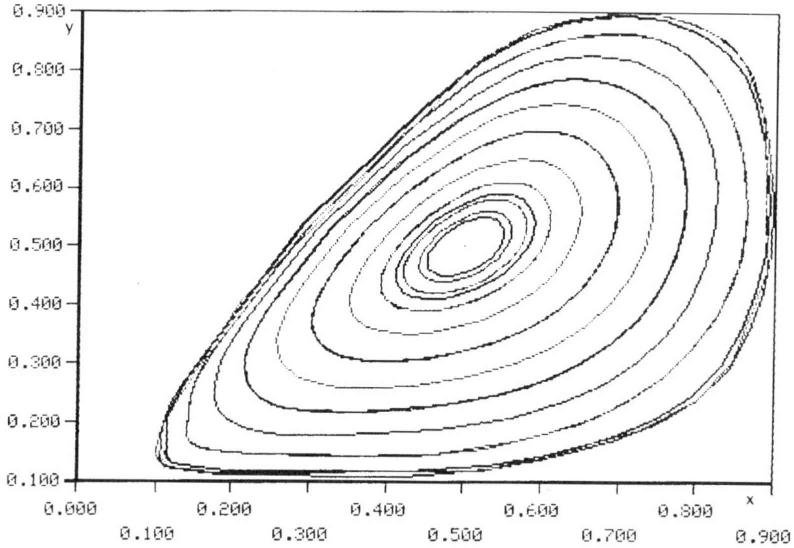

FIG. 4 *Invariant circles in the delayed logistic equation*

4. Numerical examples

4.1. A map: the delayed logistic equation. This is the discrete dynamical system defined by:

(28) $$y^{k+1} = \mu y^{k}(1-y^{k-1}),$$

where $\mu > 0$ is a parameter. Let's recall that the well known logistic equation is:

(29) $$y^{k+1} = \mu y^{k}(1-y^{k}).$$

Whereas the logistic equation is a model for the growth of population in which generations do not overlap, they do overlap in the delayed logistic equation.

Then (28) is clearly equivalent to the system:

(30) $$\begin{pmatrix} x^{k+1} \\ y^{k+1} \end{pmatrix} = F_{\mu}\begin{pmatrix} x^{k} \\ y^{k} \end{pmatrix},$$

where

(31) $$F_\mu\begin{pmatrix} x \\ y \end{pmatrix} = \begin{pmatrix} y \\ \mu y(1-x) \end{pmatrix}.$$

This system has a very complex behaviour as we can see in [1]. This is not surprising when we think about the famous behaviour of the simpler logistic map studied by Fegenbaum [13].

Clearly one can see that the steady states are:

(32) $$\begin{pmatrix} 0 \\ 0 \end{pmatrix} \text{ and } \begin{pmatrix} (\mu-1)/\mu \\ (\mu-1)/\mu \end{pmatrix}$$

and with an easy calculation of the Jacobian that the second branch of solutions is stable for $1 < \mu < 2$ and that there is a torus bifurcation for $\mu = 2$ with an exchange of stability. (In fact the two eigenvalues pass out off the unit circle via two sixth roots of the unity: $\frac{1}{2}(1 \pm i\sqrt{3})$: it's a weak resonance).The computed invariant circles are shown in FIG. 4 in the x-y plane:

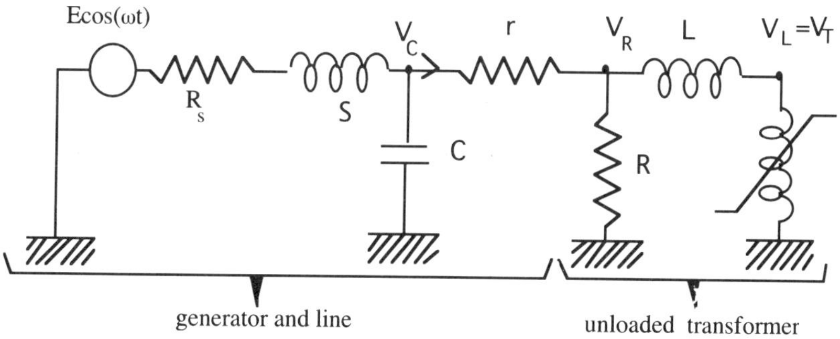

FIG. 5 *A simple model of an electric line*

4.2. A differential equation: behaviour of a non linear network (ferroresonance). Usually a Power Network is a three-phase network with distributed parameters in lines and is governed by the telegraphic equation with some linear and non-linear boundary conditions. We restrict here to a very simple situation: the single phase case with no distributed parameter (one step discretization). In fact, in the beginning, the algorithm has been devised to work with such models. The one we use is shown in FIG. 5 .

The non linear element is the transformer with the following experimental so-called **saturation law**:

(33) $$i = g(\varphi) = k_1\varphi + (k_2\varphi)^n$$

where i and f are respectively the inductance current and the flux inside the transformer and where n is odd and k_1, k_2 are positive. In the linear case $k_2 = 0$ and

$1/k_1$ is the inductance. This non linear behaviour is related to the magnetic saturation of soft iron cores. This is the ferroresonance phenomena on electric lines. We can have multiple periodic solutions and sub harmonic ones, moreover some stable solutions could be destructive...

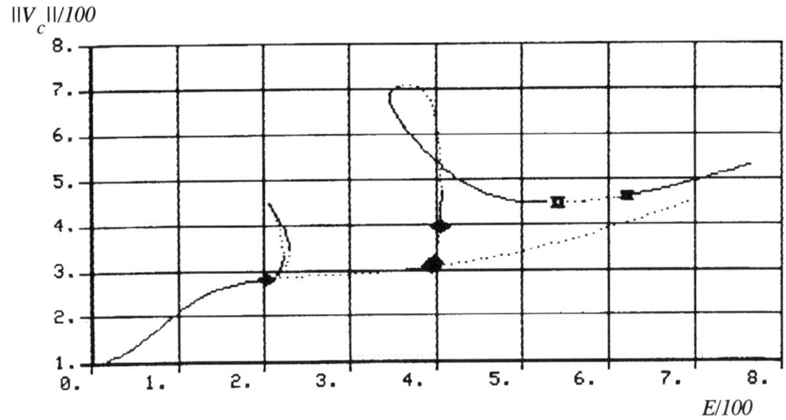

FIG. 6 *Bifurcation Diagram*

We can now write the equations and after elementary simplifications and with standard notations, we thus obtain the following three dimensional system:

$$
(34) \qquad \left\{
\begin{aligned}
\frac{d\varphi}{dt} &= \frac{(1-\frac{r}{R+r})}{1+L*g_{\varphi}(\varphi)}(V_c - rg(\varphi)) \\
\frac{dI_s}{dt} &= \frac{1}{S}(E\cos(\omega t) - V_c - R_s I_s) \\
\frac{dV_c}{dt} &= \frac{1}{C}(I_s - \frac{1}{R+r}V_c - (1-\frac{r}{(R+r)})g(\varphi))
\end{aligned}
\right.
$$

We use the following values:

$n = 9,\quad k_1 = 0.01\,H^{-1},\quad k_2 = 0.7168SI..,\quad S = 3H,\quad \omega = 100\pi$ *(i.e. 50 Hz),*

$r = 20\,\Omega,\qquad 1/R = 0S,\qquad C = 2.10^{-6}\,F,\qquad L = 0.1H.$

$E = 0$ *to* $1000V$ (continuation parameter).

Indeed this is a miniature model of a Power network; on real ones, voltage will range from *100kV* to *400kV*. A similar model has been investigated in [10].

As E increases from the zero value, starting from the null solution, we can follow the periodic states using the AUTO software ([7]) and detect two torus bifurcations on the branch and other singularities. We get the interesting bifurcation diagram shown in FIG. 6 with details in FIG.7 and FIG.8.

We have to give some explanations: solid and dashed curves represent stable and unstable solutions respectively. There is a torus bifurcation near the value *200V* and there is another one near *400V*. Very close to the last one there is a pitchfork

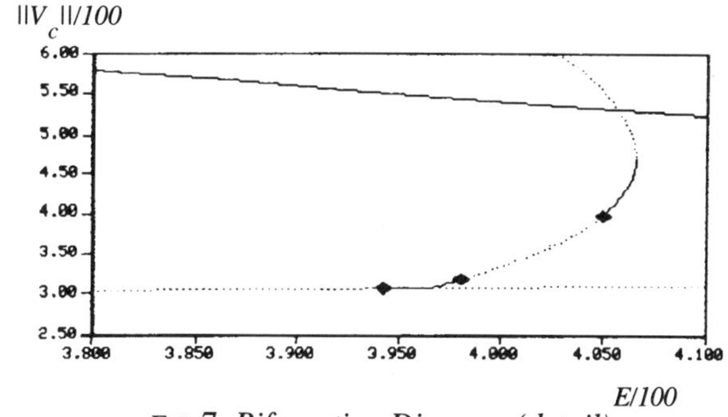

FIG 7 *Bifurcation Diagram (detail)*

bifurcation. On the bifurcated branch there are four points of period doubling (flip bifurcations). Using a Poincaré map we did compute the branch of invariant tori. The quasi-periodic solutions have been correctly computed until the occurrence of a frequency locking of order one third. Near this point the program failed. In fact we did

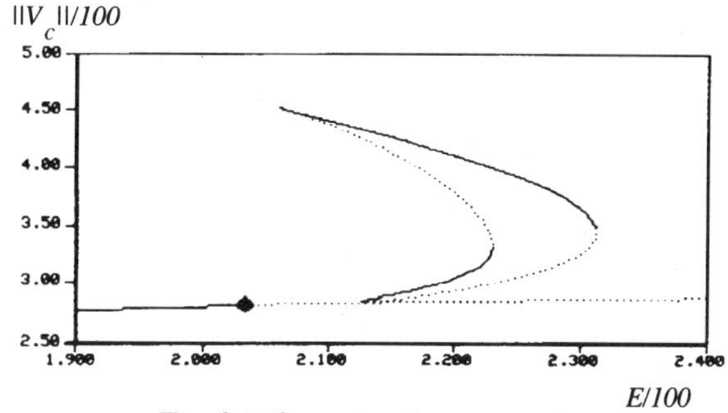

FIG 8 *Bifurcation Diagram (detail))*

observe a quick degradation of the condition number of the Jacobian. But starting from this point it has been possible to follow an order 3 sub-harmonic isola of solutions using AUTO [7]. We can see it in FIG 8. As far as we know it is the first time that such a connection has been observed on an electric line. The computed circles are shown in FIG. 9 in the flux-voltage plane.

5. Conclusion

This new algorithm for the numerical computation of invariant circles gives interesting results. At this point, it would be presumptuous and premature to compare it with other methods. In fact, we do not have yet a precise convergence proof for this algorithm and it would be interesting to know if such a proof can be achieved without hyperbolicity.

From a general and qualitative point of view, we can point out that in its formulation this method is not limited by the existence hypothesis of a particular kind of parameterization like [6]. Indeed in [6], an explicit global parameterization is used. It's true that using general coordinate transformations it's possible to generalize the algorithm and indeed the authors planned to do it. But this is a difficult task: (i) it's not easy to choose an atlas for the unknown manifold, (ii) numerically it could be difficult to deal with overlapping maps.

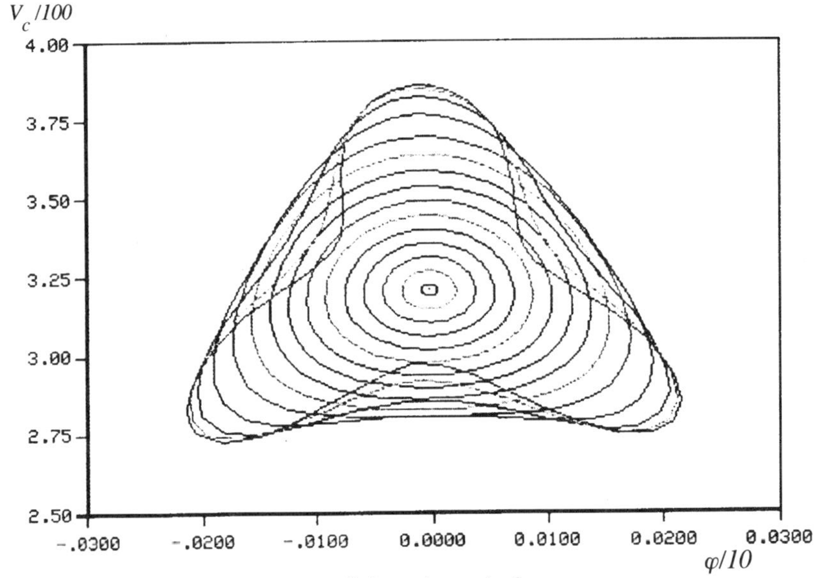

FIG. 9 *Invariant circles*

As we have said, we have trouble when the rotation number is rational with a small denominator; but we expect that a careful choice of an adaptative mesh strategy will limit the problems; thus the algorithm does not seem restricted to an irrational rotation number like [3].

Numerically the algorithm has been succesfully tested on unstable hyperbolic invariant curves. As we can expect, numeric convergence is a bit more difficult but it does work. On the other hand the principle of [12] relies upon the asymptotic attractivity of the invariant curve.

Moreover, the algorithm can start at a Hopf point without difficulties. In comparison the earlier algorithm [5] gave only a coarse approximation of the invariant circles near the Hopf point (a computation of a first order analytical approximation) and the purpose of [11] was merely to insure the existence of an invariant curve numerically.

However when we deal with flows, we have to find a Poincaré section -that could be a difficult task - and we have to compute the Poincaré map, in this case, the PDE approach of [6] seems more appropriate. But it should be pointed out that we have a natural Poincaré section when we deal with periodically forced differential equations. We can state too that our algorithm is useless when the invariant curve is not at least C^2 but the one described in [12] can deal with curves which are only C^0.

From a numerical point of view, it would be necessary to compare the different algorithms on several examples in similar conditions with no "manual tuning" of the computation. This would require we have achieved efficient implementations of the different methods. Such a comparison would have to be done very carefully and would be an heavy task. The comparison with the algorithm [12] would be the easiest.[7]

But we can look at several improvements for this algorithm. Let's give for instance two of them. The first one would be to use a C^3 and local approximation - cubic splines are not local and are only C^2. The second one would be to not use directly a Poincaré mapping when we deal with periodically forced differential equations but to use a technique similar to multiple shooting, the T^2 torus being then represented by several sections. In that way we would get a bigger problem but we would have to integrate the differential equation over a shorter time; furthermore it would be possible to have an interesting and geometric mesh strategy. We can expect then to get a problem with a better condition number. In this case, the method could become very similar to the PDE approach.

6. References

[1] ARONSON D. G., CHORY M. A., HALL G. R. AND MCGEHEE R.P. [1979] *A discrete dynamical system with subtly wild behaviour* , in "New approaches to non-linear problems in dynamics" SIAM 1980 (Proceedings of Asilomar Conference grounds, Pacific Grove California December 1979) pp. 339-359.

[2] BAI-LIN HAO[1989] *Elementary Symbolic Dynamics and chaos in dissipative systems*, World Scientific.

[3] CHAN TZE NGON [1983] *Numerical bifurcation analysis of simple dynamical systems* , Thesis Report, Concordia University Montreal Quebec Canada.

[4] DEBRAUX L. [1990] Thesis: *Analyse et contrôle de l'équation de Duffing, et des phénomènes de ferrorésonnance dans les lignes électriques.- Calcul des bassins d'attraction. -Continuation des tores invariants et des solutions quasipériodiques.*, Université de Technologie de Compiègne France.

[5] DELFILIPPI M.AND LATIL J.C. [1977] *AGENOR programme de calcul de bifurcation en un tore invariant pour un système différentiel 2π périodique.*, CNRS laboratoire de mécanique et d'acoustique.

[6] DIECI L., LORENZ J. AND RUSSELL R. [1991] *Numerical calculation of invariant Tori*, SIAM J. Sci. Statist. Comput. pp. 607-647.

[7] DOEDEL E. AND KERNEVEZ J.P. [1986] *AUTO : Software for continuation and bifurcation problems in ordinary differential equations*, Applied mathematics reports, California Institute of Technology.

[7] One difficulty if we want to compare this method or [12] with [6] is that we have to take care about the required accuracy in the computation of the Poincaré mapping and it's first derivative and we generally use general purpose routines for ordinary differential equations.

[8] HALE J. [1969] *Ordinary Differential Equations*, Wiley Interscience (John wiley & sons), pure and applied mathematics vol. XXI.

[9] MARSDEN J.E. AND MCCRACKEN M. [1976] *The Hopf Bifurcation and its Applications*, Springer Verlag Applied mathematical science 19.

[10] QUIVY L. AND KIENY C. *Pseudo-periodic ferroresonnant solutions stability in a power network*, IMACS TCI'90.

[11] THOULOUZE-PRATT E. [1980] *Existence Theorem of an invariant torus of solutions to a periodic differential system*, Laboratoire de Mécanique et d'Acoustique - CNRS - Marseille. Publications de l'Université de Provence.

[12] VAN VELDHUIZEN M. [1987] *A new algorithm for the numerical approximation of an invariant curve*, SIAM J. Sci. Statist. Comput. pp.951-962.

[13] FEIGENBAUM M.J.[1978] *Quantitative universality for a class of non linear transformations*, J. Stat. Phys, pp. 25-52.

LAURENT DEBRAUX.

Université de Reims, Département de Mathématiques, B.P. 347,

51062 Reims cedex , FRANCE.

e-mail: debraux@math-1.univ-reims.fr or ldebraux@dma.univ-compiegne.fr.

Contemporary Mathematics
Volume **172**, 1994

Numerics of Invariant Manifolds and Attractors

JENS LORENZ

October November 16, 1993

ABSTRACT. We consider autonomous systems of ODEs discretized by one-step methods with constant stepsize h. If the given system is dissipative, for example, then it often has a global attractor $\mathcal{A}$ and the one-step method has a nearby attractor $\mathcal{A}_h$. General results about structural similarity of $\mathcal{A}$ and $\mathcal{A}_h$ seem to be unknown, but we show common upper bounds for the Hausdorff dimension of $\mathcal{A}$ and $\mathcal{A}_h$.

If the ODE system has an invariant manifold M, then — under suitable assumptions — we show that the one-step method has a diffeomorphic invariant manifold M_h converging to M with the order of the numerical method.

1. Introduction

In this paper we consider autonomous dynamical systems of the form

$$\dot{x} = \Phi(x), \quad x(t) \in \mathbb{R}^m , \tag{1.1}$$

which are discretized by one-step methods

$$x^{n+1} = x^n + h\Phi_h(x^n) =: F_h(x^n), \quad n = 0, 1, 2, \dots \tag{1.2}$$

with constant stepsize $h > 0$. For simplicity of presentation, we will assume that $\Phi : \mathbb{R}^m \to \mathbb{R}^m$ is a C^1 vectorfield and that (1.1) together with any initial condition

$$x(0) = x^0, \quad x^0 \in \mathbb{R}^m \tag{1.3}$$

determines a unique solution

$$x(t) = F^t x^0, \quad t \geq 0 ,$$

1991 *Mathematics Subject Classification*. Primary 65L05 ; Secondary 34A50 .

The author was supported in part by NSF Grant #DMS-9107612 and DOE Grant #DE-FG03-92ER25128.

This paper is in final form and no version of it will be submitted for publication elsewhere.

existing for all time $t \geq 0$. We call $\{F^t, t \geq 0\}$ the continuous-time semi-group determined by (1.1). Concerning the one-step method (1.2), we make the usual assumption of p-th order convergence in finite time intervals, where the convergence is uniform in bounded sets. Also, it will be assumed that the linearized discrete solution operator

$$(F_h^n)'(x^0)$$

approximates the corresponding linearized continuous solution operator

$$(F^{hn})'(x^0)$$

in finite time intervals, uniformly in bounded sets. These assumptions are valid for all practically used methods.

In this paper we discuss two topics outlined below. First, under quite general assumptions one can show that the continuous-time semi-group $\{F^t, t \geq 0\}$ has a global attractor $\mathcal{A} \subset \mathbb{R}^m$ and that the discrete-time semi-group $\{F_h^n : n = 0, 1, \ldots\}$ has a nearby attractor $\mathcal{A}_h$ for all sufficiently small $h > 0$. What can be said about the relation between $\mathcal{A}$ and $\mathcal{A}_h$? To describe first a known result, we use the following notations:

For $x \in \mathbb{R}^m$ let $|x|$ denote the Euclidean norm of x. If $B \subset \mathbb{R}^m$ is a nonempty compact set and $x \in \mathbb{R}^m$ then let

$$\text{dist}\,(x, B) = \min\{|x - b| : b \in B\}\,.$$

If $A, B \subset \mathbb{R}^m$ are nonempty compact sets then let

$$\text{dist}\,(A, B) = \max\{\,\text{dist}\,(a, B) : a \in A\}\,.$$

It is not difficult to show that

$$\text{dist}(A, B) = 0 \Leftrightarrow A \subset B$$

and $\text{dist}(A, B) < \epsilon$ means that $A \subset N_\epsilon(B)$ where

$$N_\epsilon(B) = \{x : \ \text{dist}\,(x, B) < \epsilon\}$$

is the ϵ-neighborhood of B.

One can show under quite general assumption that

$$\text{dist}\,(\mathcal{A}_h, \mathcal{A}) \to 0 \text{ as } h \to 0\,. \tag{1.4}$$

See [**6, 7, 8**], for example. The relation

$$\text{dist}\,(\mathcal{A}, \mathcal{A}_h) \to 0 \text{ as } h \to 0 \tag{1.5}$$

requires stronger assumptions since large parts of $\mathcal{A}$ may disappear under small perturbations. Sufficient conditions for (1.5) are given in [**7**], for example.

Unfortunately, neither (1.4) nor (1.5) nor both relations together imply that the sets $\mathcal{A}_h$ are structurally similar to $\mathcal{A}$ for small $h > 0$. (Here we use the term of structural similarity in a loose way.)

A concept that has often been used for attractors is their *dimension*. To be specific, we will consider here only the Hausdorff dimension of compact sets $\mathcal{K}$, denoted by $dim\mathcal{K}$, and ask how $dim\mathcal{A}_h$ is related to $dim\mathcal{A}$. It is clear that (1.4), (1.5) imply no relation at all between these dimensions. For example, if

$$\mathcal{A} = \{0\}$$

and

$$\mathcal{A}_h = \{x \in \mathbb{R}^m : |x| \le h\}, \quad h > 0,$$

then (1.4) and (1.5) hold but

$$dim\mathcal{A} = 0, \qquad dim\mathcal{A}_h = m .$$

It does not seem to be known, in general, how the dimensions of the numerical attractors $\mathcal{A}_h$ are related to the dimension of the true attractor $\mathcal{A}$. Below we will describe a simple result which can loosely be formulated as follows: If a certain mathematical technique provides a bound

$$dim\mathcal{A} \le d$$

then one also has

$$dim\mathcal{A}_h \le d$$

for all sufficiently small $h > 0$. To give an example, consider the Lorenz system

$$\begin{aligned} \dot{x} &= \sigma(y - x) \\ \dot{y} &= rx - y - xz \\ \dot{z} &= xy - bz \end{aligned} \tag{1.6}$$

If one chooses the standard parameter values

$$\sigma = 10, \qquad b = 8/3, \qquad r = 28 ,$$

then the arguments in [**3**] show

$$\dim \mathcal{A} \le 2.468 .$$

Our result implies that also

$$\dim \mathcal{A}_h \le 2.468$$

for all sufficiently small $h > 0$.

Our second subject are compact attracting invariant manifolds M of (1.1). Under suitable assumptions the numerical operators F_h have nearby invariant manifolds M_h, which are diffeomorphic to M and which are $\mathcal{O}(h^p)$ close to M. This means that the manifolds M_h converge to M with the order of the numerical method. We give a proof of this result in a special setting and only sketch how the proof can be generalized.

A related question is how one should *compute* numerical approximations to an invariant manifold M of (1.1). For a treatment of this topic we refer to [**4**], where parameter dependent systems

$$\dot{x} = \Phi(x, \lambda)\,, \quad a \leq \lambda \leq b,$$

are considered and an algorithm is given to compute a branch of invariant manifolds $M(\lambda)$.

Here we want to explain only the basic difficulty. Consider the situation for a fixed value $\lambda = \lambda_1$ and suppose that $\tilde{M}_h$ is a *finite* set of vectors in $\mathbb{R}^m$ close to $M = M(\lambda_1)$. We assume that $\tilde{M}_h$ gives a fairly good numerical representation of the exact (unknown) manifold M and ask how we can improve the approximation. (In a continuation process, $\tilde{M}_h$ may be thought of as the numerically accepted approximation for $M(\lambda_0)$, and we try to continue from λ_0 to $\lambda_1 = \lambda_0 + \Delta\lambda$.)

If one forms the set

$$F_h^n \tilde{M}_h, \quad n \text{ large}\,,$$

will one obtain a better approximation of M? In general, this is not the case, because the finite set $F_h^n \tilde{M}_h$ usually approaches a *strongly attracting part* of M as n increases, and will not give a good representation of the whole manifold M.

To overcome this difficulty, it is suggested in [**4**] to solve suitable ordinary *boundary* value problems; simple repeated application of F_h corresponds to solving *initial* value problems. The resulting algorithm to approximate $M = M(\lambda)$ in a path following process is described and analyzed in [**4**].

2. Attractors $\mathcal{A}$ and $\mathcal{A}_h$

We first present some background material on global attractors for illustration. Then we show that a frequently used technique which yields upper bounds for $dim\mathcal{A}$ immediately implies the same bound for $dim\mathcal{A}_h$ for all sufficiently small $h > 0$.

Let $< \cdot, \cdot >$ denote the Euclidean scalar product on $\mathbb{R}^m$ with corresponding norm $|\cdot|$. Closed balls centered at the origin are denoted by

$$B_R = \{x \in \mathbb{R}^m : |x| \leq R\}\,.$$

A structural assumption which is satisfied for many (forced or unforced) dissipative systems is the following.

ASSUMPTION 2.1. *There are constants $\alpha > 0, \beta > 0$ such that*

$$< x, \Phi(x) > \leq -\alpha|x|^2 + \beta \tag{2.1}$$

for all $x \in \mathbb{R}^m$.

EXAMPLE 2.1. Consider the Lorenz system (1.6) with positive parameters σ, b, r. Using the variables

$$x, y, \tilde{z} = z - r - \sigma \,,$$

one obtains

$$\frac{d}{dt}\begin{pmatrix} x \\ y \\ \tilde{z} \end{pmatrix} = \begin{pmatrix} -\sigma & \sigma & 0 \\ -\sigma & -1 & -x \\ 0 & x & -b \end{pmatrix}\begin{pmatrix} x \\ y \\ \tilde{z} \end{pmatrix} + \begin{pmatrix} 0 \\ 0 \\ -b(r+\sigma) \end{pmatrix} \tag{2.2}$$

We denote the right-hand side by $\tilde{\Phi}(u),\ u = (x, y, \tilde{z})^T$. Then we obtain

$$< u, \tilde{\Phi}(u) > \; = \; -\sigma x^2 - y^2 - b\tilde{z}^2 - \tilde{z}b(r+\sigma) \; .$$

Let $\alpha = \min\{\sigma, 1, \frac{b}{2}\}$ and note that

$$|\tilde{z}b(r+\sigma)| \;\leq\; \frac{1}{2}b\tilde{z}^2 + \frac{1}{2}b(r+\sigma)^2$$

since $|pq| \leq \frac{1}{2}(p^2 + q^2)$. This yields the estimate

$$\begin{aligned} < u, \tilde{\Phi}(u) > \quad &\leq -\sigma x^2 - y^2 - \frac{b}{2}\tilde{z}^2 + \frac{1}{2}b(r+\sigma)^2 \\ &\leq -\alpha|u|^2 + \frac{1}{2}b(r+\sigma)^2 \, . \end{aligned}$$

Consequently, the system (2.2) satisfies Assumption 2.1.

To illustrate the use of Assumption 2.1, let $x(t),\ t \geq 0$, denote a solution of (1.1). Then we have by (2.1)

$$\begin{aligned} \frac{1}{2}\frac{d}{dt}|x(t)|^2 \; &= < x(t), \Phi(x(t)) > \\ &\leq \; -\alpha|x(t)|^2 + \beta \, , \end{aligned}$$

and therefore

$$|x(t)|^2 \leq \frac{\beta}{\alpha} + (|x(0)|^2 - \frac{\beta}{\alpha})e^{-2\alpha t}, \qquad t \geq 0 \, .$$

This estimate shows that any ball B_{R_0} where $R_0 > \sqrt{\beta/\alpha}$ is an absorbing set. Here one defines

DEFINITION 2.1. *A compact set $B \subset \mathbb{R}^m$ is called absorbing for $\{F^t, t \geq 0\}$ if the following two properties hold*
P1) *$F^t B \subset B$ for all $t \geq 0$;*
P2) *for all $R > 0$ there exists $t_R \geq 0$ so that*

$$F^t B_R \subset B \quad \textit{for} \quad t \geq t_R \, .$$

If $\{F^t, t \geq 0\}$ has the compact absorbing set B, then it is elementary to show that the set

$$\mathcal{A} := \bigcap_{t \geq 0} F^t B$$

has the following three properties:
P3) $\mathcal{A}$ is compact;
P4) $F^t \mathcal{A} = \mathcal{A}$ for all $t \geq 0$;
P5) for all $R > 0$,

$$\text{dist}\,(F^t B_R, \mathcal{A}) \to 0 \text{ as } t \to \infty\,.$$

Here P5 expresses global attractivity of $\mathcal{A}$ and P4 implies that $\mathcal{A}$ contains no transients. Since it is easy to see that for any semi-group $\{F^t, t \geq 0\}$ there is *at most* one set $\mathcal{A}$ with the properties P3, P4, P5, the following definition is justified.

DEFINITION 2.2. *A set $\mathcal{A} \subset \mathbb{R}^m$ is called the global attractor for $\{F^t, t \geq 0\}$ if* P3, P4, P5 *hold.*

These well-known arguments demonstrate that any dynamical system (1.1) which satisfies Assumptions 2.1 has a unique global attractor $\mathcal{A}$.
Generalizing Definition 2.2, we define the notion of an attractor as follows.

DEFINITION 2.3. *A set $\mathcal{A} \subset \mathbb{R}^m$ is called an attractor for $\{F^t, t \geq 0\}$ if* P3, P4, *and the following condition* P6 *hold;*
P6) *There is an open set $\mathcal{X} \supset \mathcal{A}$ so that for all $\epsilon > 0$ there is a time $t_\epsilon \geq 0$ with*

$$F^t \mathcal{X} \subset N_\epsilon(\mathcal{A}) \quad \textit{if} \quad t \geq t_\epsilon.$$

Concerning one-step methods, we will make the following — slightly restrictive — assumption for ease of presentation.

ASSUMPTION 2.2. *There is an $h_0 > 0$ such that $F_h : \mathbb{R}^m \to \mathbb{R}^m$ is a C^1 operator for $0 < h \leq h_0$. The operator F_h determines the numerical trajectory $x^n = F_h^n x^0, n = 1, 2, \ldots$ approximating the solution $x(t) = F^t x^0$ of the initial value problem*

$$\dot{x} = \Phi(x), \quad x(0) = x^0,$$

and has the following property:
P7) *For all $R > 0$ and all $T > 0$ there is a constant $C(R, T)$ with*

$$|F_h^n x^0 - F^{hn}(x^0)| \leq C(R, T) h^p \tag{2.3}$$

$$|(F_h^n)'(x^0) - (F^{nh})'(x^0)| \leq C(R, T) h \tag{2.4}$$

if $0 \leq nh \leq T,\ |x^0| \leq R,\ 0 < h \leq h_0$.

REMARK 2.1. *If one considers implicit one-step methods, then F_h is usually not defined on all of $\mathbb{R}^m$. This technical problem can be overcome as follows. We choose a large R so that B_R contains the region of interest and then choose a C^∞ cut-off function*

$$c : [0, \infty) \to [0, 1]$$

with $c(r) = 1$ for $0 \le r \le R^2$ and $c(r) = 0$ for $r \ge R^2 + 1$. If we replace (1.1) *by*

$$\dot{x} = \tilde{\Phi}(x) := c(|x|^2)\Phi(x) ,$$

then $\tilde{\Phi}(x) = \Phi(x)$ for $x \in B_R$, and the dynamics is not altered in the region of interest. Also, $|(\tilde{\Phi})'(x)|$ is bounded on $\mathbb{R}^m$. Therefore, if one applies the one-step method $x^{n+1} = \tilde{F}_h(x^n)$ to the altered system, then Assumption 2.2 *is reasonable. In a set slightly smaller than B_R the operators F_h and $\tilde{F}_h$ agree.*

Assuming that $\{F^t, t \ge 0\}$ has an attractor $\mathcal{A}$, it is elementary to prove that the discrete semi-groups $\{F_h^n, n = 0, 1, \ldots\}$ have nearby attractors $\mathcal{A}_h$ for all sufficiently small $h > 0$, and

$$\operatorname{dist}(\mathcal{A}_h, \mathcal{A}) \to 0 \quad \text{as} \quad h \to 0. \tag{2.5}$$

(For such a result, the estimate (2.4) is not needed.) Here the notion of an attractor for F_h^n is defined in complete analogy to Definition 2.3. We refer to [**7, 8**] for establishing the existence of $\mathcal{A}_h$ and convergence (2.5). See also [**6**] for a similar result on perturbed semigroups.

We now turn to estimating the Hausdorff dimensions of $\mathcal{A}$ and $\mathcal{A}_h$. (For a definition of the Hausdorff dimension, see [**11**], for example.) Let

$$\mathcal{K} \subset \mathcal{X} \subset \mathbb{R}^m$$

where $\mathcal{K}$ is compact and $\mathcal{X}$ is open. Suppose that $S : \mathcal{X} \to \mathbb{R}^m$ is a C^1 map with $S\mathcal{K} = \mathcal{K}$. Then one can obtain upper bounds for $dim\mathcal{K}$ using the singular values of the matrices $S'(x), x \in \mathcal{K}$. We will apply such a result for

$$\mathcal{K} = \mathcal{A}, \quad S = F^T, \quad T \text{ large}$$

and

$$\mathcal{K} = \mathcal{A}_h, \quad S = F_h^n, \quad n \text{ large}$$

NOTATION. Let $L \in \mathbb{R}^{m \times m}$ be nonsingular and denote the singular values of L by

$$\alpha_1(L) \ge \alpha_2(L) \ge \ldots \ge \alpha_m(L)$$

(These are the eigenvalues of $(L^T L)^{1/2}$.) Then we define the products

$$\omega_k(L) = \alpha_1(L) \cdots \alpha_k(L), \quad k = 1, \dots m$$

and interpolate as follows: If $1 \le d \le m$ we write

$$d = k + s, \quad k \in \{1, \dots, m-1\}, \quad 0 \le s \le 1$$

and set

$$\omega_d(L) = \omega_k(L)^{1-s} \omega_{k+1}(L)^s.$$

For a proof of the result formulated in the following theorem we refer to [**3**]. (The results in [**3**] are proved in an infinite dimensional setting and some simplifications occur in the case considered here. Also, we can weaken our assumptions slightly, but the result described below is sufficient for our applications and can be directly deduced from [**3**, Theorem 3.1].)

THEOREM 2.1. *Let $\mathcal{K} \subset \mathcal{X} \subset \mathbb{R}^m$ where $\mathcal{K}$ is compact and $\mathcal{X}$ is open. Assume that $S : \mathcal{X} \to \mathbb{R}^m$ is a C^1 map with $S\mathcal{K} = \mathcal{K}$ and $\det S'(x) \ne 0$ for all $x \in \mathcal{X}$. If there is a number d with $1 \le d \le m$ and*

$$\omega_d(S'(x)) < 1 \quad \textit{for all} \quad x \in \mathcal{K}$$

then $\dim \mathcal{K} \le d$.

To apply the result, we make the following assumption motivated by our discussion of attractors above.

ASSUMPTION 2.3. *There are compact sets $\mathcal{A}$ and $\mathcal{A}_h$ (for sufficiently small $h > 0$) with*

$$dist(\mathcal{A}_h, \mathcal{A}) \to 0 \quad \textit{as} \quad h \to 0$$

and

$$F^t \mathcal{A} = \mathcal{A}, \ t \ge 0, \qquad F_h \mathcal{A}_h = \mathcal{A}_h, \ 0 < h \le h_1.$$

THEOREM 2.2. *Let* Assumptions 2.2 *and* 2.3 *be valid and suppose that there is a time $T > 0$ and a number d with $1 \le d \le m$ and*

$$\omega_d((F^T)'(x)) < 1 \quad \textit{for all} \quad x \in \mathcal{A}.$$

Then

$$\dim \mathcal{A} \le d \quad \textit{and} \quad \dim \mathcal{A}_h \le d$$

for all sufficiently small $h > 0$.

PROOF. The estimate $dim\mathcal{A} \leq d$ follows directly from Theorem 2.1. Continuity of the function

$$x \rightarrow \omega_d((F^T)'(x))$$

yields existence of numbers $\delta > 0, \epsilon > 0$ with

$$\omega_d((F^T)'(x)) \leq 1 - \delta \quad \text{for all} \quad x \in N_\epsilon(\mathcal{A}).$$

Note that we have $\mathcal{A}_h \subset N_\epsilon(\mathcal{A})$ for $0 < h \leq h_1$. For given h we determine an integer $N = N(h)$ so that $h(N-1) < T \leq hN$ and obtain

$$|(F_h^N)'(x) - (F^{hN})'(x)| \leq Ch$$

and

$$|(F^{hN})'(x) - (F^T)'(x)| \leq Ch$$

for $x \in N_\epsilon(\mathcal{A})$. These inequalities together with continuity of the function

$$L \rightarrow \omega_d(L)$$

yield

$$\omega_d((F_h^N)'(x)) \leq 1 - \frac{\delta}{2} \quad \text{for all} \quad x \in \mathcal{A}_h$$

if $0 < h \leq h_2$. Now we can apply Theorem 2.1 again (with $\mathcal{K} = \mathcal{A}_h, S = F_h^N$) and obtain the desired estimate $dim\mathcal{A}_h \leq d$. □

If one wants to apply the above theorem, one has to derive estimates for the singular values of the linearized solution operator $(F^T)'(x), x \in \mathcal{A}$. Techniques for doing this are described in [**1, 3, 11**], for example.

3. Invariant Manifolds M and M_h

Consider a dynamical system $\dot{x} = \Phi(x)$ and assume that M is a compact attracting invariant manifold without boundary for the semi-group $\{F^t, t \geq 0\}$. Again, we discretize the dynamical system by a one-step method $x^{n+1} = F_h x^n$ and make Assumption 2.2. Under suitable conditions on generalized Lyapunov type numbers — which we will formulate below — one can prove that M *persists* under small (in the C^1 sense) perturbations of the vectorfield Φ. More precisely, if we perturb Φ slightly then the perturbed dynamical system also has an invariant manifold close to M, which is diffeomorphic to M. Such a result is proved by in [**5, 9, 10**], for example. We want to show here a similar result for *perturbations by discretization*, i.e., we want to show that F_h has an invariant manifold M_h diffeomorphic to M, which converges to M with the order of the numerical method. We will not show the result in its most general form, however. It is our aim to give a proof which is readily accessible to numerical analysts since there is a growing interest in topics like invariant manifolds from the computational point of view. — For a similar local result in case of a center manifold of a fixed point, see [**2**].

Let $T^p = (\mathbb{R} mod 2\pi)^p$ denote the standard p-torus. (The number p here is not related to the order p of the numerical method.) To present the ideas in their simplest form, we will first consider a partitioned system

$$\dot{\phi} = f(\phi, r), \quad \dot{r} = g(\phi, r) \tag{3.1}$$

where $\phi(t) \in T^p, r(t) \in \mathbb{R}^q$. The functions f and g are in C^1. Also, we assume that $g(\phi, 0) \equiv 0$, so that

$$M = \{(\phi, 0) : \phi \in T^p\}$$

is an invariant p-torus for (3.1). At the end of the section we will sketch how to treat more general invariant p-tori and more general invariant manifolds.

The solution of (3.1) with initial condition

$$x(0) = x^0 = (\phi^0, r^0)$$

is denoted by

$$x(t) = (\phi(t), r(t)) = F^t x^0 = F^t(\phi^0, r^0).$$

It is convenient to use the notations

$$\phi(t) = F_1^t x^0, \quad r(t) = F_2^t x^0;$$

i.e., F_1^t and F_2^t denote the ϕ-part and the r-part of the solution operator F^t, respectively. The Jacobian of $F^t(\cdot)$ has the block form

$$(F^t)' = \begin{pmatrix} F_{1\phi}^t & F_{1r}^t \\ F_{2\phi}^t & F_{2r}^t \end{pmatrix} \tag{3.2}$$

where the indices ϕ and r indicate differentiation with respect to the ϕ-variables and r-variables, respectively. It is important to note that our assumption $g(\phi, 0) \equiv 0$ implies that

$$F_2^t(\phi, 0) \equiv 0$$

and therefore the block $F_{2\phi}^t$ is zero on M. The growth or decay of the diagonal blocks $F_{1\phi}^t$ and F_{2r}^t determines if there is attractivity of the dynamics *within* M and *towards* M, respectively. To quantify this we follow [**5**] and define the generalized Lyapunov type numbers

$$\begin{aligned} \nu(x) &= \inf\{a > 0 : \frac{1}{a^t}|F_{2r}^t(F^{-t}x)| \to 0 \text{ as } t \to \infty\} \\ \sigma(x) &= \inf\{s > 0 : |F_{1\phi}^{-t}x||F_{2r}^t(F^{-t}x)|^s \to 0 \text{ as } t \to \infty\} \end{aligned}$$

for $x \in M$. For illustration, consider a trajectory $x(t) = F^t x^0$ in M and assume that for large $t > 0$

$$\begin{aligned} |F_{1\phi}^{-t} x^0| &\sim e^{\alpha t}, \quad \alpha > 0 , \\ |F_{2r}^t (x(-t))| &\sim e^{-\beta t}, \quad \beta > 0 . \end{aligned}$$

Thus there is attractivity — in forward time — within and towards M. We have

$$\nu(x^0) = e^{-\beta} < 1$$

and

$$\sigma(x^0) = \frac{\alpha}{\beta} > 0.$$

We now make the following assumption.

ASSUMPTION 3.1. *For all $x \in M$ it holds that*

$$\nu(x) < 1 \quad \textit{and} \quad \sigma(x) < 1.$$

REMARK 3.1. *The condition $\sigma(x) < 1$ expresses that the attractivity of the flow towards M is stronger than the attractivity within M.* Assumption 3.1 *will lead to a C^1 perturbation theorem: For sufficiently small $h > 0$ the operator F_h has a (locally unique) invariant C^1 manifold M_h, diffeomorphic to M and converging to M with the order of the numerical method. As in* [**5**], *the result can be extended to a corresponding C^k perturbation theorem for any integer $k \geq 1$ if one assumes that $f, g \in C^k$, that*

$$\nu(x) < 1 \quad \textit{and} \quad k\sigma(x) < 1$$

for all $x \in M$, and that $x \to F_h x$ is a C^k operator.

To give a precise formulation of our result we introduce a class of Lipschitz functions from T^p to $\mathbb{R}^q$. This requires to measure distances in T^p. To this end, we first define a distance on $\mathbb{R} mod 2\pi$ as follows: If $\alpha, \beta \in \mathbb{R} mod 2\pi$ then let

$$|\alpha - \beta| = |\gamma|$$

where $\gamma \in \mathbb{R}$ is the unique representative in $[-\pi, \pi)$ of $\alpha - \beta$, and $|\gamma|$ is its absolute value. If $\phi = (\phi_1, \ldots, \phi_p), \psi = (\psi_1, \ldots, \psi_p)$ are in T^p, then we set

$$|\phi - \psi| = |\phi_1 - \psi_1| + \cdots + |\phi_p - \psi_p|.$$

With this distance, T^p is a complete metric space. Next we introduce a class of Lipschitz functions from T^p to $\mathbb{R}^q$.

DEFINITION 3.1. *For $\epsilon > 0, \delta > 0$ let $Lip_{\epsilon,\delta}$ denote the set of all functions $R : T^p \to \mathbb{R}^q$ with*

$$|R\phi| \le \epsilon \quad \textit{and} \quad |R\phi - R\psi| \le \delta|\phi - \psi|$$

for all $\phi, \psi \in T^p$.

(Here $|R\phi|$ denotes the Euclidean norm of $R\phi \in \mathbb{R}^q$, for definiteness.) The set $Lip_{\epsilon,\delta}$ becomes a complete metric space for the distance

$$|R - S|_\infty = \max\{|R\phi - S\phi| : \phi \in T^p\}.$$

The following theorem, based on ideas in [**5**], will be proved in several steps.

THEOREM 3.1. *Consider a system* (3.1) *with $f, g \in C^1, g(\phi, 0) \equiv 0$ and* Assumption 3.1. *Assume that the one-step method $x^{n+1} = F_h x^n$ is applied and that* Assumption 2.2 *(with $\mathbb{R}^m$ replaced by $T^p \times \mathbb{R}^q$) is fulfilled. Then there exist $\epsilon > 0, \delta > 0, h_0 > 0$ so that for $0 < h \le h_0$ there is a unique function $R_h \in Lip_{\epsilon,\delta}$ with the property that*

$$M_h = \{(\phi, R_h\phi) : \phi \in T^p\}$$

is invariant under F_h, $F_h M_h = M_h$. Furthermore, $R_h \in C^1$ and

$$|R_h|_\infty = \mathcal{O}(h^p),$$

where p is the order of the numerical method.

1. The proof is given in several steps. First note that

$$F_{1\phi}^{-t}x = \left(F_{1\phi}^{t}(F^{-t}x)\right)^{-1}, \quad x \in M,$$

and therefore the expression in the definition of $\sigma(x)$ can be written as

$$|F_{1\phi}^{-t}x||F_{2r}^{t}(F^{-t}x)|^s = |(F_{1\phi}^{t}y)^{-1}||F_{2r}^{t}y|^s, \quad y = F^{-t}x\ .$$

Then the Uniformity Lemma in [**5**] implies that there exists a sufficiently large time $T > 0$ so that

$$|F_{2r}^{T}x| \le \frac{1}{4}, \quad |(F_{1\phi}^{T}x)^{-1}||F_{2r}^{T}x| \le \frac{1}{4}$$

for all $x \in M$. (Here the bound $\frac{1}{4}$ could be replaced by any other positive number.)

By continuity, there exist $\epsilon > 0, \Delta > 0$ so that

$$|F_{2r}^{t}x| \le \frac{1}{3}, \quad |(F_{1\phi}^{t}x)^{-1}||F_{2r}^{t}x| \le \frac{1}{3} \tag{3.3}$$

for all $x = (\phi, r)$ with $|r| \leq \epsilon$ and all t with $T \leq t \leq T + \Delta$. Recalling the equation $F_2^t x = 0, x \in M$, we also obtain

$$(3.4) \qquad |F_2^t x| \leq \frac{\epsilon}{3}$$

for all x, t as above. Now let $0 < h \leq \Delta$ and let $N = N(h)$ denote the smallest integer with

$$T \leq Nh \leq T + \Delta \ .$$

We have

$$|F_h^j x - F^{jh} x| \leq Ch^p, \quad |(F_h^j)' x - (F^{jh})' x| \leq Ch,$$

for $0 \leq jh \leq T + \Delta$ and all x in the (bounded) region of interest. In particular, the above estimates (3.3), (3.4) are valid for sufficiently small h with F^t replaced by F_h^N if the bounds are increased from $\frac{1}{3}$ to $\frac{1}{2}$, say. This will be used below.

2. Let $R \in Lip_{\epsilon,\delta}$ where $\epsilon > 0$ and $\delta > 0$ will be determined below. We fix h and $N = N(h)$ as above and consider an integer j with $1 \leq j \leq N$. For any given $\alpha \in T^p$ we want to solve the equation

$$(3.5) \qquad (F_h^j)_1(\phi, R\phi) = \alpha$$

for $\phi \in T^p$. (With $(F_h^j)_1$ and $(F_h^j)_2$ we denote the ϕ-component and r-component of F_h^j, respectively.) The following elementary lemma — based on contraction — is proved in [**4**, Lemma 2.5].

LEMMA 3.1. *Let A, B denote mappings from T^p into itself. Assume that A is $1-1$ and onto, and that A^{-1} is Lipschitz:*

$$|A^{-1}\alpha - A^{-1}\beta| \leq L|\alpha - \beta| \quad \forall \alpha, \beta \in T^p \ .$$

If B is close to A in the sense that

$$(3.6) \qquad |A\phi - A\psi - (B\phi - B\psi)| \leq q|\phi - \psi| \quad \forall \phi, \psi \in T^p$$

with $qL < 1$, then B is also $1-1$ and onto, and

$$|B^{-1}\alpha - B^{-1}\beta| \leq \frac{L}{1 - qL}|\alpha - \beta| \quad \forall \alpha, \beta \in T^p \ .$$

To apply the lemma, we set

$$A\phi = F_1^{jh}(\phi, 0), \quad B\phi = (F_h^j)_1(\phi, R\phi), \quad \phi \in T^p.$$

Note that A satisfies the assumptions of the lemma by elementary ODE arguments and the invariance of M. To show (3.6), we let $\phi, \psi \in T^p$ be arbitrary and set

$$\kappa_s = s\phi + (1-s)\psi, \quad \mu_s = (\kappa_s, sR\phi + (1-s)R\psi), \quad 0 \le s \le 1 .$$

Then we have

$$\begin{aligned} A\phi - A\psi &= \Big(\int_0^1 F_{1\phi}^{jh}(\kappa_s, 0)ds\Big)(\phi - \psi) , \\ B\phi - B\psi &= \Big(\int_0^1 (F_h^j)_{1\phi}(\mu_s)ds\Big)(\phi - \psi) + \Big(\int_0^1 (F_h^j)_{1r}(\mu_s)ds\Big)(R\phi - R\psi) . \end{aligned}$$

Noting that $|R\phi - R\psi| \le \delta|\phi - \psi|, (F_h^j)' = (F^{jh})' + \mathcal{O}(h)$ and $|R|_\infty \le \epsilon$, one obtains

$$|A\phi - A\psi - (B\phi - B\psi)| \le C(\epsilon + h + \delta)|\phi - \psi| .$$

This shows that we can apply Lemma 3.1 for sufficiently small ϵ, δ, h, and obtain unique solvability of (3.5).

3. We will use (3.5) for $j = 1$ and $j = N$. First let $j = N$ and recall $T \le Nh \le T + \Delta$. The smallness assumptions (3.3), (3.4) carry over to F_h^N. This is the basis for proving Lemma 3.2 below. We let $R \in Lip_{\epsilon,\delta}$ and assume that ϵ, δ, h are so small that (3.5) (with $j = N$) has a unique solution $\phi \in T^p$ for any given $\alpha \in T^p$. Then we set

$$(G_h R)(\alpha) := (F_h^N)_2(\phi, R\phi) . \tag{3.7}$$

LEMMA 3.2. *There exist sufficiently small $\epsilon > 0, \delta > 0, h_0 > 0$ so that the following holds:*
(i) Whenever $0 < h \le h_0$ and $R \in Lip_{\epsilon,\delta}$, then $G_h R$ is well-defined and

$$G_h R \in Lip_{\epsilon/2,\delta/2} .$$

(ii) There is a contraction constant $0 \le q < 1$, independent of $0 < h \le h_0$, so that

$$|G_h R - G_h S|_\infty \le q|R - S|_\infty$$

for all $R, S \in Lip_{\epsilon,\delta}$.

This lemma can be proved in the same way as Lemma 2.8 and Lemma 2.9 of [**4**]. The proof is therefore omitted.

4. Let R_h denote the unique fixed point of G_h in $Lip_{\epsilon,\delta}$. We want to show that $|R_h|_\infty = \mathcal{O}(h^p)$. To this end, let $\underline{0}$ denote the zero function in $Lip_{\epsilon,\delta}$. For given $\alpha \in T^p$ the value $(G_h\underline{0})(\alpha)$ is determined as follows: Solve

$$(F_h^N)_1(\phi, 0) = \alpha$$

for $\phi \in T^p$ and set

$$(G_h \underline{0})(\alpha) = (F_h^N)_2(\phi, 0) \ .$$

Since $F_2^{hN}(\phi, 0) = 0$ we have $(F_h^N)_2(\phi, 0) = \mathcal{O}(h^p)$ by Assumption 2.2. This shows that

$$|G_h \underline{0}|_\infty \leq Ch^p \ .$$

Since G_h is a contraction, we obtain

$$|G_h \underline{0} - R_h|_\infty = |G_h \underline{0} - G_h R_h|_\infty \leq q|\underline{0} - R_h|_\infty$$

and therefore

$$|R_h|_\infty \leq q|R_h|_\infty + \mathcal{O}(h^p) \ .$$

Since $0 \leq q < 1$ is independent of h, we have shown $|R_h|_\infty = \mathcal{O}(h^p)$.

5. We prove next that

$$M_h = \{(\phi, R_h\phi) : \phi \in T^p\}$$

is invariant under F_h. (Recall that F_h is the operator determining the one-step method.) First note that $R_h \in Lip_{\epsilon/2, \delta/2}$ by Lemma 3.2. We now use (3.5) for $j = 1$ and define an operator H_h analogously to G_h, with $j = N$ replaced by $j = 1$: For $R \in Lip_{\epsilon,\delta}$ and $\alpha \in T^p$ let

$$(H_h R)(\alpha) = (F_h)_2(\phi, R\phi)$$

where $\phi \in T^p$ is the unique solution of (3.5) for $j = 1$. Using the fact that $F_h^N F_h = F_h F_h^N$ it is easy to show that

$$G_h H_h R = H_h G_h R$$

as long as

$$R, \ H_h R, \ G_h R \in Lip_{\epsilon,\delta} \ .$$

We apply this relation with $R = R_h$, the fixed point of G_h. (Note that $H_h R_h \in Lip_{\epsilon,\delta}$ since $R_h \in Lip_{\epsilon/2,\delta/2}$ and h is small.) Therefore,

$$G_h H_h R_h = H_h G_h R_h = H_h R_h$$

and *uniqueness* of the fixed point of G_h implies $H_h R_h = R_h$.

Now let $(\phi, R_h\phi) \in M_h$ be arbitrary and set

$$(F_h)_1(\phi, R_h\phi) = \alpha \ . \tag{3.8}$$

Then

$$(F_h)_2(\phi, R_h\phi) = (H_h R_h)(\alpha) = R_h\alpha \ .$$

This shows that

$$F_h(\phi, R_h\phi) = (\alpha, R_h\alpha) \in M_h \ ,$$

i.e., $F_h M_h \subset M_h$. We can also start with an arbitrary point $(\alpha, R_h\alpha) \in M_h$, define ϕ by (3.8) and then argue as above. This shows that $F_h M_h = M_h$.

By similar arguments one can prove that if

$$\tilde{M}_h = \{(\phi, S_h\phi) : \phi \in T^p\}$$

is invariant under F_h and $S_h \in Lip_{\epsilon,\delta}$, then $H_h S_h = S_h$, and therefore $G_h S_h = S_h$. Uniqueness of the fixed point of G_h in $Lip_{\epsilon,\delta}$ yields $S_h = R_h$. This shows that F_h has a *unique* invariant manifold in the specified class.

6. So far we have shown that the function R_h, which determines the invariant manifold M_h of F_h, is Lipschitz. For a proof that $R_h \in C^1$, we refer to [**5**]. The arguments given there can also be used to prove that $R_h \in C^k$ under the stronger assumptions described in Remark 3.1.

Let us now sketch how the effect of discretization can be analyzed in more general situations. To this end, let $\dot{x} = \Phi(x), x(t) \in \mathbb{R}^m$, denote a dynamical system discretized by a one-step method, as before. We assume that a smooth function $w : T^p \to \mathbb{R}^m$ parametrizes a p-torus

$$M = \{w(\phi) : \phi \in T^p\} \ ,$$

which is invariant under the dynamics of $\dot{x} = \Phi(x)$. In a sufficiently small neighborhood $N_\rho(M)$ of M, $\rho > 0$, one can define a coordinate transformation

$$\mathcal{T}x = (\phi, r) = y, \quad x \in N_\rho(M) \ ,$$

where $\phi \in T^p, r \in \mathcal{O}_\rho$. Here

$$\mathcal{O}_\rho = \{r \in \mathbb{R}^q : |r| < \rho\}$$

is a neighborhood of the origin in $\mathbb{R}^q, q = m - p$. This transformation $\mathcal{T}$ can be obtained as follows: Let $n_1(\phi), \dots , n_q(\phi)$ denote an orthonormal basis of the normal space to M at the point $w(\phi)$; under suitable smoothness assumptions on w, one can choose the $n_j(\phi)$ as smooth functions of $\phi \in T^p$. (For a construction of the $n_j(\phi)$ using a modified Householder transformation and numerical applications, we refer to [**4**].) If x is sufficiently close to M and $w(\phi)$ is the point in M closest to x, then one can write

$$x = w(\phi) + \sum_{j=1}^{q} r_j n_j(\phi), \quad r \in \mathbb{R}^q . \tag{3.9}$$

This determines the transformation

$$\mathcal{T} x = (\phi, r) .$$

(Analytically, it is easier to study $\mathcal{T}$ by defining $x = \mathcal{T}^{-1}(\phi, r)$ using (3.9).) Note that M becomes

$$\tilde{M} = \{(\phi, 0) : \phi \in T^p\}$$

in the new coordinates. Transforming the given system $\dot{x} = \Phi(x)$ to (ϕ, r)-coordinates, we arrive at a partitioned system (3.1) in $T^p \times \mathcal{O}_\rho$. The one-step method $x^{n+1} = F_h x^n$ can also be transformed,

$$y^{n+1} = \mathcal{T} x^{n+1} = \mathcal{T} F_h \mathcal{T}^{-1} y^n .$$

Assumption 2.2 for F_h leads to corresponding (local) properties of

$$\tilde{F}_h = \mathcal{T} F_h \mathcal{T}^{-1} .$$

In this way, we can reduce the analysis of general (smooth) p-tori to the situation of Theorem 3.1.

To study general invariant manifolds M (instead of p-tori), one has to work with a finite set of local coordinate charts (instead of one parametrization $w : T^p \to \mathbb{R}^q$). The technical details are carried out in [**5**] for the case of perturbations of the vectorfield Φ. As in the case of Theorem 3.1, all arguments can be carried over to perturbations by discretization. One only has to observe that (a localized) Assumption 2.2 is all that is needed to carry over the proofs. The fact that F_h^j approximates F^t only at discrete times is irrelevant.

REFERENCES

1. A. V. Babin and M. I. Vishik, *Attractors of partial differential equations and estimates of their dimension,* Uspekhi Mat. Nauk **38** (1983), 133–187 (in Russian); Russian Math. Surveys **38** (1983), 151–213.
2. W.-J. Beyn and J. Lorenz, *Center manifolds of dynamical systems under discretization,* Numer. Funct. Anal. and Optimiz. **9** (1987), 381–414.
3. P. Constantin, C. Foias, and R. Temam, *Attractors representing turbulent flows,* Memoirs of the Amer. Math. Soc. **53**, No. 314, 1985.
4. L. Dieci, J. Lorenz, *Computation of invariant tori by the method of characteristics,* SIAM J. Numer. Anal. (submitted for publication).
5. N. Fenichel, *Persistence and smoothness of invariant manifolds for flows,* Indiana Univ. Math. J. **21** (1971), 193–226.
6. J. K. Hale, *Asymptotic Behavior of Dissipative Systems,* Mathematical Surveys and Monographs, vol. 25, Amer. Math. Soc., Providence R. I., 1988.
7. A. Humphries and A. Stuart, *Runge-Kutta methods for dissipative and gradient dynamical systems,* SIAM J. Numer. Anal. (to appear).

8. P. Kloeden and J. Lorenz, *Stable attracting sets in dynamical systems and in their one-step discretizations,* SIAM J. Numer. Anal. **23** (1986), 986–995.
9. R. Sacker, *A new approach to the perturbation theory of invariant surfaces,* Comm. Pure and Applied Math. **18** (1965), 717–732.
10. R. Sacker, *A perturbation theorem for invariant manifolds and Hölder continuity,* J. Math. and Mech. **18** (1969), 705–762.
11. R. Temam, *Infinite-dimensional dynamical systems in mechanics and physics,* Springer–Verlag, New York, 1988.

Department of Mathematics and Statistics, The University of New Mexico, Albuquerque, NM 87131

E-mail address: lorenz@altona.unm.edu

Contemporary Mathematics
Volume **172**, 1994

Interval stochastic matrices and simulation of chaotic dynamics

P. Diamond, P. Kloeden and A. Pokrovskii

Abstract

The relationship between a chaotic dynamical system and approximate spatial discretizations is discussed. Collapsing effects are the major difficulty. These are investigated and some strategies to suppress them are analyzed. The paper presents a unified approach to the problem based on the systematic use of a new mathematical tool, *interval stochastic matrices*.

1 Introduction

A dynamical system will be regarded as generated by a mapping f of a compact metric space Ω into itself. This mapping is assumed to be a continuous or discontinuous Borel mapping. That is, if $\mathcal{B}$ is the class of Borel subsets of Ω, then $f^{-1}(S) \in \mathcal{B}$ for all $S \in \mathcal{B}$. For chaotic systems, analyzing specific trajectories such as equilibrium points or cycles is less important than investigating richer and more complex attractors and various global stochastic–like properties. The simplest such property is invariance of a measure. A probability measure μ on Ω is called *invariant* for f if for any Borel set $S \subseteq \Omega$ the volume of S with respect to the measure μ coincides with the μ-volume of its full preimage $f^{-1}(S)$. For example, the Dirac measure concentrated at an equilibrium point of f is always invariant. So too is the atomic measure uniformly distributed on any cycle of the system f. Invariant measure is thus a natural generalization of equilibrium points and cycles. For systems with chaotic behaviour usually there exist other invariant measures; for instance, absolutely continuous measures or measures with support on a Cantor set. The method of invariant measures is appropriate for the analysis of chaotic systems.

Classical results state that a smooth dynamical system preserves some of its properties under small *smooth perturbations*. Some general results which spring to

1991 *Mathematics Subject Classification.* Primary 58F13, 65G10; Secondary 60J20.
This research was supported by the Australian Research Council Grant A 89132609.

mind include the Hartman–Grobman Theorem, Stable Manifold Theorem, Shadowing Lemma and structural stability theorems (see, for example, [12, 14]). The situation is more difficult if continuous but nonsmooth perturbations are considered. Results about structural stability are no longer true in general, but, under natural assumptions, nondegenerate attractors and invariant measures are reasonably robust.

The situation changes, however, if we consider perturbations which are produced by *spatial discretization* of a system. It is well known that even the most intricate attractors and completely mixing invariant measures can collapse into, say, a single fixed point and its corresponding atomic measure. Such effects pose problems for the theoretical analysis of internal computer representations of systems. The aim of this paper is to discuss some of these problems and to introduce a useful mathematical tool: *interval stochastic matrices.*

The structure of the paper is as follows: Section 2 contains simple examples and Theorem 1, which gives sufficient conditions for collapse to occur. In Section 3 three "anti–collapsing" strategies are discussed and some related theoretical problems are outlined. In Section 4 the main technical tools are introduced and Section 5 applies these to specific problems.

2 Nature of collapsing effects

Because a digital computer is an automaton with a finite number of states, a typical computer realization of a dynamical system is a mapping φ of a finite subset $\mathbf{L} \subseteq \Omega$ into itself for which the graph $\mathbf{Gr}(\varphi)$ is close in a natural sense to $\mathbf{Gr}(f)$. That is, the typical realization is a dynamical system on a finite state space.

As mentioned in the Introduction, even close realizations φ can be quite different, from the point of view of dynamical system theory, from the original system f. Let us mention a classical example. Consider the tent map f on the interval $[0, 1]$:

$$f(x) = \begin{cases} 2x & \text{if} \quad x \leq 1/2, \\ 2 - 2x & \text{if} \quad x \geq 1/2. \end{cases} \tag{1}$$

It is chaotic: it has a unique absolutely continuous invariant measure and cycles of all orders, is completely mixing and so on. Consider as a discrete model φ of f the restriction of f on the N-digital binary lattice

$$\mathbf{L}_N = \left\{0, \frac{1}{2^N}, \frac{2}{2^N}, \frac{3}{2^N}, \ldots, \frac{2^N - 1}{2^N}\right\}. \tag{2}$$

Clearly, φ is asymptotically trivial: $\varphi^k \equiv 0$ if $k \geq N$. Certainly, φ has only the zero cycle and only one invariant measure concentrated at zero.

Such collapsing effects are in some sense always possible. Let us formulate a simple result in this direction. Let μ_0 be an absolutely continuous measure on $\mathcal{B}$.

Define the sequence of measures $\{\mu_n\}$ by

$$\mu_{n+1}(S) = \mu_n\left(f^{-1}(S)\right) \ , \ S \in \mathcal{B} \ , \ n = 0, 1, 2, \dots \ .$$

A measure μ_* is said to be *attractive* if all such sequences $\{\mu_n\}$ converge Cesàro weakly [1] to μ_*. That is, the Cesàro means

$$m_n = \frac{\mu_0 + \mu_1 + \ldots + \mu_{n-1}}{n}$$

converge weakly to μ_* for all choices of absolutely continuous μ_0. For wide classes of systems with chaotic behavior there exists an attractive measure which is absolutely continuous on all of Ω for expansive systems, or with support on a Cantor set for hyperbolic systems. Usually, this measure is regarded as the most interesting invariant measure [15].

Consider a given system f with an attractive measure μ_* with support $\mathbf{Supp}(\mu_*)$. Recall that the *support* of a probability measure μ is the *minimal closed subset* $S \subseteq \Omega$ with $\mu(S) = 1$.

We will characterize the density of the lattice $\mathbf{L}$ by its spatial step

$$h(\mathbf{L}) = \sup_x \inf_\xi \{\rho(\xi, x) \ : \ \xi \in \mathbf{L}, \ x \in \Omega\} \ . \tag{3}$$

Theorem 1. *Let f satisfy the Lipschitz condition $\rho(f(x), f(y)) \le \lambda \rho(x, y)$ and let S be an f-invariant subset of $\mathbf{Supp}(\mu_*)$. Then for any $\varepsilon > 0$ and sufficiently small $h(\mathbf{L})$ there exists a discretization $\varphi : \mathbf{L} \to \mathbf{L}$ of f satisfying*

$$\mathbf{Sep}(\mathbf{Gr}(\varphi), \mathbf{Gr}(f)) \le (\lambda^2 + \lambda + 1) h(\mathbf{L}), \tag{4}$$

such that the support of each φ-invariant measure is in an ε-neighborhood of S.

In (4) $\mathbf{Sep}(A, B)$ denotes the Hausdorff separation in $\Omega \times \Omega$ with respect to the metric on the space $\Omega \times \Omega$ defined by

$$\rho_1\left((x_1, x_2), (y_1, y_2)\right) = \max\left\{\rho(x_1, y_1), \rho(x_2, y_2)\right\}.$$

The proof of Theorem 1 is analogous to the proof of Theorem 5 in [7], which deals with attractors.

Consequently, a measure μ_* is attractive for spatial discretizations *only if its support* $\mathbf{Supp}(\mu_*)$ *has no invariant proper subsets*. It is well known that the actual situation is often different. In particular, the support of an attractive measure of a hyperbolic system may contain periodic points. Hence there are invariant proper subsets consisting of cycles in $\mathbf{Supp}(\mu_*)$, so the system could collapse under discretization onto any of these cycles. Some other details about the behaviour of attractors of dynamical systems under discretization are given in [5].

3 Strategies for collapsing effects

What can be done to avoid such effects? There are some general strategies, including the following three:

S1. To analyze properties of some specific discretization which is carefully chosen from a certain set of appropriate discretizations.

S2. To analyze a discretization which also includes a random process; this is usually done by replacing φ by a Markov chain on the lattice $\mathbf{L}$.

S3. To sweep the problem under the carpet.

All three of these strategies are quite reasonable though the fundamental theoretical questions concerning each of them are still open.

Nevertheless, the issues concerning the first strategy are quite clear.

Q1. *To give recommendations, at least general, for choosing a discretization which guarantees that collapsing effects are avoided.*

There are some deep theoretical results (see for instance [3]), mainly connected with using the Stetter discretization [16], but these are not always applicable to the situation under consideration.

In addition, there are a number of brilliant theoretical results associated with the second strategy, especially general theorems of Kifer [13] and Blank [1], but, again, the main question is open:

Q2. *What is an appropriate level of randomness?*

If the stochastic component in the second strategy is too large then the dynamics of the model will differ markedly from those of the original system, but if it is not strong enough, then collapsing effects will be present. Some details are discussed at the end of this section.

A theoretical question connected with the strategy S3 is the question

Q3. *When and why the strategy of ignoring the problem is justified.*

That is, when are collapsing effects unlikely (although still possible by virtue of Theorem 1).

While analyzing Strategy S2, we have investigated the combined effect of *discretization* and *introduction of a random component* for the tent map (1). Some details were unexpected.

Consider the sequence of i.i.d. random variables ξ_n , $n = 1, 2, \ldots$ defined as follows: let $p \in [0, 1/2)$, and let

$$\xi_n = \begin{cases} 0 & \text{with probability } 2p \ , \\ -1 & \text{with probability } q = 1/2 - p \ , \\ +1 & \text{with probability } q \ . \end{cases} \tag{5}$$

For $N = 1, 2, \ldots$ and each $\eta_0 \in \mathbf{L}_N$, define the random sequence $\eta_n \in \mathbf{L}_N$, by

$$\eta_n = f(\eta_{n-1} + \xi_n 2^{-N}), \quad n = 1, 2, \ldots \, . \tag{6}$$

where f is the tent map (1) and $\mathbf{L}_N$ is defined by (2). This sequence represents the computer simulation of the map f in the finite arithmetic on $\mathbf{L}_N$, with roundoff modeled by the random error $\xi_n 2^{-N}$ in the last place of machine accuracy.

It is fairly straightforward to show

Lemma 1. *For any integer $N \geq 1$ and $a \in [0, 1)$, the limit*

$$F_N(a) = \lim_{n \to \infty} n^{-1} \text{card}\{j : \eta_j \in [0, a) \ , \ 1 \leq j \leq n\}$$

exists almost surely, and is independent of η_0. Each function F_N is the distribution of a measure μ_N on $\mathbf{L}_N$. The measures μ_N converge weakly to a probability measure $\mu_* = \mu_*(q)$ on $[0,1]$ as $N \to \infty$, and $\mu_*(q)$ is an invariant measure of the tent map (1), where q is the probability defined in (5).

Denote by x^j_N , $j = 1, 2, \ldots, N$ the j-th symbol in the $\mathbf{L}_N$ binary representation of $x \in I$. Note that the symbol x^j_N is not defined if $N < j$. Where no confusion is possible, we simply write x^j, suppressing N.

Theorem 2. *The measure $\mu_*(q)$ coincides with Lebesgue measure $m(\cdot)$ at $q = 1/4$ and at $q = 1/2$. For any $q \in (0, 1/2) \setminus \{1/4\}$ the measure $\mu_*(q)$ is singular; it has no atoms and the support is $\mathbf{Supp}(\mu_*(q)) = [0, 1]$. For all $q \in (0, 1/2]$, the following hold:*

$$\mu_*(q)\{x : x^j = 0\} = \mu_*(q)\{x : x^j = 1\} = 1/2 \ , \quad j = 1, 2, \ldots \tag{7}$$

and

$$\mu_*(q)\{x : x^j + x^{j+1} = 1\} = 3q - 4q^2 \ , \quad j = 1, 2, \ldots \, . \tag{8}$$

This assertion can be proven analogously to the Theorem 1 from [6]. It is worth noting that the measure $\mu_*(q)$ does not depend monotonically on the probability q. The parameter q could be thought of as characterizing the intensity of perturbation arising from computer roundoff. As q increases to $q = 1/4$, the randomized map

$$\eta \mapsto f(\eta + 2^{-N}\xi) \ , \quad \eta \in I_N \ ,$$

becomes completely mixing at q=1/4. Another form of regularity typified by relations like (8), then appears and disappears only at q=1/2.

4 Interval stochastic matrices and multi–valued discretizations

The idea which will be described below is useful when using either the first or the second strategy and would possibly also be useful in a rigorous analysis of the third strategy. What is especially important is that it demonstrates that all these strategies are related.

Let $\mathcal{M}_d$ denote the totality of all real square $d \times d$ matrices $A = (a_{ij})$ with nonnegative entries $a_{ij} \geq 0$. Vectors $v \in \Re^d$ will be treated as columns and matrix multiplication Av will be on the left. $\mathcal{M}_d$ has a natural partial order given by

$$A \leq B \Leftrightarrow a_{ij} \leq b_{ij} \ , \ i,j = 1, \ldots, d \ ,$$

where $A = (a_{ij}), B = (b_{ij})$ are elements of $\mathcal{M}_d$. Recall that a matrix $C = (c_{ij})$ is a *stochastic matrix* if

$$\sum_{i=1}^{d} c_{ij} = 1 \ , \ j = 1, \ldots, d \ .$$

The class of all stochastic matrices in $\mathcal{M}_d$ will be denoted by $\mathcal{S}_d$.

Let $\mathcal{M}_d^-$ be the set of all matrices $A = (a_{ij}) \in \mathcal{M}_d$ satisfying

$$\sum_{i=1}^{d} a_{ij} \ \leq \ 1 \ , \ j = 1, \ldots, d$$

and $\mathcal{M}_d^+$ the set of all matrices $B = (b_{ij}) \in \mathcal{M}_d$ satisfying

$$\sum_{i=1}^{d} b_{ij} \ \geq \ 1 \ , \ j = 1, \ldots, d \ .$$

For any two matrices $A \in \mathcal{M}_d^-$ and $B \in \mathcal{M}_d^+$ such that $A \leq B$, let $\widehat{AB}$ denote the set of all stochastic matrices between A and B, that is

$$\widehat{AB} = \{C \in \mathcal{S}_d : A \leq C \leq B\} \ .$$

The set $\widehat{AB}$ will be called the *interval stochastic matrix with boundaries* A and B.

Finally, if σ_d is the standard simplex in $\Re^d$, that is,

$$\sigma_d = \left\{ (p_1, \ldots, p_d) \in \Re^d \ : \ p_i \geq 0, \ i = 1, \ldots, d \quad \text{and} \quad \sum_{i=1}^{d} p_i = 1 \right\} ,$$

then for any vector $p \in \sigma_d$ and for any interval stochastic matrix $\widehat{AB}$ define

$$\widehat{AB}\, p = \{Cp : \ C \in \widehat{AB}\}. \tag{9}$$

Our principal result is an explicit representation of the set (9). Let $\mathcal{I}_d$ be the class of all subsets of $\{1, 2, \ldots, d\}$. For any $j \in \{1, \ldots, d\}$ and $I \in \mathcal{I}_d$ define the *upper* (j, I)*−flow*

$$H_j(I, \widehat{AB}) = \min\{\sum_{i \in I} b_{ij}, \ 1 - \sum_{i \notin I} a_{ij}\}.$$

Theorem 3. *The set* $\widehat{AB}\,p$ *is precisely the set of all vectors* $q \in \sigma_d$ *satisfying*

$$\sum_{j=1}^{d} p_j H_j(I, \widehat{AB}) \geq \sum_{i \in I} q_i \quad \text{for all} \quad I \in \mathcal{I}_d \,.$$

The proof is presented in the Appendix.

An important corollary to this theorem can be formulated as follows. A vector $p \in \sigma_d$ is said to be *semi–invariant* for the interval stochastic matrix $\widehat{AB}$ if for each $I \in \mathcal{I}_d$

$$\sum_{j=1}^{d} p_j H_j(I, \widehat{AB}) \geq \sum_{i \in I} p_i \quad \text{for all} \quad I \in \mathcal{I}_d \,.$$

For any stochastic matrix C, let $\text{Fix}(C)$ denote the set of all vectors $x \in \sigma_d$ such that $Cx = x$.

Corollary 1. *The set*

$$\bigcup_{C \in \widehat{AB}} \text{Fix}(C)$$

is precisely the set of all semi–invariant vectors of the interval stochastic matrix $\widehat{AB}$.

Let us return to dynamical systems and their spatial discretizations. Consider for a moment a multi–valued discretization $\Phi : \mathbf{L} \to 2^{\mathbf{L}}$. There are some natural classes of measures on $\mathbf{L}$ which are candidates for invariant measures of Φ, two of which will be considered here.

Markov chains. A measure is Markov Φ–invariant if it is an invariant for some Markov chain C on $\mathbf{L}$ satisfying $p(\xi, \eta) = 0$ if $\eta \notin \Phi(\xi)$. Here, $p(\xi, \eta)$ is the transition probability of passing from state $\xi \in \mathbf{L}$ of C to the state $\eta \in \mathbf{L}$.

Flows. A measure μ is flow Φ–invariant if, for any subset $\mathbf{L}_* \subseteq \mathbf{L}$, $\mu(\mathbf{L}_*) \leq \mu(\Phi^{-1}(\mathbf{L}_*))$.

Denote by **m-inv**(Φ) the totality of Markov Φ–invariant measures and by **f-inv**(Φ) the totality of flow invariant measures. From Theorem 3 it follows that

Theorem 4. **m-inv**(Φ) = **f-inv**(Φ).

So both ideas are equivalent and it is possible to use the term *invariant measure* for multi–valued approximations.

5 Simulation of chaotic systems

Now we will show how the theorems 3 and 4 work together in analysis of spatial discretizations.

EXAMPLE 1. The basic estimate

Let $\mathcal{P}$ denote the totality of Borel probability measures on Ω. The *Prokhorov metric* ρ_P on $\mathcal{P}$ can be defined by

$$\rho_P(\mu_1, \mu_2) = \inf\{\varepsilon > 0 : \mu_1(\mathcal{O}_\varepsilon(S)) \le \mu_2(S) - \varepsilon, \ \forall S \in \mathcal{B}\}. \tag{10}$$

where $\mathcal{O}_\varepsilon(S)$ is the ε–neighbourhood of S. This metric is a standard in many branches of probability theory — see, for instance, [11].

Choose a fixed multi–valued discretization Φ_f satisfying

$$\mathbf{Gr}(\Phi_f) = \{(\xi, \eta) : \rho_1((\xi, \eta), \mathbf{Gr}(f)) \le h(\mathbf{L})\},$$

where h is defined by (3). Points from the graph of Φ_f are points from the $h(\mathbf{L})$–tube around the graph of f. In a sense it is a reasonable minimal multi–valued realization of f.

Theorem 5. *For any invariant measure μ of f there exists an invariant measure μ_* of Φ_f satisfying the inequality*

$$\rho_P(\mu_*, \mu) \le h(\mathbf{L}). \tag{11}$$

This theorem estimates the minimal level of "multi-valuedness" which is sufficient to suppress collapsing effects.

EXAMPLE 2. Application to the strategy S2

By virtue of Theorem 5 and the Markov characterization of invariant measures of the multi–valued mapping Φ_f we immediately obtain

Corollary 2. *For any invariant measure μ of f there exists a Markov chain on $\mathbf{L}$ with a stationary measure μ_* satisfying (11) and with transient probabilities $p(\xi, \eta)$ satisfying*

$$p(\xi, \eta) = 0 \quad \text{for} \quad \rho((\xi, \eta), \mathbf{Gr}(f)) > h(\mathbf{L}).$$

This corollary provides an estimate of the minimal level of stochasticity which is sufficient to suppress collapsing effects. This level is surprisingly low. In an obvious sense it coincides with the spatial step of the discretization under consideration. Certainly, it does not provide a full answer to the question Q2 from Section 3, but it is a step in this direction. We mention also that assertions similar to Corollary 2 were useful in analyzing some algorithms to find interesting invariant measures [8].

EXAMPLE 3. Reduction to a linear programming problem

By Theorem 5 and the flow characterization of invariant measures of multi–valued mappings we get another corollary:

Corollary 3. *For any invariant measure μ of f there exists a measure μ_* on* $\mathbf{L}$, *satisfying* (11), *and*

$$\mu_*(\mathbf{L}_*) \leq \mu_*(\Phi_*^{-1}(\mathbf{L}_*)) \qquad \text{for} \qquad \mathbf{L}_* \subseteq \mathbf{L}.$$

This is interesting because often an invariant measure with certain extremal properties is of interest. Corollary 3 reduces the numerical search of such a measure to a linear programming problem.

EXAMPLE 4. Application to the strategy S1

Let $\mathbf{T^d}$ be the standard d–dimensional torus, let $f : \mathbf{T^d} \mapsto \mathbf{T^d}$ be a mapping with invariant Lebesgue measure and let $\mathbf{L}_\nu$ be a lattice on $\mathbf{T^d}$ induced by the standard uniform $1/\nu$ lattice on the cube.

Theorem 6. *For every $\varepsilon > 0$ there exists a permutation π of* $\mathbf{L}$ *satisfying*

$$\mathbf{Sep}(\mathbf{Gr}(\pi), \mathbf{Gr}(f)) \leq \varepsilon.$$

This theorem [10] is somewhat surprising because the mapping f is not even assumed to be injective. It can be regarded as an example of how to choose a good approximation. Clearly, mappings such as permutations avoid collapsing effects. It is important that there are rapid algorithms to find such permutations corresponding to f constructively. They can be realised as simple computer programs.

6 Conclusion

The ideas of previous sections seem especially useful in the analysis of discontinuous systems f, which are very important today, with applications to control theory, number theory, non–smooth mechanics, and so on.

Let $\overline{\mathbf{Gr}(f)}$ be the closure of the set $\mathbf{Gr}(f)$ and for any $S \in \mathcal{B}$ denote by $\overline{f^{-1}}(S)$ the set

$$\overline{f^{-1}}(S) = \left\{ x \in \Omega : \text{ there exists } y \in \overline{S} \text{ such that } (x, y) \in \overline{\mathbf{Gr}(f)} \right\}.$$

A measure $\mu \in \mathcal{P}$ is said to be *semi–invariant* for the system f if

$$\mu(S) \leq \mu(\overline{f^{-1}}(S)) \quad \text{for} \quad S \subseteq \Omega.$$

The methods outlined above allow the development of a rich theory of semi–invariant measures [7, 8, 9, 10]. Note only a few results in this direction:

- For continuous systems μ is invariant if and only if it is semi–invariant.
- $\mathbf{f\text{-}inv}(f)$ is nonempty, convex and closed; it depends on f upper semicontinuously.

- $\mu \in \mathbf{f\text{-}inv}(f)$ if and only if it is a fixed point of weak closure of f^*, where f^* is defined by $(f^*\mu)(S) = \mu(f(S))$.
- There are natural analogues of classical theorems such as the Poisson theorem on nonwandering points; the Birkhoff—Khinchin ergodic theorem, etc.
- Flow invariant measures are exactly the measures which can be approximated by natural computer experiments.

Appendix. Proof of Theorem 1

Let p be a fixed vector from σ_d and let $Q(p)$ be the set of all vectors $q \in \sigma_d$ satisfying

$$\sum_{j=1}^{d} p_j H_j(I, \widehat{AB}) \geq \sum_{i \in I} q_i \quad \text{for all} \quad I \in \mathcal{I}_d \, .$$

The theorem is proved by establishing $Q(p) \supseteq \widehat{AB}p$ and $Q(p) \subseteq \widehat{AB}p$.

Step 1. Let $q \in \widehat{AB}p \cap \sigma_d$. We prove that $q \in Q(p)$. From the definition there exists a matrix $C = (c_{ij}) \in \widehat{AB}$ such that $q = Cp$. Hence, for any $I \in \mathcal{I}_d$

$$\sum_{i \in I} \sum_{j=1}^{d} c_{ij} p_j = \sum_{i \in I} q_i \, . \tag{12}$$

Moreover, for any $j \in \{1, \ldots, d\}$ and $I \in \mathcal{I}_d$

$$\sum_{i \in I} a_{ij} \leq \sum_{i \in I} c_{ij} \leq \sum_{i \in I} b_{ij} \, ,$$

from which follows

$$\sum_{i \in I} c_{ij} \leq \min\{\sum_{i \in I} b_i^j, 1 - \sum_{i \notin I} a_i^j\} \leq H_j(I, \widehat{AB}). \tag{13}$$

From (12) and (13), $q \in Q(p)$ follows.

Step 2. Let $q \in Q(p) \cap \sigma_d$ be fixed, that is,

$$\sum_{j=1}^{d} p_j H_j(I, \widehat{AB}) \geq \sum_{i \in I} q_i \quad \text{for all} \quad I \in \mathcal{I}_d \, . \tag{14}$$

We show that $q \in \widehat{AB}p$. For this it is sufficient to establish the existence of a matrix $C \in \widehat{AB}$ such that $Cp = q$. Suppose that no such matrix exists. For each $C \in \widehat{AB}$ define

$$\delta(C) = \|Cp - q\|^2.$$

The quantity $\delta(C)$ depends continuously on the matrix C. Since the set $\widehat{AB}$ is compact there exists a matrix $D \in \widehat{AB}$ such that

$$\delta(D) = \min_{C \in \widehat{AB}} \delta(C) \; > \; 0 \, . \tag{15}$$

Since $p \in \sigma_d$, so also is $Dp \in \sigma_d$. Since $Dp \neq q$, by virtue of (15) the set

$$I_0 = \{i \in \{1, \ldots, d\} : \; (Dp)_i < q_i\}$$

is nonempty. Hence

$$\sum_{i \in I_0} \sum_{j=1}^{d} d_{ij} p_j \; < \; \sum_{i \in I_0} q_i$$

or, equivalently,

$$\sum_{j=1}^{d} p_j \sum_{i \in I_0} d_{ij} \; < \; \sum_{i \in I_0} q_i . \tag{16}$$

To obtain the contradiction with the inequality (14) it is sufficient to prove

Lemma 2. *For each $k \in \{1, \ldots, d\}$, $p_k \sum_{i \in I_0} d_{ik} \; \geq \; p_k H_k(I_0, \widehat{AB})$.*

PROOF: Suppose that for some $k \in \{1, \ldots, d\}$ the lemma is untrue. That is, $p_k > 0$ and

$$\sum_{i \in I_0} d_{ik} \; < \; H_k(I_0, \widehat{AB}). \tag{17}$$

From (17) and the definition of upper (k, I)–flows, it must follow that

$$\sum_{i \in I_0} d_{ik} \; < \; \sum_{i \in I_0} b_{ik} \; , \tag{18}$$

$$\sum_{i \in I_0} d_{ik} \; < \; 1 - \sum_{i \notin I_0} a_{ik} . \tag{19}$$

Since D is a stochastic matrix, (19) can be rewritten in the form

$$\sum_{i \notin I_0} d_{ik} \; > \; \sum_{i \notin I_0} a_{ik} . \tag{20}$$

From (18) and (20) there exist indices $i_1 \in I_0$ and $i_2 \notin I_0$ such that for some sufficiently small $\varepsilon_0 > 0$

$$d_{i_1 k} \; < \; b_{i_1 k} - \varepsilon_0 \quad \text{and} \quad d_{i_2 k} \; > \; a_{i_2} + \varepsilon_0 .$$

For any $\varepsilon \in (0, \varepsilon_0]$, consider the matrix $D(\varepsilon) = (d(\varepsilon)_{ij})$ defined by

$$d(\varepsilon)_{ij} = \begin{cases} d_{i_1 k} + \varepsilon & \text{if } i = i_1 \text{ and } j = k \\ d_{i_2 k} - \varepsilon & \text{if } i = i_2 \text{ and } j = k \\ d_{ij} & \text{if } j \neq k \text{ or } (i \neq i_1 \text{ and } i \neq i_2) \; . \end{cases}$$

Then for each $\varepsilon \in (0, \varepsilon_0]$, since D is a stochastic matrix, the matrix $D(\varepsilon)$ is also a stochastic matrix and satisfies $A \leq D(\varepsilon) \leq B$. Hence

$$D(\varepsilon) \in \widehat{AB}\,, \quad 0 < \varepsilon \leq \varepsilon_0\,. \tag{21}$$

Now choose $\varepsilon_* \in (0, \varepsilon_0]$ such that

$$0 > (D(\varepsilon_*)p)_{i_1} - q_{i_1} > (Dp)_{i_1} - q_{i_1}\,; \tag{22}$$

such an ε_* exists because $i_1 \in I_0$ and $p_k \neq 0$. From (22) it follows that

$$\left((D(\varepsilon_*)p)_{i_1} - q_{i_1}\right)^2 = (|(Dp)_{i_1} - q_{i_1}| - \varepsilon_* p_k)^2 < ((Dp)_{i_1} - q_{i_1})^2 - (\varepsilon_* p_k)^2 \tag{23}$$

But since $i_2 \notin I_0$,

$$0 \leq (Dp)_{i_2} - q_{i_2}\,,$$

and

$$\left((D(\varepsilon_*)p)_{i_2} - q_{i_2}\right)^2 = (|(Dp)_{i_2} - q_{i_2}| - \varepsilon_* p_k)^2 \leq ((Dp)_{i_1} - q_{i_1})^2 + (\varepsilon_* p_k)^2\,. \tag{24}$$

Moreover,

$$\left((D(\varepsilon_*)p)_{i_2} - q_{i_2}\right)^2 = ((Dp)_i - q_i)^2 \quad i \neq i_1, i_2\,. \tag{25}$$

The relations (23) – (25) imply

$$\delta\left(D(\varepsilon_*)\right) < \delta(D)\,. \tag{26}$$

But $D(\varepsilon_*) \in \widehat{AB}$ since (21) and $\varepsilon_* \in (0, \varepsilon_0]$. Therefore, (26) contradicts the construction of D in (15). The lemma is proved and hence so too is the theorem. □

References

[1] Blank M., *Small perturbations of chaotic dynamical systems*, Russian Math. Soc. Surveys, **44**, No 6 (1989), 1–33.

[2] Boyarsky A., *Randomness implies order*, J. Math.Anal. Applns., **76** (1986), 483–497.

[3] Boyarsky A. and Gora P., *Why computers like Lebesgue measure*, Comp. Math. Applns, **16** (1988), 321–329.

[4] Boyarsky A. and Scarowsky M., *Long periodic orbits of the triangle map*, Proc. Amer. Math. Soc., **97** (1986), 247–254.

[5] Diamond P. and Kloeden P. *Spatial discretization of mappings*, Comp. Math. Applns., **25** (1993), 85–94.

[6] Diamond P., Kloeden P. and Pokrovskii A., *An invariant measure arising in computer simulation of a chaotic dynamical system*, J. Nonlinear Sciences, to appear.

[7] Diamond P., Kloeden P. and Pokrovskii A., *Weakly chain recurrent points and spatial discretizations of dynamical systems*, Random & Computational Dynamics, to appear.

[8] Diamond P., Kloeden P. and Pokrovskii A., *Analysis of an Algorithm for Computing Invariant Measures*, Nonlin. Anal. TMA, to appear.

[9] Diamond P., Kloeden P. and Pokrovskii A., *Interval stochastic matrices and the computation of invariant measures*, submitted for publication.

[10] Diamond P., Kloeden P. and Pokrovskii A., *Numerical modeling of toral dynamical systems with invariant Lebesgue measure*, submitted for publication.

[11] Ethier S. and Kurtz T., "Markov Processes: Characterization and Convergence", Wiley, New York, 1986.

[12] Guckenheimer J. and Holmes P., "Nonlinear Oscillations, Dynamical Systems and Bifurcation of Vector Fields", Springer-Verlag, New–York, 1983.

[13] Kifer Yu., "Random Perturbations of Dynamical Systems", Birkhäuser, Boston, 1988.

[14] Shub M., "Global Analysis of Dynamical Systems", Springer–Verlag, New York, 1987.

[15] Sinai Ya.G., *The stochasticity of dynamical systems,* Selecta Math. Soviet. **1** (1981),100–119.

[16] Stetter H.J., "Analysis of Discretization Methods for Ordinary Differential Equations", Springer–Verlag, Berlin, 1976.

Department of Mathematics, The University of Queensland, 4072 Australia

Department of Mathematics, Deakin University, Geelong, 3217 Australia

Department of Mathematics, The University of Queensland, 4072 Australia
Permanent address: Institute of Information Transmission Problems, Russian Academy of Sciences

Contemporary Mathematics
Volume **172**, 1994

MATHEMATICAL AND NUMERICAL ANALYSIS OF A MEAN-FIELD EQUATION FOR THE ISING MODEL WITH GLAUBER DYNAMICS

C.M. Elliott, A.R. Gardiner, I. Kostin and Bainian Lu

Abstract. In this paper, we consider a mean-field equation of motion, derived by Penrose [7] for the dynamic Ising model with Glauber dynamics. The properties of the global dynamics are studied for the equation and an explicit Euler discretisation. A study of the bifurcation problem on the periodic one dimensional lattice is made. Some numerical computations are presented.

1. Introduction

We consider the following mean–field equation of motion for the dynamic Ising model on a periodic lattice Λ:

$$\text{(1.1a,b,c)} \qquad \begin{cases} \mathbf{u}_t + \mathbf{u} = \tanh(\beta \mathbf{A}\mathbf{u}) \ t > 0 \\ \mathbf{u}(0) = \mathbf{u}_0 \in V_\Lambda \\ \mathbf{u}_{a+N\mathbf{e}^i} = \mathbf{u} \qquad\qquad a \in \Lambda, 1 \le i \le d \end{cases}$$

where Λ denotes the lattice of $\mathbb{Z}^d$ with N^d sites defined by

$$\Lambda := \{a : a = \sum_{i=1}^{d} a_i \mathbf{e}^i,\ a_i \in \mathbb{Z},\ 1 \le a_i \le N\}$$

with $\{\mathbf{e}^i\}$ being the standard unit vectors of $\mathbb{Z}^d$. We say that Λ is a d-dimensional lattice. We denote by V_Λ the N^d dimensional space of lattice vectors $\mathbf{v} = (v_a)_{a\in\Lambda^*}$ satisfying $v_{a+N\mathbf{e}^i} = v_a$. Here $\mathbf{u} = (u_a)_{a\in\Lambda}$ and u_a denotes the expectation of the spin at site a of the lattice and Λ^* is defined by

$$\{a : a = \sum_{i=1}^{d} a_i \mathbf{e}^i,\ a_i \in \mathbb{Z}\}.$$

The $N^d \times N^d$ symmetric matrix $\mathbf{A}$ is defined by, for $v \in V_\Lambda$

$$\text{(1.2)} \qquad \{\mathbf{A}\mathbf{v}\}_a := \sum_{b\in\Lambda} E_{ab} v_b$$

1991 *Mathematics Subject Classification.* 65C20 35B35

The work of CME and ARG was supported by the SERC grants GR/F85659 and GR/H61445. The work of IK was supported by the Royal Society. The work of BL was supported by the British Council.

where E_{ab} is the Ising interaction between sites a and b satisfying, for all $a, b \in \Lambda$

$$(1.3) \qquad (i) \quad 1 \geq E_{ab} \geq 0 \qquad (ii) \quad E_{ab} > 0 \Longleftrightarrow b \in N(a).$$

Here $N(a)$ denotes the neighbourhood of the site a defined by

$$N(a) = \{b : \sum_{i=1}^{d} |a_i - b_i| = 1\}.$$

The parameter $\beta = 1/\theta$, where $\theta(> 0)$ is the scaled absolute temperature. Furthermore throughout the paper we use the convention that for any lattice vector $\mathbf{u}$, the component at site a is $(\mathbf{u})_a = u_a$ and for any $f : I\!R \to I\!R$, $\{f(\mathbf{u})\}_a = f(u_a)$. The dynamical system was derived by Penrose [7] from an Ising model on the lattice Λ. It approximately represents the behaviour in the mean of the Ising model with Glauber (spin–flip) stochastic dynamics, Glauber [4].

It is convenient for us to introduce some new notation in order to both analyse the equation and also to compare with other models. We define θ_a, θ_c by

$$(1.4) \qquad \begin{aligned} \theta_a &:= \sum_{b \in N(a)} E_{ab} \\ \theta_c &:= \max_{a \in \Lambda} \theta_a = \|\mathbf{A}\|_\infty, \end{aligned}$$

where $\|\mathbf{A}\|_\infty$ is the infinity norm of the matrix $\mathbf{A}$ given by

$$\|\mathbf{A}\|_\infty := \max_{a \in \Lambda} \sum_{b \in N(a)} E_{ab}.$$

We define an $N^d \times N^d$ symmetric matrix $\mathbf{L}$ by

$$(1.5) \qquad \mathbf{L} := \mathbf{A} - \theta_c \mathbf{I}.$$

It follows from (1.5) that $-\mathbf{L}$ is diagonally dominant, has positive diagonal elements and so is positive semi–definite. The inverse of $\tanh(\bullet)$ is denoted by $\phi(\bullet)$ so that

$$\phi(r) = \frac{1}{2} \ln \frac{1+r}{1-r}.$$

Hence (1.1a) may be rewritten as

$$(1.6) \qquad \theta\phi(\mathbf{u}_t + \mathbf{u}) = \mathbf{L}\mathbf{u} + \theta_c \mathbf{u}.$$

Note that for (1.1a) to have a meaning $\|\mathbf{u}_t + \mathbf{u}\|_\infty < 1$, where $\|\mathbf{v}\|_\infty = \max_{a \in \Lambda} |v_a|$, $\mathbf{v} \in V_\Lambda$. Furthermore from the physical point of view it is reasonable to suppose that $\|\mathbf{u}(t)\|_\infty \leq 1$ for all t. Indeed we shall prove in section 2 that if $\|\mathbf{u}_0\|_\infty \leq 1$ then $\|\mathbf{u}(t)\|_\infty < 1$, $\forall t > 0$. We shall also make some remarks concerning the case

$\|\mathbf{u}_0\|_\infty > 1$. But let us suppose that for $t > 0$, $\|\mathbf{u}(t)\|_\infty < 1$ and introduce the homogeneous 'free energy' functions

$$\psi_0(r) := \frac{\theta}{2}((1+r)\ln(1+r) + (1-r)\ln(1-r)) \tag{1.7}$$
$$\psi(r) := \psi_0(r) - \frac{\theta_c}{2}r^2$$

for $r \in (-1,1)$. Then as noted by Penrose [7], an important feature of the system (1.1) is the existence of a Lyapunov functional given in our notation by

$$I(\mathbf{u}) := -\frac{1}{2}(\mathbf{Lu}, \mathbf{u}) + (\mathbf{e}, \psi(\mathbf{u})) \tag{1.8}$$

where $(\bullet, \bullet)$ is the discrete L^2 inner product

$$(\mathbf{u}, \mathbf{v}) = \sum_{a\in\Lambda} u_a v_a \qquad \forall \mathbf{u}, \mathbf{v} \in V_\Lambda,$$

and $\{\mathbf{e}\}_a = 1$, $\forall a \in \Lambda$. We set $\|\mathbf{v}\| = (\mathbf{v}, \mathbf{v})^{1/2}$, $\forall \mathbf{v} \in V_\Lambda$.

This can be seen by rewriting (1.6) as

$$\theta(\phi(\mathbf{u}_t + \mathbf{u}) - \phi(\mathbf{u})) = \mathbf{Lu} - \psi'(\mathbf{u}) \tag{1.9}$$

and taking the inner product with $\mathbf{u}_t$. Since $\phi'(r) \geq 1$ for all $r \in (-1,1)$ it follows that

$$\theta\|\mathbf{u}_t\|^2 + \frac{d}{dt}I(\mathbf{u}) \leq 0 \tag{1.10}$$

which implies that $I(\mathbf{u}(t))$ is decreasing with time.

Remarks.

1. θ_c is said to be the critical temperature, This is because for $\theta > \theta_c$ the function $\psi(\bullet)$ is convex with a unique minimum at $r = 0$ whereas for $\theta < \theta_c$, $\psi(\bullet)$ is a double equal well potential with minimum at the binodal points $r = \pm u^m$ where $u^m < 1$ is the positive root of

$$\frac{2\theta_c}{\theta} = \ln\left(\frac{1+u^m}{1-u^m}\right)/u^m \tag{1.11}$$

and $\psi''(r) < 0$ is the spinodal interval $r \in (-u^s, u^s)$ where $u^s = (1-\theta/\theta_c)^{1/2} < u^m$. For $\theta \geq \theta_c$ there is a unique minimiser of ψ at $u = u^m = 0$. It turns out that because of (1.10) the dynamical system (1.1) drives $\mathbf{u}(t)$ to a steady state $\mathbf{u}^*$ as $t \to \infty$. For $\theta > \theta_c$ there is a unique equilibrium $\mathbf{u}^* = 0$ whereas for $\theta < \theta_c$ there are multiple equilibria.

2. Observe that if

$$E_{ab} = 1 \qquad b \in N(a) \tag{1.12}$$

then $\theta_c = 2d$ and $\mathbf{L}$ is a finite difference approximation of the Laplacian on the lattice Λ with the spatial distance between adjacent points being 1. Thus (1.6) can be regarded as a discretisation of a parabolic equation. Furthermore setting

$$\mathbf{z} := \mathbf{u}_t + \mathbf{u} \tag{1.13}$$

we obtain

$$\theta\phi(\mathbf{z})_t = \mathbf{L}\mathbf{z} - \psi'(\mathbf{z}). \tag{1.14}$$

This equation can be seen to be related to the following spatial discretisation of the Allen–Cahn equation (Elliott and Stuart [3]).

$$\theta\mathbf{z}_t = \mathbf{L}\mathbf{z} - \psi'(\mathbf{z}). \tag{1.15}$$

3. It is convenient to note here that

$$\begin{aligned} |(\mathbf{v}, \mathbf{A}\mathbf{v})| &\leq \sum_{a\in\Lambda}\sum_{b\in\Lambda} E_{ab}|v_a||v_b| \\ &\leq \frac{1}{2}\sum_{a\in\Lambda}\sum_{b\in\Lambda} E_{ab}(v_a^2 + v_b^2) \\ &\leq \theta_c\|\mathbf{v}\|^2. \end{aligned} \tag{1.16}$$

4. Related discrete lattice models for phase transitions which in some cases are discretisations of the Allen-Cahn and Cahn-Hilliard equations have been studied by Cahn, Chow and Van Vleck [1].

2. Equilibrium, Global Existence and Attractor

2.1 Equilibrium. The steady state problem is :

$$\mathbf{u} \in V_\Lambda : \ \mathbf{u} = \tanh(\beta\mathbf{A}\mathbf{u}). \tag{2.1}$$

We have the following proposition concerning $\mathcal{E}$, the set of equilibria

Proposition 2.1.

(1) $\mathbf{0} \in \mathcal{E}$ *and if* $\theta_c < \theta$ *then* $\mathcal{E} = \{\mathbf{0}\}$.
(2) $\|\mathbf{u}\|_\infty \leq u^m \quad \forall\mathbf{u} \in \mathcal{E}$
(3) *Suppose* $\theta_a > \theta \ \ \forall a \in \Lambda$. *It follows that* $\pm u^m\mathbf{e} \in \mathcal{E}$ *if and only if* θ_a *is independent of* a.

Proof. (1) Clearly $\mathbf{0}$ solves (2.1). Furthermore it holds that

$$\| \tanh(\beta \mathbf{A}\mathbf{v}) - \tanh(\beta \mathbf{A}\mathbf{w})\|_\infty \le \beta \|\mathbf{A}(\mathbf{v}-\mathbf{w})\|_\infty \le \frac{\theta_c}{\theta}\|\mathbf{v}-\mathbf{w}\|_\infty. \tag{2.2}$$

Uniqueness follows for $\theta_c/\theta < 1$.

(2) Suppose $|u_b| = \|\mathbf{u}\|_\infty$ for some $b \in \Lambda$. It follows that

$$|u_b| = |\tanh(\beta \sum_{a\in N(b)} E_{ab}u_a)| \le \tanh((\theta_c/\theta)|u_b|). \tag{2.3}$$

Setting $f(x) := x - \tanh(\theta_c x/\theta)$ we have that $f(x)$ is monotone increasing on $(u^s, 1)$ and $f(|u_b|) \le f(u^m) = 0$. Hence we have that

$$\|\mathbf{u}\|_\infty = |u_b| \le u^m. \tag{2.4}$$

(3) If $\theta_a = \theta_c > \theta \quad \forall a \in \Lambda$ then it is obvious that $\pm u^m\mathbf{e}$ solves (2.1). On the other hand if $x_0\mathbf{e}$ solves (2.1) then for each $a \in \Lambda$

$$x_0 = \tanh(\theta_a x_0/\theta). \tag{2.5}$$

Since this equation has only one positive and negative solution for $\theta_a > \theta$ we have θ_a is independent of a. □

Remark:. Clearly if $d = 1$ then associated with each $\mathbf{u} \in \mathcal{E}$ there are N other solutions obtained from $\mathbf{u}$ by translation i.e. $\mathbf{v} = \{v_a\} = \{u_{a+1}\}$ also solves (2.1).

2.2 Global Existence, Absorbing Set and Attractor.

Lemma 2.2. *Let $f \in C^1(0,\infty)$ satisfy*

$$\frac{df}{dt} \le C - Df(t), \ \forall t \ge 0$$

for constants $C \ge 0$ and $D > 0$. Then for all $t \ge 0$

$$f(t) \le \max(f(0), C/D).$$

Proof. Integrating the differential equation yields

$$f(t) \le \exp(-Dt)f(0) + (1 - \exp(-Dt))C/D$$

which gives the required result. □

Theorem 2.3. *For every $\mathbf{u}_0 \in V_\Lambda$ there exists a unique solution $\mathbf{u} \in C^\infty(0,\infty;V_\Lambda)$ of (1.1). The mapping $\mathbf{u}_0 \to \mathbf{u}(t)$ is continuous in V_Λ for each $t > 0$. Hence the family of solution operators $\{S(t)\}_{t\geq 0}$, defined by $S(t)\mathbf{u}_0 \equiv \mathbf{u}(t)$, forms a continuous semi group on V_Λ .*

Proof. Existence theory for ordinary differential equation yields a unique C^∞ local solution $\mathbf{u}(t)$ such that for each $a \in \Lambda$,

$$\frac{1}{2}\frac{d}{dt}|u_a(t)|^2 + |u_a(t)|^2 = u_a(t)\tanh(\beta \sum_{b\in N(a)} E_{ab}u_b(t)) \leq \frac{1}{2} + \frac{1}{2}|u_a(t)|^2.$$

Applying Lemma 2.2 we obtain

$$\|\mathbf{u}(t)\|_\infty \leq \max(1, \|\mathbf{u}_0\|_\infty), \tag{2.6}$$

and, hence, a unique global solution exists.

For two solutions $\mathbf{u}(t)$ and $\mathbf{v}(t)$ of (2.1) we set $\mathbf{e}(t) = \mathbf{u}(t) - \mathbf{v}(t)$ so that

$$\frac{d}{dt}\mathbf{e} + \mathbf{e} = \tanh(\beta\mathbf{A}\mathbf{u}) - \tanh(\beta\mathbf{A}\mathbf{v}).$$

Continuity with respect to the initial data follows by taking the inner product with $\mathbf{e}$ and using Gronwall's inequality since $\tanh(\bullet)$ is globally Lipschitz continuous. $\square$

Lemma 2.4. *For each $\mathbf{u}_0 \in V_\Lambda$ there exists $T_1 = T_1(\mathbf{u}_0) > 0$ such that the solution of (2.3) satisfies*

$$\|\mathbf{u}(t)\|_\infty < 1 \quad \forall t > T_1(\mathbf{u}_0).$$

Proof. Using (2.3) we find that

$$\|\tanh(\beta\mathbf{A}\mathbf{u}(t))\|_\infty \leq \tanh((\theta_c/\theta)\max(1, \|\mathbf{u}_0\|_\infty)) =: c_0(\mathbf{u}_0) < 1.$$

It follows that for each $a \in \Lambda$,

$$\frac{1}{2}\frac{d}{dt}|u_a|^2 + |u_a|^2 = u_a \tanh(\beta \sum_{b\in N(a)} E_{ab}u_b) \leq \frac{c_0^2}{2} + \frac{|u_a|^2}{2}.$$

Therefore for each $a \in \Lambda$,

$$|u_a(t)|^2 \leq \exp(-t)(|u_a(0)|^2 - c_0^2) + c_0^2,$$

which yields the desired result. $\square$

Lemma 2.5. *Let $\theta_c > \theta$ and $\|\mathbf{u}_0\|_\infty \leq u_m$. Then*

$$\|\mathbf{u}(t)\|_\infty \leq u_m \quad \forall t$$

Proof. Suppose $u_a(t) = u_m$ and $\|\mathbf{u}(t)\|_\infty = u_m$. It follows that

$$\frac{du_a(t)}{dt} \leq \tanh\frac{\theta_c}{\theta}u_m - u_m = 0.$$

Similarly if $u_a(t) = -u_m$ and $\|\mathbf{u}(t)\|_\infty = +u_m$ then

$$\frac{du_a(t)}{dt} \geq 0.$$

$\square$

Theorem 2.6. *The set* $B = \{\mathbf{v} \in V_\Lambda : \|\mathbf{v}\|_\infty < 1\}$ *is an absorbing set for* $\{S(t)\}$ *on* V_Λ. *There exists a global attractor* $\mathcal{A} \subset V_\Lambda$ *for* $\{S(t)\}$ *given by* $\omega(B)$ *the* ω *– limit set of* B. *The functional* $I(\bullet)$ *is a Lyapunov functional on* B *for the semi–group* $\{S(t)\}$ *and, in addition,* $\omega(\mathbf{u}_0) \subset \mathcal{E}$ *for each* $\mathbf{u}_0 \in V_\Lambda$.

Proof. These are consequences of the previous lemmas and the elementary theory of dissipative dynamical systems, (cf Hale[5], Ladyzhenskaya [6], Temam [8].) □

3. Bifurcaton Problem for the One Dimensional Lattice

In this section we consider the stationary problem (2.1) on a one dimensional lattice with $E_{ab} = 1$, $b \in N(a)$, which can be rewritten as

$$\beta(u_{k+1} + u_{k-1}) = \phi(u_k), \qquad k = 1, \dots, N, \tag{3.1}$$

where $\mathbf{u} \equiv \{u_k\}_{k=1}^N \in \mathbb{R}^N$, and $u_0 \equiv u_N, u_{N+1} \equiv u_1$. We shall assume that N is an odd number, $N = 1 + 2m$, $m \in \mathbb{N}$. Obviously, $\mathbf{u} = \mathbf{0}$ always solves this equation. Here we study small non-zero solutions of (3.1).

As above, by $\mathbf{A}$ we denote the linear operator $\mathbf{A} : V_\Lambda \to V_\Lambda$,

$$(\mathbf{A}\mathbf{v})_k = (v_{k+1} + v_{k-1}), \qquad \mathbf{v} = \{v_k\}_{k=1}^N.$$

It is easy to see that $\mathbf{A}$ is a self-adjoint operator in V_Λ. It has one one-dimensional eigensubspace $X_0 = \operatorname{span}\{\mathbf{z}_0\}$, $(\mathbf{z}_0)_k = 1$, with corresponding eigenvalue $\lambda_0 = 2$ and m two-dimensional eigensubspaces X_j,

$$X_j = \operatorname{span}\{\mathbf{z}_j^+, \mathbf{z}_j^-\}, \qquad j = 1, \dots, m,$$
$$(\mathbf{z}_j^+)_k = \sin\frac{2\pi jk}{N}, \qquad (\mathbf{z}_j^-)_k = \cos\frac{2\pi jk}{N}, \qquad k = 1, \dots, N,$$

with corresponding eigenvalues $\lambda_j = 2\cos(2\pi j/N)$. For every $r > 0$ denote by U_r the open ball

$$U_r = \{\mathbf{v} \in \mathbb{R}^N : \|\mathbf{v}\| < r\}$$

and define the non-linear operator $\Phi : U_1 \to \mathbb{R}^N$, by

$$(\Phi(\mathbf{v}))_k = \phi(v_k) - v_k.$$

The mapping Φ is smooth and

$$\sup_{\mathbf{v}\in U_r} \|\Phi'(\mathbf{v})\| = \frac{r^2}{1 - r^2}, \qquad \sup_{\mathbf{v}\in U_r} \|\Phi''(\mathbf{v})\| = \frac{2r}{(1 - r^2)^2}. \tag{3.2}$$

Now we can rewrite (3.1) as

$$\mathbf{A}\mathbf{u} = \frac{1}{\beta}(\Phi(\mathbf{u}) + \mathbf{u}). \tag{3.3}$$

We first show that there are no small non-zero solutions of (3.3) when $1/\beta$ is bounded away from the spectrum of $\mathbf{A}$.

Lemma 3.1. *Let for some $\delta > 0$*

$$|1/\beta - \lambda_j| \geq \delta, \qquad \forall j.$$

Then the ball U_r contains no non-zero solutions of (3.3) *provided*

$$r^2 < \left(1 + 1/(\delta\beta)\right)^{-1}.$$

Proof. By the assumptions of the lemma, the operator $\mathbf{A} - 1/\beta$ is invertible. Hence, (3.3) is equivalent to

$$\mathbf{u} = \mathcal{D}(\mathbf{u}) \tag{3.4}$$

where the non-linear operator $\mathcal{D} : U_1 \to \mathbb{R}^N$ is defined by

$$\mathcal{D}(\mathbf{u}) = (\mathbf{A} - 1/\beta)^{-1}\frac{1}{\beta}\Phi(\mathbf{u}).$$

For all $\mathbf{u} \in U_r$ the norm of the operator $\mathcal{D}'(\mathbf{u})$ does not exceed $\delta^{-1}\beta^{-1}r^2(1-r^2)^{-1} < 1$. Along with the obvious equality $\mathcal{D}(0) = 0$ this proves that $\mathcal{D}$ maps U_r into itself and contracts it. Therefore, (3.4) has a unique solution in U_r, which has to be zero. □

We now fix some positive eigenvalue λ_i of the operator $\mathbf{A}$ and consider the situation when $1/\beta$ is close to λ_i. Thus we introduce the new parameter

$$\gamma = \beta\lambda_i - 1$$

and assume it to be small. Equation (3.3) now reads

$$(\mathbf{A} - \lambda_i)\mathbf{u} = \frac{\lambda_i}{\gamma + 1}(\Phi(\mathbf{u}) - \gamma\mathbf{u}). \tag{3.5}$$

For the rest of this section P and Q denote the orthogonal projections onto X_i and $\mathbb{R}^N \ominus X_i$, respectively. The neighbourhood of zero under consideration is now

$$\mathcal{O}_r = \{\mathbf{v} \in \mathbb{R}^N : \|P\mathbf{v}\| \leq r, \|Q\mathbf{v}\| \leq r\}.$$

Lemma 3.2. *Denote*

$$\mu \equiv \mu(\gamma, r) = \frac{\lambda_i}{\gamma + 1} \cdot \sin^{-2}\frac{\pi}{N} \cdot \left(\frac{r^2}{1 - r^2} + |\gamma|\right) \tag{3.6}$$

and assume $|\gamma|$ and r to be sufficiently small for μ to satisfy $\mu(\gamma, r) \leq 1/\sqrt{2}$. Then equation (3.5) *is equivalent in $\mathcal{O}_r$ to*

$$\begin{aligned} \mathbf{u} &= p + S_\gamma(p), \\ \gamma p &= P\Phi(p + S_\gamma(p)), \quad p \in P\mathcal{O}_r, \end{aligned} \tag{3.7}$$

for some smooth map $S_\gamma : PO_r \to QO_r$ *satisfying* $S_\gamma(0) = 0$ *and* $\|S'_\gamma(p)\| \le \mu(1-\mu^2)^{-1/2}$.

Proof. Multiply (3.5) by P and by Q, and denote $p = P\mathbf{u}$, $q = Q\mathbf{u}$. This results in the system

$$\mathbf{u} = p + q, \tag{3.8}$$

$$\gamma p = P\Phi(p+q), \tag{3.9}$$

$$(\mathbf{A} - \lambda_i)q = \frac{\lambda_i}{\gamma+1}(Q\Phi(p+q) - \gamma q) \tag{3.10}$$

We first prove solvability of (3.10) for a given p, using the same trick as in the proof of Lemma 3.1. The operator $(\mathbf{A} - \lambda_i)$ is invertible on $Q\mathbb{R}^N$ and

$$\|(\mathbf{A}|_{Q\mathbb{R}^N} - \lambda_i)^{-1}\| \le \sin^{-2}\frac{\pi}{N}, \qquad \forall i. \tag{3.11}$$

Therefore (3.10) is equivalent to

$$q = \mathcal{D}_\gamma(p+q) \tag{3.12}$$

with $\mathcal{D}_\gamma : \mathcal{O}_r \to Q\mathbb{R}^N$ defined by

$$\mathcal{D}_\gamma(\mathbf{v}) = \frac{\lambda_i}{\gamma+1}(\mathbf{A} - \lambda_i)^{-1}(Q\Phi(\mathbf{v}) - \gamma Q\mathbf{v}).$$

Making use of (3.2) and (3.11), one easily shows that

$$\|\mathcal{D}'_\gamma(\mathbf{v})\| \le \mu \le 1/\sqrt{2}, \qquad v \in \mathcal{O}_r, \tag{3.13}$$

Along with the fact that $\mathcal{D}_\gamma$ takes zero to zero this implies that for every $p \in P\mathcal{O}_r$, the operator $\mathcal{D}_\gamma(p + \cdot\,)$ maps the ball $Q\mathcal{O}_r$ into itself and contracts it. Hence, for every $p \in P\mathcal{O}_r$ equation (3.12) possesses a unique solution $q \in Q\mathcal{O}_r$, which we denote by $S_\gamma(p)$. The mapping S_γ is smooth and $S_\gamma(0) = 0$. Substituting q by $S_\gamma(p)$ in (3.8) and (3.9), one obtains the desired system. To complete the proof it remains to estimate the norm of the derivative of S_γ. Differentiating (3.12) in p yields

$$S'_\gamma(p) = \mathcal{D}'_\gamma(p + S_\gamma(p))(P + S'_\gamma(p)).$$

Take the norm of both sides of this equality and apply (3.13). Taking into account that the ranges of P and $S'_\gamma(p)$ are orthogonal, one obtains

$$\|S'_\gamma(p)\|^2 \le \mu^2(1 + \|S'_\gamma(p)\|^2),$$

which is equivalent to the claimed estimate. □

Solvability of equation (3.7) is easy to study for $i = 0$, since in this case p is a one-dimensional variable. For $i \ge 1$ equation (3.7) is two-dimensional but nevertheless can be reduced to a one-dimensional one in some sense.

Lemma 3.3. *The subspace PV_Λ contains at least N different one-dimensional subspaces, which we denote $Y_1, \dots, Y_N$, such that $P\Phi(\,\cdot + S_\gamma(\,\cdot\,))$ maps $\mathcal{O}_r \cap Y_n$ to Y_n for each n.*

Proof. Let $\bar{Y}_n \in V_\Lambda$ be the subspace of all vectors $\mathbf{v} = \{v_k\}_{k=1}^N$ whose periodic extensions $\{v_k\}_{k=-\infty}^{+\infty}$ onto $\mathbb{Z}$ are even with respect to n, i.e.,

$$v_{n+k} = v_{n-k}, \qquad k \in \mathbb{Z}.$$

As is easy to see, $\mathbf{A}$, P, Q, and Φ map $\mathcal{O}_r \cap \bar{Y}_n$ into $\bar{Y}_n$. Hence, so do $\mathcal{D}_\gamma$, S_γ, and $\Phi(\,\cdot + S_\gamma(\,\cdot\,))$. Denote $Y_n = PV_\Lambda \cap \bar{Y}_n$. The operator $P\Phi(\,\cdot + S_\gamma(\,\cdot\,))$ then maps $\mathcal{O}_r \cap Y_n$ into Y_n. The explicit discription

$$Y_n = \operatorname{span}\{\zeta_n\}, \qquad (\zeta_n)_k = \cos\frac{2\pi i(k-n)}{N}$$

shows, in particular, that all the subspaces Y_n are different. □

We proceed now with the study of solvability of (3.7). By Z we denote X_0 when $i = 0$ and one of the subspaces Y_n, $n = 1, \dots, N$, when $i \geq 1$.

Lemma 3.4. *There exist $r > 0$ and $\gamma_0 > 0$ such that equation* (3.7) *has no nonzero solutions in $Z \cap \mathcal{O}_r$ when $-\gamma_0 \leq \gamma \leq 0$ and exactly two nonzero solutions in $Z \cap \mathcal{O}_r$ when $0 < \gamma < \gamma_0$. These solutions tend to zero as γ tends to zero.*

Proof. Any solution $p \in Z$ must be of the form $p = \alpha\omega$, α being a real number and ω being the unit vector in Z. Then (3.7) is equivalent to

$$\gamma\alpha = \big(\omega,\, \Phi(\alpha\omega + S_\gamma(\alpha\omega))\big), \quad \alpha < r.$$

Define functions $f_\gamma, g : (-r, r) \to \mathbb{R}$ by

$$f_\gamma(\alpha) = \big(\omega,\, \Phi(\alpha\omega + S_\gamma(\alpha\omega)) - \Phi(\alpha\omega)\big),$$
$$g(\alpha) = \big(\omega,\, \Phi(\alpha\omega)\big),$$

to rewrite the above equation as

$$G(\alpha) \equiv g(\alpha) - \gamma\alpha + f_\gamma(\alpha) = 0. \tag{3.14}$$

To solve this equation we first need to estimate $f'_\gamma(\alpha)$. Differentiation of the definition of $f'_\gamma(\alpha)$ results in

$$f'_\gamma(\alpha) = \big(\omega,\, [\Phi'(\alpha\omega + S_\gamma(\alpha\omega)) - \Phi'(\alpha\omega)]\,\omega\big) + \big(\omega,\, \Phi'(\alpha\omega + S_\gamma(\alpha\omega))\, S'_\gamma(\alpha\omega)\,\omega\big).$$

Together with (3.2) and the estimate for the norm of operator S'_γ (see Lemma 3.2) this implies

$$|f'(\alpha)| < 16\mu\alpha^2 \tag{3.15}$$

(here we also used the inequalities $|\alpha\omega + S_\gamma(\alpha\omega)| \leq \sqrt{2}\alpha$ and $\alpha \leq r < 1/2$). With the help of the explicit formula

$$g(\alpha) = \sum_k \omega_k(\phi(\alpha\omega_k) - \alpha\omega_k)$$

one easily shows that

$$g(0) = g'(0) = g''(0) = 0, \qquad g'''(0) > 0, \tag{3.16}$$

and hence $g(\alpha) = C\alpha^3 + O(\alpha^4)$ for some $C > 0$. Choose $r > 0$ and $\gamma_1 > 0$ sufficiently small for inequalities

$$\begin{aligned} |f'_\gamma(\alpha)| &\leq g'(\alpha)/4, \qquad \forall\alpha \in [0, r],\ \forall\gamma \in [-\gamma_1, \gamma_1] \\ 2g(\alpha) &\leq \alpha g'(\alpha), \qquad \forall\alpha \in [0, r], \end{aligned} \tag{3.17}$$

to hold (by (3.15) and (3.16) this is always possible). This in particular implies $|f_\gamma(\alpha)| \leq g(\alpha)/4$. Now fix $\gamma_0 \in (0, \gamma_1]$ such that

$$\frac{3}{4}g(r) - \gamma_0 r > 0 \tag{3.18}$$

If $\gamma \leq 0$, then $G' > 0$ on $(0, r)$. This means that $G > 0$ on $(0, r)$ since $G(0) = 0$. So there are no positive solutions of (3.14). By the symmetry of the original problem (3.1), there are no negative solutions, either. Assume now $\gamma > 0$. Then $G'(0) < 0$, which implies $G(\alpha) < 0$ for all α sufficiently small. But by (3.18), $G(r) > 0$, so equation (3.14) has at least one solution α^* in $(0, r)$. One can easily see that $\alpha_1 \leq \alpha^* \leq \alpha_2$, where $\alpha_l = \alpha_l(\gamma)$ solve the equations

$$\begin{aligned} \frac{5}{4}g(\alpha_1) - \gamma\alpha_1 = 0, \\ \frac{3}{4}g(\alpha_2) - \gamma\alpha_2 = 0. \end{aligned}$$

The estimate

$$G'(\alpha) = g'(\alpha) - \gamma + f'_\gamma(\alpha) \geq \frac{3}{4}g'(\alpha) - \gamma$$

together with (3.17) shows that G' is positive on $[\alpha_1, \alpha_2]$, and therefore α^* is the unique solution on this interval. It tends to zero as $\gamma \to 0$ because so does α_2. By the symmetry of problem (3.1) $-\alpha^*$ is also a solution of (14). □

We summarize the obtained results in the following

Theorem 3.5. *There exists a neighbourhood $\mathcal{O} \subset \mathbb{R}^N$ of zero and $\delta > 0$ such that*

1. *$\mathcal{O}$ contains no non-zero solutions of* (3.1) *if $|1/\beta - \lambda_j| \geq \delta$, $\forall j$;*
2. *$\mathcal{O}$ contains no non-zero solutions of* (3.1) *if $|1/\beta - \lambda_j| < \delta$ for some j and $1/\beta \geq \lambda_j$;*
3. *$\mathcal{O}$ contains exactly two non-zero solutions of* (3.1) *if $|1/\beta - \lambda_0| < \delta$ and $1/\beta < \lambda_0$; these solutions tend to zero as $1/\beta \to \lambda_0$;*
4. *$\mathcal{O}$ contains at least $2N$ non-zero solutions of* (3.1) *if $|1/\beta - \lambda_j| < \delta$ for some $j \neq 0$ and $1/\beta < \lambda_j$; these solutions tend to zero as $1/\beta \to \lambda_j$.* □

Remark. For all $\beta > 1/2$ equation (3.1) has two constant solutions $\mathbf{u}^+$ and $\mathbf{u}^-$, $u_k^{\pm} = \pm u$, u being the positive solution of the scalar equation

$$2\beta u = \phi(u). \tag{3.19}$$

For small $|\beta - 1/2|$ they must coincide with ones whose existence is guaranteed by statement (3) of Theorem 3.5. □

In the rest of the section we study stability of the constructed stationary solutions. The linearization of equation (1.1a) near a stationary solution $\mathbf{u}$ reads

$$\mathbf{v}_t = -\mathbf{v} + (1 - \tanh^2(\beta \mathbf{A}\mathbf{u}))\, \beta \mathbf{A}\mathbf{v},$$

or

$$\mathbf{v}_t = -\mathbf{v} + (1 - \mathbf{u}^2)\, \beta \mathbf{A}\mathbf{v}. \tag{3.20}$$

The scalar product of (3.20) with $\mathbf{v}$ yields

$$\frac{1}{2}\frac{d}{dt}\|\mathbf{v}\|^2 = -b_{\mathbf{u}}[\mathbf{v}, \mathbf{v}],$$

where the quadratic form $b_{\mathbf{u}}[\mathbf{v}, \mathbf{v}]$ is defined by

$$b_{\mathbf{u}}[\mathbf{v}, \mathbf{v}] = \sum_k \left(v_k^2 - (1 - u_k^2)\, \beta (\mathbf{A}\mathbf{v})_k v_k\right).$$

The stationary solution $\mathbf{u}$ is stable if the form $b_{\mathbf{u}}$ is positive definite. On the other hand, if $b[\mathbf{v}, \mathbf{v}] < 0$ for some $\mathbf{v}$, then $\mathbf{u}$ is unstable.

Lemma 3.6. *The zero stationary solution of* (1.1a) *is stable for* $\beta < 1/2$ *and unstable for* $\beta > 1/2$.

Proof. The first claim of the lemma is obvious if one takes into account that $\|\mathbf{A}\| = 2$. If $\beta > 1/2$, then $b_{\mathbf{u}}[\mathbf{v}, \mathbf{v}] < 0$ for $(\mathbf{v})_k = 1$, $\forall k$. This proves the second claim. □

Lemma 3.7. *Non-zero constant stationary solutions of* (1.1) *are always stable.*

Proof. Let $u_k = u$, where $u \in (-1, 1) \setminus \{0\}$ is a solution of (3.19). Taking into account that $\|\mathbf{A}\| = 2$, we have

$$b_{\mathbf{u}}[\mathbf{v}, \mathbf{v}] = \sum_k \left(v_k^2 - (1 - u^2)\, \beta (\mathbf{A}\mathbf{v})_k v_k\right) \geq \|\mathbf{v}\|^2 \left(1 - 2\beta(1 - u^2)\right).$$

Together with (3.19) this yields

$$b_{\mathbf{u}}[\mathbf{v}, \mathbf{v}] \geq \|\mathbf{v}\|^2 \left(1 - \frac{\phi(u)}{u}(1 - u^2)\right) = \|\mathbf{v}\|^2 \left(1 - \frac{\phi(u)}{u\phi'(u)}\right)$$

Finally notice that $\phi(u) < u\phi'(u)$ for $u \neq 0$, and so the form $b_{\mathbf{u}}[\mathbf{v}, \mathbf{v}]$ is positive definite. □

Lemma 3.8. *Small non-constant stationary solutions of* (1.1) *are unstable.*

Proof. First notice that by Theorem 3.5 for small non-constant solutions to exist the parameter β must satisfy the inequality

$$1/\beta < \lambda_1 = 2\cos(2\pi/N).$$

We estimate the form $b_{\mathbf{u}}[\mathbf{v}, \mathbf{v}]$ for $v_k = 1$. Obviously, $(\mathbf{A}\mathbf{v})_k = 2$, and so

$$b_{\mathbf{u}}[\mathbf{v}, \mathbf{v}] = \sum_k \left(1 - 2\beta(1 - u_k^2)\right) < \sum_k \left(1 - 2/\lambda_1(1 - u_k^2)\right) = N(1 - 2/\lambda_1) + 2/\lambda_1\, \|\mathbf{u}\|^2.$$

Therefore, $b_{\mathbf{u}}[\mathbf{v}, \mathbf{v}] < 0$ provided $\|\mathbf{u}\|^2 < N(1 - \lambda_1/2)$. □

4. The Explicit Euler Scheme

In this section we consider the explicit Euler scheme and discuss its long time behaviour. The scheme is given by

(4.1a) $$\partial \mathbf{u}^n + \mathbf{u}^n = \tanh(\beta \mathbf{A} \mathbf{u}^n), \ n = 1, 2, \ldots$$

(4.1b) $$\mathbf{u}^0 = \mathbf{u}_0$$

where $\mathbf{u}^n \in V_\Lambda \ \ \forall n \geq 0$ and

$$\partial f^n = (f^{n+1} - f^n)/\Delta t$$

and Δt is the time step. It is convenient to note the following lemma.

Lemma 4.1. *Suppose $f^n \in I\!R$, $n \geq 0$ and*

(4.2a) $$\partial f^n \leq C - D f^{n+1}, \ f^n \geq 0$$

where C and D are positive constants. Then

(4.2b) $$f^{n+1} \leq C/D + (f^0 - C/D)(1 + D\Delta t)^{-(n+1)}, \ n \geq 0$$

Lemma 4.2. *Let $\{\mathbf{u}^n\}_{n \geq 0}$ solve (4.1). Then for any ϵ_1, $\epsilon_2 (> 0)$, $0 < \delta_0 < 1$ and*

$$\Delta t < \delta_1 = \min\{\delta_0, 1 - \epsilon_2 - c_0^2\}$$

there exists $n_0(\|\mathbf{u}_0\|_\infty, \Delta t)$ such that

$$\|\mathbf{u}^n\|_\infty^2 < K^2 := (c_0^2 + \epsilon_2)/(1 - \Delta t) < 1, \ n > n_0$$

where $c_0 = \tanh(\theta_c q/\theta)$ and $q^2 = (1 + \epsilon_1)/(1 - \delta_0)$.

Proof. Clearly

$$\frac{1}{2}\partial |u_a^n|^2 + \frac{\Delta t}{2}|\partial u_a^n|^2 + |u_a^{n+1}|^2 = \Delta t u_a^{n+1} \partial u_a^n + u_a^{n+1} \tanh(\beta \mathbf{A} \mathbf{u}^n)_a$$

which implies

$$\partial |u_a^n|^2 \leq |\tanh(\theta_c \|\mathbf{u}^{n+1}\|_\infty/\theta)|^2 - (1 - \Delta t)|u_a^{n+1}|^2.$$

First we note by Lemma 4.1 that

$$\|\mathbf{u}^n\|_\infty \leq \max\{\|\mathbf{u}_0\|_\infty, (1 - \delta_0)^{-1/2}\},$$

and that for any ϵ_1 there exists $n_1 = n_1(\mathbf{u}_0, \delta t)$ such that

$$\|\mathbf{u}^n\|_\infty^2 \leq (1 + \epsilon_1)/(1 - \delta_0), \qquad n \geq n_1.$$

Furthermore we have

$$\partial |u_a^n|^2 \leq c_0^2 - (1 - \Delta t)|u_a^{n+1}|^2, \qquad n \geq n_1.$$

and again applying Lemma 4.1 it follows that there exists $n_0 \geq n_1$ such that for $n \geq n_0$

$$\|\mathbf{u}^n\|_\infty^2 < (c_0^2 + \epsilon_2)/(1 - \Delta t).$$

Choosing $\Delta t < \delta_1$, yields the result . □

Remark. Since $q > 1$ it follows that $K^2 = (c_0^2 + \epsilon_2)/(1 - \Delta t) > (u^m)^2$. So Lemma 4.2 does not contradict the expectation that $\mathbf{u}^n$ converges as $n \to \infty$ to an element of $\mathcal{E}$ which is bounded by u^m in the maximum norm.

Lemma 4.3. *Let $\delta \in (0,1)$, and $\Delta t \leq 2(1-\delta)\theta/(\theta_c + \phi'(K)\theta)$. Then provided $\|\mathbf{u}^n\| < K$ for all n it holds that*

$$\delta\theta\Delta t\|\partial\mathbf{u}^n\|^2 + I(\mathbf{u}^{n+1}) \leq I(\mathbf{u}^n), \qquad \forall n \geq 0. \tag{4.4}$$

Proof. We rewrite the explicit Euler scheme (4.1a) as

$$\theta\phi(\partial\mathbf{u}^n + \mathbf{u}^n) = A\mathbf{u}^n.$$

It follows that

$$\theta(\phi(\partial\mathbf{u}^n + \mathbf{u}^n) - \phi(\mathbf{u}^n)) = A\mathbf{u}^n - \theta\phi(\mathbf{u}^n). \tag{4.5}$$

Observe that since $\phi'(r) \geq 1$ for $r \in (-1,1)$, we have

$$\theta(\phi(\partial\mathbf{u}^n + \mathbf{u}^n) - \phi(\mathbf{u}^n), \partial\mathbf{u}^n) \geq \theta\|\partial\mathbf{u}^n\|^2 \tag{4.6}$$

and

$$\begin{aligned} \theta\phi(r)(r-s) &= \psi_0(r) - \psi_0(s) + \frac{\theta}{2}\phi'(\xi)(r-s)^2 \\ &\leq \psi_0(r) - \psi_0(s) + |\phi'(K)|\frac{\theta}{2}(r-s)^2. \end{aligned} \tag{4.7}$$

Thus taking the inner product of (4.5) with $\Delta t\partial\mathbf{u}^n$ yields

$$\begin{aligned} \theta\Delta t\|\partial\mathbf{u}^n\|^2 \leq &-\frac{1}{2}(\mathbf{A}\mathbf{u}^n, \mathbf{u}^n) + \frac{1}{2}(\mathbf{A}\mathbf{u}^{n+1}, \mathbf{u}^{n+1}) - \frac{1}{2}\Delta t^2(\mathbf{A}\partial\mathbf{u}^n, \partial\mathbf{u}^n) \\ &+ (\psi_0(\mathbf{u}^n) - \psi_0(\mathbf{u}^{n+1}), \mathbf{e}) + \frac{\theta\Delta t^2}{2}\phi'(K)\|\partial\mathbf{u}^n\|^2. \end{aligned}$$

By (1.16) and the definition of $I(\bullet)$ we have

$$\Delta t\theta\|\partial\mathbf{u}^n\|^2 + I(\mathbf{u}^{n+1}) \leq I(\mathbf{u}^n) + \frac{(\theta_c + \phi'(K)\theta)}{2}\Delta t^2\|\partial\mathbf{u}^n\|^2$$

which yields the required result. $\square$

For any $\mathbf{u}_0 \in V_\Lambda$ we can solve (4.1) for all n. Furthermore it is easy to see that the map $\mathbf{u}_0 \to \mathbf{u}^n$ is continuous for each n. Therefore the family of solution operators $\{S^n\}_{n\geq 0}$ defined by $S^n\mathbf{u}_0 \equiv \mathbf{u}^n$ forms a continuous semi-group on V_Λ. The following theorem is now an immediate consequence of Lemmas 4.2 and 4.3 and the elementary theory of discrete dissipative dynamical systems. (cf Elliott and Stuart [3], Hale [5], Ladyzhenskaya [6]).

Remark:(1). The time step restrictions of Lemmas 4.2 and 4.3 are independent of the lattice size. This is in contrast to explicit Euler discretisations of the Allen-Cahn equation.

Remark:(2). The scheme

$$\partial \mathbf{u}^n + \mathbf{u}^{n=1} = \tanh(\beta \mathbf{u}^n), \ n = 0, 1, ... \tag{4.8a}$$

$$\mathbf{u}^0 = \mathbf{u}_0 \tag{4.8b}$$

is just as simple to implement as the explicit Euler method (4.1). Applying the argument of lemma 4.3 we have that

$$\theta(\phi(\partial \mathbf{u}^n + \mathbf{u}^n) - \phi(\mathbf{u}^{n+1}), \partial \mathbf{u}^n) \geq \theta \|\partial \mathbf{u}^n\|^2$$

and

$$\theta\phi(r)(r-s) \geq \psi_0(r) - \psi_0(s) + \frac{\theta}{2}(r-s)^2.$$

Rewriting (4.8a) in the form

$$\theta(\phi(\partial \mathbf{u}^n + \mathbf{u}^n) - \phi(\mathbf{u}^{n+1})) = A\mathbf{u}^n - \theta\phi(\mathbf{u}^{n+1})$$

and taking the inner product with $\Delta t \partial \mathbf{u}$ yields

$$\theta \Delta t \|\partial \mathbf{u}^n\|^2 \leq -\frac{1}{2}(\mathbf{A}\mathbf{u}^n, \mathbf{u}^n) + \frac{1}{2}(\mathbf{A}\mathbf{u}^{n+1}, \mathbf{u}^{n+1}) - \frac{1}{2}\Delta t^2 (\mathbf{A}\partial \mathbf{u}^n, \partial \mathbf{u}^n)$$
$$+ (\psi_0(\mathbf{u}^n) - \psi_0(\mathbf{u}^{n+1}), \mathbf{e}) - \frac{\theta \Delta t^2}{2}\|\partial \mathbf{u}^n\|^2.$$

Therefore

$$\Delta t \theta \|\partial \mathbf{u}^n\|^2 + I(\mathbf{u}^{n+1}) \leq I(\mathbf{u}^n) + \frac{(\theta_c - \theta)}{2}\Delta t^2 \|\partial \mathbf{u}^n\|^2,$$

if $\theta \geq \theta_c$ then Δt is unrestricted while if $\theta < \theta_c$ and $\delta \in (0, 1)$ then (4.4) holds if $\Delta t \leq 2(1-\delta)\theta/(\theta_c - \theta)$.

Theorem 4.4. *Let Δt be sufficiently small. The ball $B = \{\mathbf{v} \in V_\Lambda : \|\mathbf{v}\|_\infty < K\}$ is an absorbing set for $\{S^n\}$. There exists a global attractor $\mathcal{A}_{\Delta t} \subset V_\Lambda$ for $\{S^n\}$ given by $\omega(B)$ the $\omega-$limit set of B. The functional $I(\bullet)$ is a Lyapunov function on B and $\omega(\mathbf{u}_0) \subset \mathcal{E}$ for each $\mathbf{u}_0 \in V_\Lambda$.*

5. Numerical Computations

In this section we present the results of numerical simulations of the mean-field equations using the explicit Euler scheme (4.1) in one and two dimensions. In all computations E_{ab} is taken to be 1. It follows that $\theta_c = 2d$. Because of the instability of the non constant equilibrium solutions one generally expects as $t \to +\infty$ the trajectories should converge to $u^m\mathbf{e}$ or $-u^m\mathbf{e}$.

5.1 1D Computations. To obtain a non zero equilibrium solution $\beta > 0.5$.

Experiment 1. We take $\beta = 0.51$, $\Delta t = 0.8$ and use 200 lattice points with the initial data

$$u_j(0) = 0.25 \sin(6\pi j/200) + \delta_j \qquad j = 0, 2, \ldots, 199$$

where δ_j is random noise which is less that 0.0125 in magnitude. The evolution of the solution is shown in Figure 1. A pattern of phase domains and transition layers shown in Figure 1a rapidly develops. The solution at this point appears to converge to a steady state. However, the profile is undergoing very slow evolution and at $t \approx 20000$ the positive phase domain to the right of the lattice collapses. The remaining positive domains remain essentially unchanged during this. This is shown in Figure 1b. This pattern of very slow evolution and the relatively quick collapse of domains repeats itself as is shown in Figures 1c,d with the negative equilibrium state finally being attained at around $t = 80000$. This behaviour is most clearly shown in the plot of the discrete Lyapunov functional (1.8) against time shown in Figure 2. After the initial drop in $I(u)$ as the positive and negative phase domains form the energy remains fairly constant for long periods of time, dropping to a lower level at the times where the various positive domains shown in Figure 1 collapse.

Experiment 2. We now take initial data as a random perturbation about the state $\mathbf{u} = 0$ with values distributed uniformly between ± 0.05. We again use 200 lattice points along with $\beta = 0.54$ and $\Delta t = 0.75$. The results are illustrated in Figures 3 and 4. The results show similar behaviour to those outlined in experiment 1. The solution rapidly evolving into the pattern of positive and negative domains which are shown in Figure 3a. The solution then undergoes long periods of very slow evolution interspersed with relatively rapid change as the various positive domains collapse, as shown in Figures 3b,c,d. The solution finally attains its negative equilibrium value at $t \approx 69000$. The Lyapunov energy for this problem shown in Figure 4 again illustrates the cascading effect as the transition layers are destroyed.

Using the Euler method for numerical computations in one dimension it is found, in general, that if $\beta > 0.6$ then $\max_j |u_j^{n+1} - u_j^n| < 10^{-17}$. Since $\|\mathbf{u}\|_\infty$ is of O(1) then these calculations cannot be performed, even using double precision arithmetic. This very slow motion is of the form discovered by Carr and Pego [2] for the Allen-Cahn equation.

5.2 2D Computations. To obtain a non zero equilibrium solution $\beta > 0.25$. All two dimensional experiments are performed on a 100×100 lattice.

Experiment 3. The initial data is taken to be a circle of radius 40.5 located at the centre of the lattice. Points inside and outside of the lattice are set to be $+0.5$ and -0.5 respectively. Taking $\beta = 0.3$ and $\Delta t = 0.5$ Figure 5 shows the solution at particular times, a white dot denoting a lattice point with a value > 0. It can clearly be seen that the radial symmetry of the problem is maintained as the positive domain shrinks. The negative equilibrium state ($\mathbf{u} = -0.6586$) is eventually attained at $t \approx 3067$. In Figure 6 are shown the Lyapunov energy, the radius of the circle and the radius squared all plotted against time. It is clear that

r^2 against time is linear which indicates that the motion of the interface between positive and negative domains is governed by mean curvature flow.

Experiment 4. Here $\mathbf{u}_0$ is taken to be

$$u(i,j) = 0.1(\sin(6\pi i/100) + \sin(6\pi j/100)), \qquad i,j = 0,1,\ldots,99,$$

giving the pattern shown in Figure 6a. We take β and Δt to be 0.8 and 0.5 respectively. The domains where $\mathbf{u}$ are positive and negative quickly move to ± 0.9966 and breakdown of the regular pattern of the initial condition can be seen to be occurring by $t = 100$. By $t = 500$ one large domain where $\mathbf{u} > 0$ has developed. which shrinks to a circle and then undergoes motion similar to that in experiment 3. This can clearly be seen in the Lyapunov energy in Figure 6, where for $t > 700$ the time dependence of the energy is similar to that obtained in Experiment 3.

Experiment 5. Finally we take a random perturbation of zero for the initial condition with values distributed uniformly between ± 0.05. In this experiment $\beta = 0.3$ and $\Delta t = 0.75$. The evolution of the solution is shown in Figure 9. Distinct phase domains quickly form. These shrink and grow depending upon the curvature of the boundary between the domains. In this case the positive equilibrium attracts the solution, with the domains where $\mathbf{u}$ is negative becoming almost circular before they finally collapse. The equilibrium state in this case is finally attained at around $t = 1125$. The Lyapunov energy shown in Figure 10 reflects this behaviour with a gradual reduction in the energy as the interfaces between positive and negative solutions move and shrink.

References

1. J.W. Cahn , S. Chow and E.S. Van Vleck, *Spatially discrete nonlinear diffusion equations*, Private communication.
2. J. Carr and R.L. Pego, *Metastable patterns in solutions of* $u_t = \epsilon^2 u_x x - f(u)$, Comm. Pure Appl. Math. **42** (1989), 523-576.
3. C. M. Elliott and A. M. Stuart, *The global dynamics of discrete semilinear parabolic equations*, SIAM J. Num. Anal. **30** (1993), 1622–1663.
4. R. J. Glauber, *Time-dependent statistics of the Ising model*, J. Math. Phys. **45** (1963), 294–307.
5. J. Hale, *Asymptotic behaviour of dissipative systems*, AMS Mathematical Surveys and Monographs **25** (1988).
6. O. A. Ladyzhenskaya, *On determination of minimal global attractors for the Navier-Stokes equations and other partial differential equations*, Uspekhi Mat. Nauk **42:6** (1987), 25-60; English translation in, Russian Math. Surveys **42:6** (1987), 27-73.
7. O. Penrose, *A mean-field equation of motion for the dynamic Ising model*, J. Stat. Phys. **63** (1991), 975–986.
8. R. Temam, *Infinite dimensional dynamical systems in mechanics and physics*, Springer-Verlag, New York, 1988.

Centre for Mathematical Analysis and its Applications, University of Sussex Falmer, Brighton, BN1 9QH, UK.
Email address:C.M. Elliott: C.M.Elliott@ uk.ac.susx
Email address:A.R. Gardiner: A.R.Gardiner@ uk.ac.susx

Current address: (Bainian Lu) Department of Mathematics, Shaanxi Normal University, Xi'an, Shaanxi, 710062, People's Republic of China

Current address: (I. Kostin) Steklov Mathematical Institute, St.Petersburg Branch, 27 Fontanka, St. Petersburg, 191011, Russia

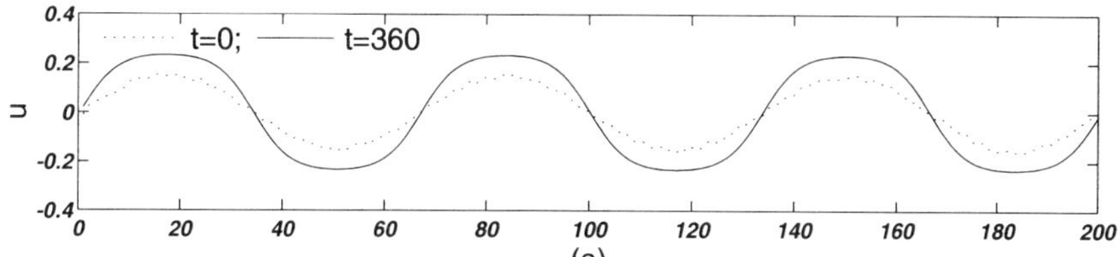

(a)

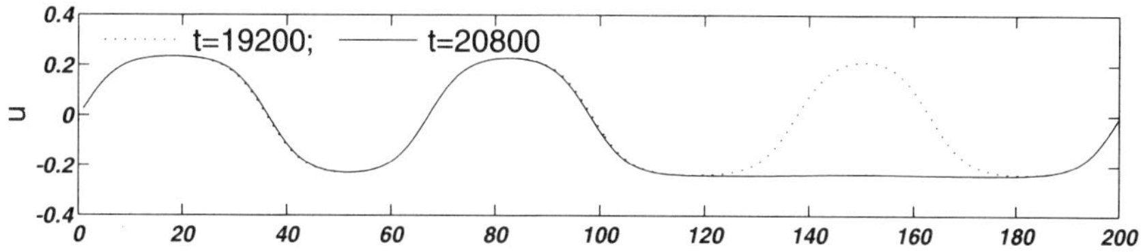

(b)

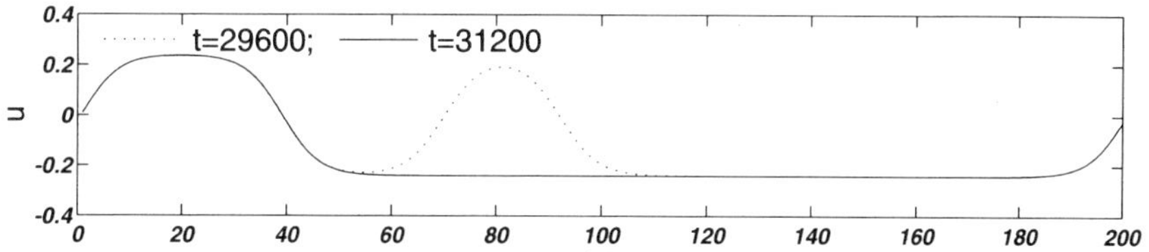

(c)

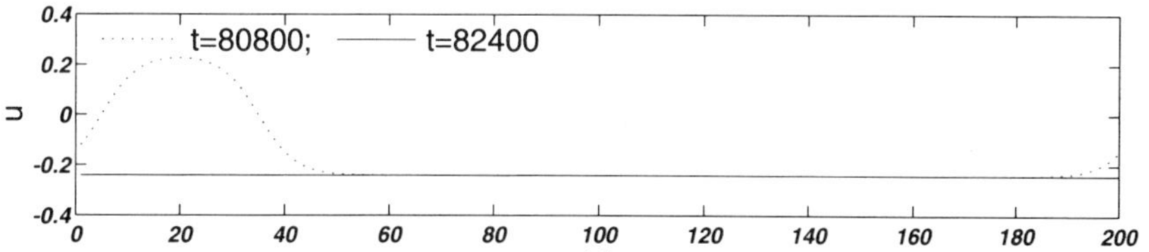

(d)

Figure 1

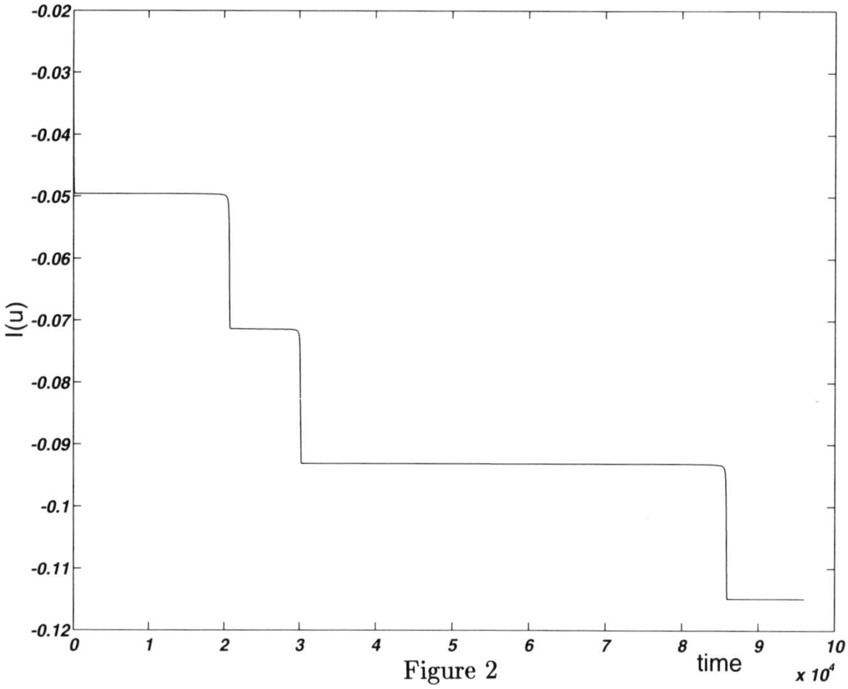
-0.02
-0.03
-0.04
-0.05
-0.06
-0.07
-0.08
-0.09
-0.1
-0.11
-0.12
I(u)
0
1
2
3
4
5
6
7
8
9
10
time
x 10^4

Figure 2

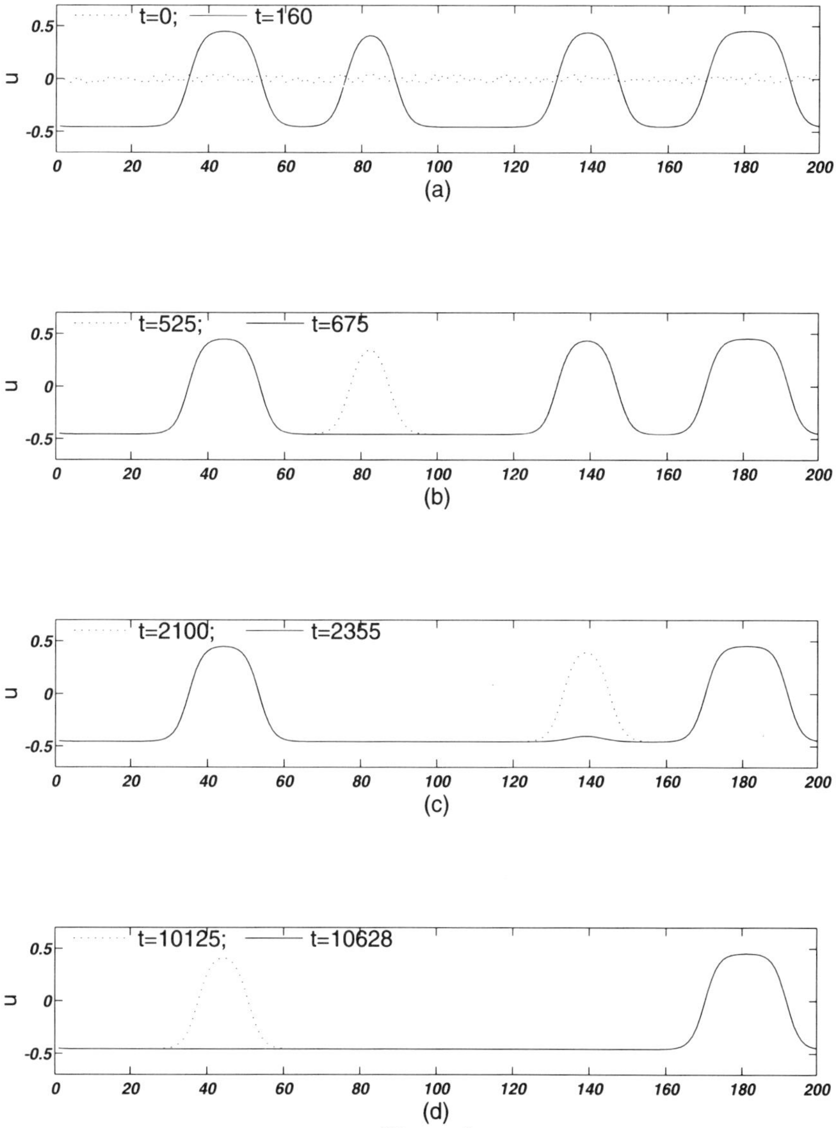
t=0; t=160
u
0.5
0
-0.5
0 20 40 60 80 100 120 140 160 180 200
(a)
t=525; t=675
u
0.5
0
-0.5
0 20 40 60 80 100 120 140 160 180 200
(b)
t=2100; t=2355
u
0.5
0
-0.5
0 20 40 60 80 100 120 140 160 180 200
(c)
t=10125; t=10628
u
0.5
0
-0.5
0 20 40 60 80 100 120 140 160 180 200
(d)

Figure 3

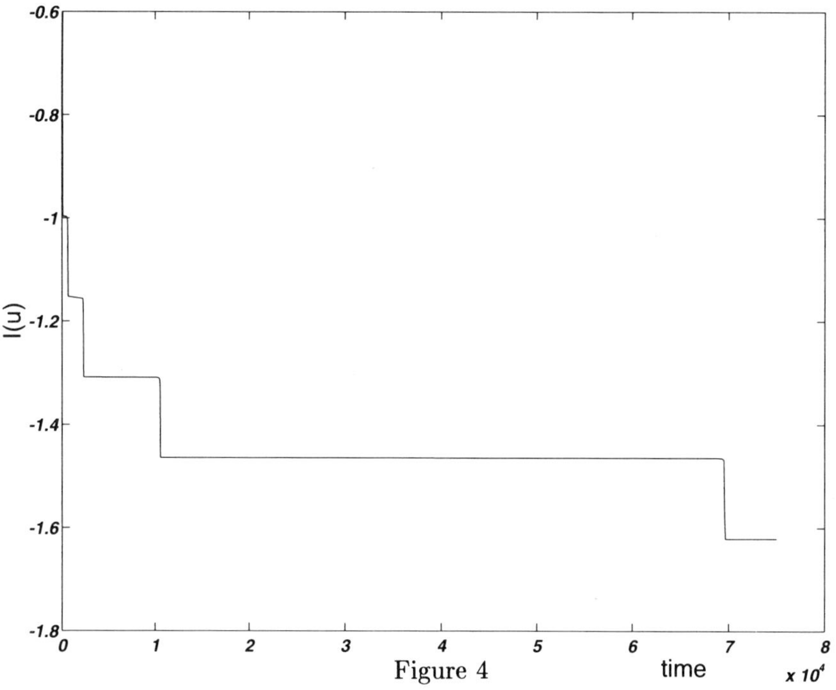

Figure 4

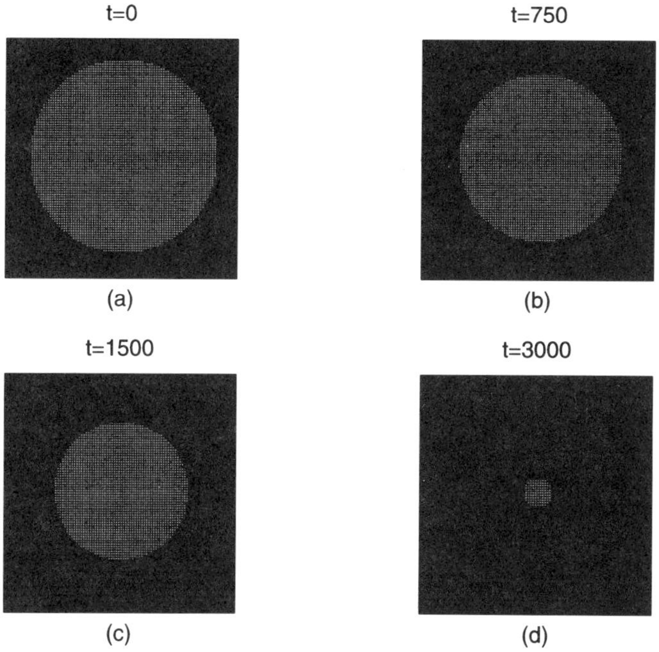

Figure 5

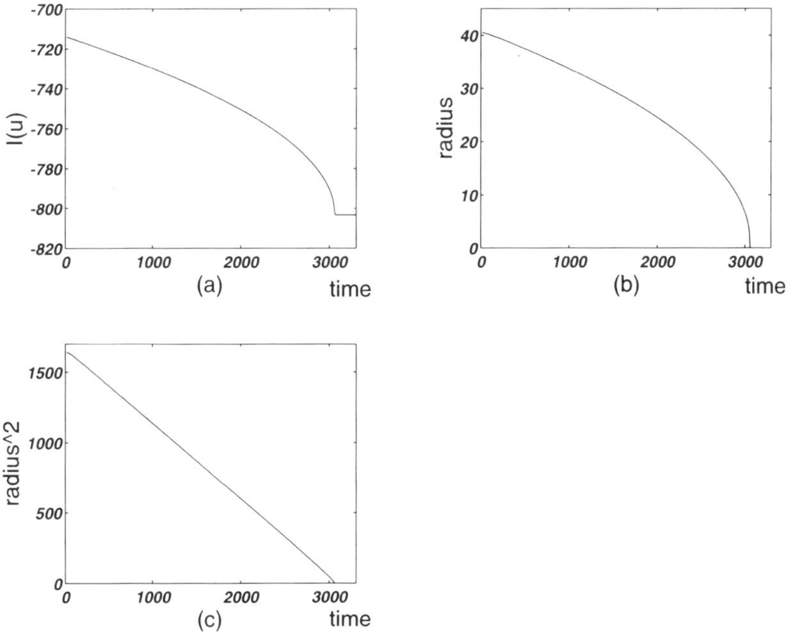

Figure 6

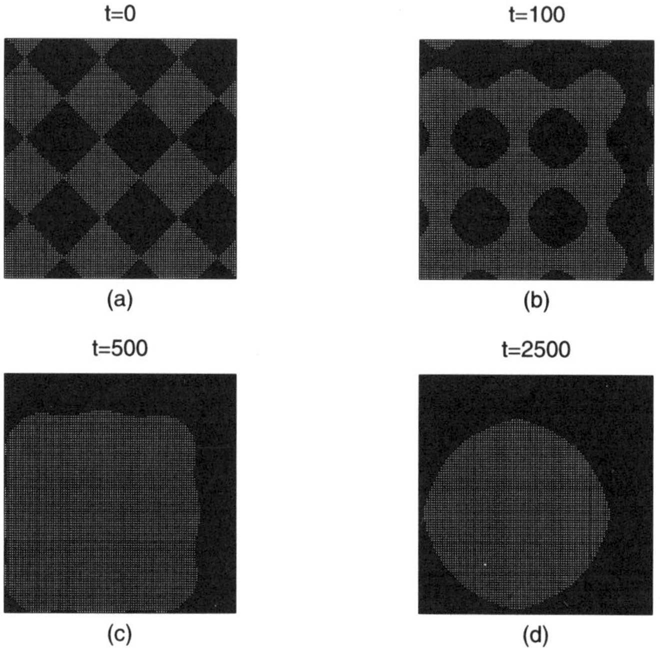

Figure 7

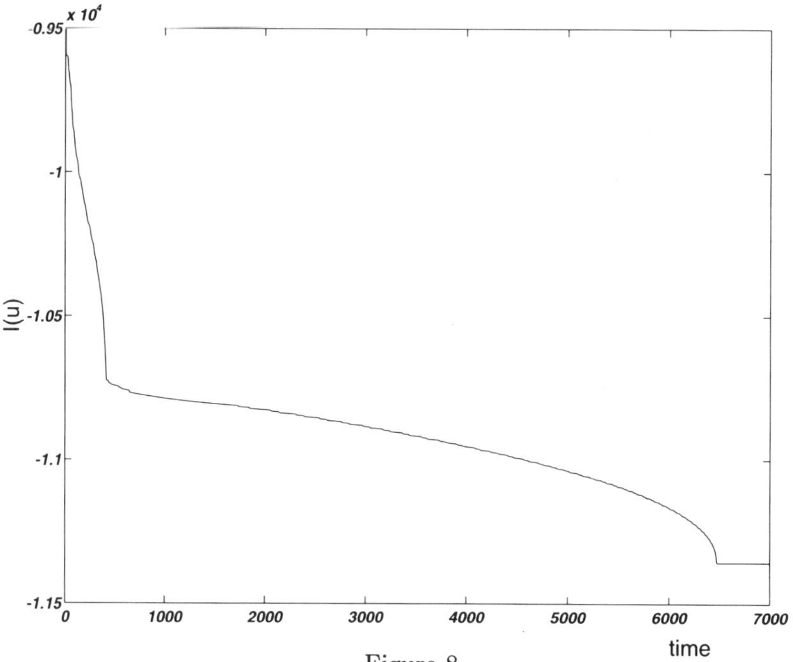

Figure 8

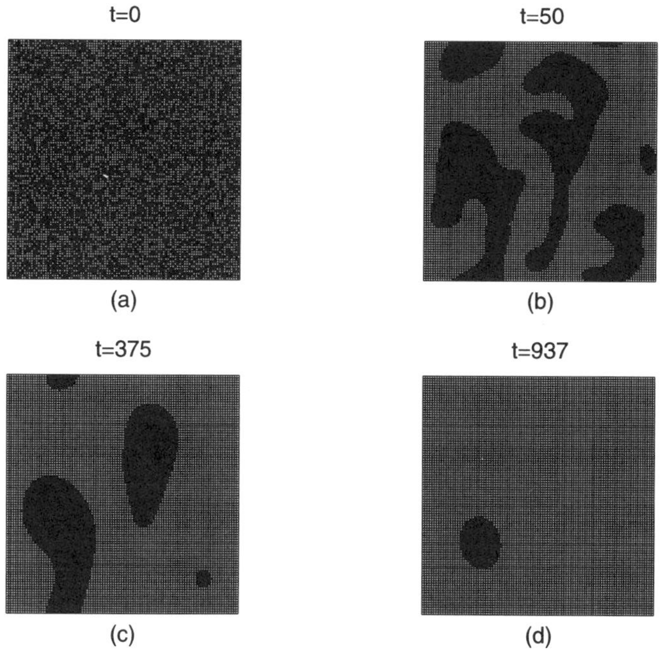

Figure 9

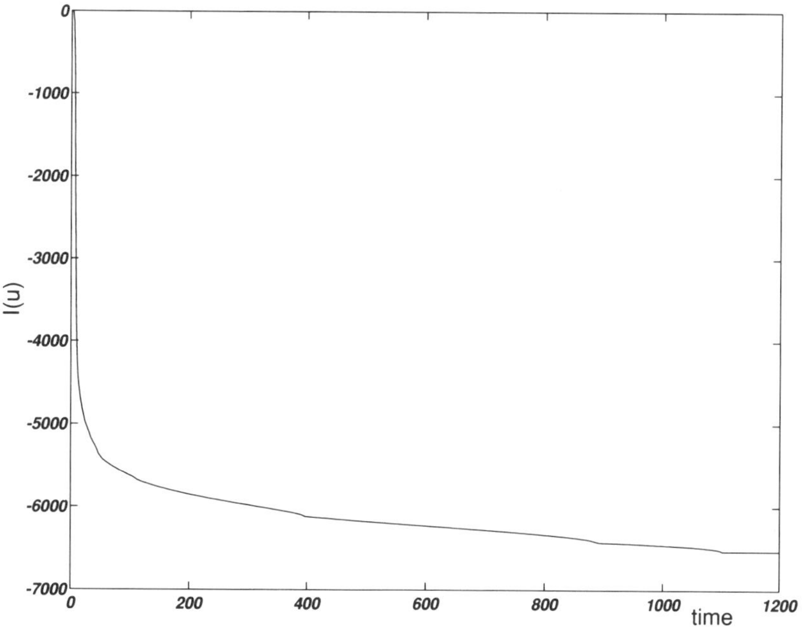

Figure 10

Contemporary Mathematics
Volume **172**, 1994

ATTRACTORS FOR WEAKLY COUPLED MAP LATTICES

VOLKER MATTHIAS GUNDLACH

ABSTRACT. We present a review of the rigorous mathematical work on coupled map lattices in the region of weak coupling focussing on the investigations of attractors. We are mainly concerned with structural stability results and the phase of spatio-temporal chaos for weak diffusively coupled map lattices. Finally we explain how a shadowing theorem could be used to relate the dynamics of coupled map lattices and reaction-diffusion equations.

1. INTRODUCTION

In the past decades the efforts to understand rather complex behaviour and complicated structures for time evolutions in many different branches of science were concentrated on the investigations of low-dimensional systems. This has led to some good understanding of phenomena exhibiting spatial coherence, while the study of systems with many degrees of freedom was more or less neglected. Proceeding on the understanding of low-dimensional systems it should be easier now to deal with spatio-temporal phenomena. This applies in particular when one restricts the attention to phenomenological studies of model systems that are constructed out of spatially distributed low-dimensional systems via coupling. Such systems known as *spatially extended systems* might give a good insight into the mechanisms determining spatio-temporal phenomena.

Amongst these spatially extended systems *coupled map lattices (CML)* are a very popular choice, mainly because they are easy to handle on parallel computers. In particular the fact that they allow a variety of immediate changes is of importance for finding all sorts of spatio-temporal phenomena. The spatial extension for this kind of system is given by a lattice. For simplicity let us consider just the case $\mathbb{Z}$. In each site of the lattice we consider a dynamical system given by some mapping f on a manifold Y in order to obtain a map F on $X = Y^{\mathbb{Z}}$ by putting $(F(\underline{x}))_j = f(x_j)$ for $\underline{x} = \{x_i\}_{i\in\mathbb{Z}} \in X$. Then a *coupled map lattice* is a mapping of the form

$$\Phi_\varepsilon = A_\varepsilon \circ F$$

1991 *Mathematics Subject Classification.* Primary 34C35, 58F11; Secondary 58F12, 58F39.

The participation in the "Chaotic Dynamics" workshop at Deakin University was made possible by a DFG grant.

where $A_\varepsilon : X \to X$ is differentiable (with respect to a chosen topology on X), depends continuously on ε and is ε-close to the identity on X in the C^1-topology. We call such an A_ε *interaction operator* or *coupling*. In the case of vector spaces Y with norm $\|.\|$ and corresponding supremum norm $\|.\|_1$ on X such that X obtains the structure of a Banach manifold the most common interaction operators are the ones of finite radius, in particular the one for the nearest neighbour coupling using standard averaging, namely

$$(A_\varepsilon(\underline{x}))_i = a_\varepsilon^{(i)}(\underline{x}) = (1-\varepsilon)x_i + \frac{\varepsilon}{2}(x_{i-1} + x_{i+1}).$$

It can be seen as a discrete version for the Laplacian. For this coupling the CML becomes $(\Phi_\varepsilon(\underline{x}))_i = (1-\varepsilon)f(x_i)+\frac{\varepsilon}{2}(f(x_{i-1})+f(x_{i+1}))$ and for this a lot of computer studies have been executed in the case that $Y = [0,1]$ and f is the logistic map.

Two essential problems in the study of coupled map lattices are the following:

(1) Which types of essentially different attractors for CML exist and hence which kinds of spatio-temporal behaviour can these mappings exhibit?
(2) How can one compare the dynamics of CML with those of more complicated higher dimensional systems, in particular PDEs, and how can one deduce analogous behaviour and phenomena for the more intricate systems?

In this paper we want to present a method based on a structural stability result that can be helpful in some cases, in particular for weak coupling in CML.

2. Structural Stability

The central results in this paper are related to the persistence of hyperbolic structures for CML. They were proven simultaneously and independently by Pesin and Sinai [14] and Gundlach and Rand [8]. While the latter presented an extended version on bundles applicable to the construction of symbolic dynamics for CML as well as to the proof of a shadowing result, Pesin and Sinai used a more suitable topology (the one induced by $\|.\|_1$) and worked without compactness assumptions. Unfortunately these results are not explicitly stated in their work. We will present in this section the adaption of the results from Gundlach and Rand [8] to the set-up of Pesin and Sinai [14], which can also be found in Campbell and Rand [5]. In all these paper the notion of hyperbolic invariant sets is crucial.

Definition 1. *If $f : E \to E$ is a differentiable mapping in a Banach manifold, $\Lambda \subset E$ is closed and $f_{|\Lambda}$ is a homeomorphism on Λ, then Λ is called hyperbolic for f if there exists a Df-invariant splitting $T_\Lambda E = E_1 \oplus E_2$ such that $Df|_{E_1}$ is an automorphism expanding E_1 and $Df|_{E_2}$ an endomorphism with norm less than 1.*

A general theorem - not just for the set-up of CML - can be formulated as follows.

Theorem 1. *Let X be a Banach manifold with induced metric d and $f : X \to X$ a differentiable mapping. Assume that there exists a set $\Lambda \subseteq X$ which is hyperbolic for f. Then for any $r > 0$ there exist numbers $\alpha = \alpha(r) > 0$, $C = C(r) > 0$ and neighbourhoods $W = W(r)$ of f in the space $C_h^1(X)$ of differentiable homeomorphisms on X and $U = U(r)$ of Λ with the following property: If $h : Y \to Y$ is a homeomorphism on a subset Y of X, $\phi \in W$ and if $i : Y \to U$ is a continuous mapping satisfying $d(ih, \varphi i) < \alpha$, then there exists a unique continuous mapping $j : Y \to X$ such that $jh = \varphi j$, $d(i,j) \le r$, $d(i,j) \le Cd(ih, \varphi i)$.*

This theorem covers the case where X is the domain of a CML f. The main applications of this result are the derivations of a shadowing lemma (see Theorem 11) and of structural stability of invariant sets (Theorem 3). For the first situation just consider Y as a pseudo-orbit and h as the corresponding shift. For the second result one needs two invariant sets close to each other, one for a mapping f, one for a mapping close to f. If the two invariant sets are hyperbolic, a conjugacy between the systems can be found. So one only needs to know the persistence of the hyperbolicity.

Theorem 2. *Suppose X is a Banach manifold and $f : X \to X$ is a differentiable mapping with a hyperbolic invariant set $\Lambda \subset X$. Then there exist neighbourhoods U of Λ and W of f in $C_h^1(X)$ with the following property: if $g \in W$ and K is a g-invariant set in U, then K is hyperbolic for g .*

3. Attractors and Stability

The structural stability results of Section 2 provide a good way to find attractors for CML in case of weak coupling. Suppose that $Y = \mathbb{R}^m$ and $f : Y \to Y$ is a diffeomorphism on bounded hyperbolic invariant subsets Λ_i of Y, $i \in J$. Let X be the Banach manifold with norm $\|.\|_1$. Assume that the interaction operator $A_\varepsilon = (a_\varepsilon^{(i)})_{i\in\mathbb{Z}}$ exhibits a dependence of $a_\varepsilon^{(i)}(\underline{x})$ on x_j which is exponentially decreasing with the distance of the sites i, j and call this coupling *diffusive.* We consider $\varepsilon \geq 0$ being so small that A_ε and its Jacobian are invertible and necessarily close to the identity. Obviously $L = \prod_{i=-\infty}^{\infty} \Lambda_{j_i} \subset X$ for $j_i \in J$ is an invariant set for $\Phi_0 = F$. It is hyperbolic. A combination of Theorem 1 and Theorem 2 applied to the described set-up yields the following result.

Theorem 3. *There exist $\varepsilon_1 > 0$ and a continuous mapping $h : [0, \varepsilon_1) \to C^0(L, X)$, the space of continuous mappings from L to X, such that*

(1) *$h(0)$ is the inclusion inc_L of L in X;*
(2) *if $\varepsilon \in [0, \varepsilon_1)$, then $L_\varepsilon = h(\varepsilon)(L) \subset X$ is an invariant set for the coupled map lattice Φ_ε which is hyperbolic;*
(3) *for $0 \leq \varepsilon < \varepsilon_1$, $h(\varepsilon)$ is a homeomorphism from L onto L_ε, and $h(\varepsilon) \circ F = \Phi_\varepsilon \circ h(\varepsilon)$;*
(4) *there exists a constant $c > 0$ such that $d(h(\varepsilon), inc_L) \leq c\varepsilon$ for $0 \leq \varepsilon < \varepsilon_1$.*

The most interesting cases of invariant sets are those with attracting neighbourhoods. Of course these so-called attractors for CML can be obtained analogously from the ones for the local dynamics. The study of the dynamical behaviour is one of the most important topics in the theory of CML, in particular as there exists a wider range of phenomena than in the low-dimensional theory. So far there have been only mathematical investigations of two kinds of attractors for the CML under consideration in this paper. One is the attractor L_ε obtained for weak coupling from $L = \Lambda^{\mathbb{Z}}$ where Λ is a chaotic attractor for f (see Section 4). The other case is the one of periodic patterns and homogeneous structures, which are of low dimension.

For weak coupling Theorem 3 states that the attractor for Φ_ε for small ε is homeomorphic to the one of F. So if f possesses more than one attractor, there will also exist more than one for Φ_ε, in fact infinitely many. The adopted one will depend on initial conditions. This phenomenon was also found in computer studies and is known as *frozen random patterns.* As the ε in Theorem 3 depends on L, the robustness of L_ε for increasing ε depends on L. This phenomenon known as *pattern selection* was experimentally observed, too. In computer studies with the logistic map defining local dynamics it was found that with increasing ε *zigzag patterns* are prefered. These are temporally and spatially periodic solutions with period 2. More generally, points $\underline{x} \in X$ with $\Phi_\varepsilon^m \underline{x} = \underline{x}$ and $x_{i+n} = x_i$ for all $i \in \mathbb{Z}$ are called (m,n)*-periodic* solutions. $(1,1)$-solutions define *homogeneous structures.*

Though it is not clear yet when attractors for these periodic solutions exist, one might ask for their stability. Afraimovich and Bunimovich [1] examined the stability of (m,n)-periodic solutions for diffusive coupling, which need not be weak. Recently they have extended this work with Chow [6] to *mosaic* solutions (occuring e.g. in discrete versions of the Cahn-Hilliard or Cahn-Allen equation) consisting of spatially distributed time-periodic solutions. A different generalisation of (m,n)-periodic solutions is given by travelling waves of speed k/p defined by $(\Phi_\varepsilon^n(\underline{x}))_i = \psi(ip+kn)$ for some function ψ. Afraimovich and Pesin [3] investigated the stability of these solutions, in particular in discrete versions of the Ginzburg-Landau equation.

Afraimovich and Pesin [2] considered CML on $X = (\mathbb{R}^m)^{\mathbb{N}}$ with interactions related to problems in hydrodynamics given by parabolic partial differential equations like the Ginzburg-Landau equation. Such *drift systems* are determined by a coupling which starts off at a site $i_0 \geq 1$ and satisfies the drift property that the influence of sites further away than i_0 stays bounded. A basic example is given by

$$\Phi_\varepsilon(\underline{x})_i = \begin{cases} f(x_i) & \text{for } 0 \leq i < i_0 \\ \Phi_\varepsilon(\underline{x})_i = (1-\varepsilon)f(x_i) - \sum_{j=1}^n \varepsilon_j f(x_{i-j}) & \text{for } i \geq i_0 \end{cases}$$

where f has to be C^3, $0 < n < i_0$, $\varepsilon_j \geq 0$ and $\sum_{j=1}^n \varepsilon_j = 1$. The form of the interaction has some influence on the differentiability of Φ_ε: instead of the Fréchet differentiability the Gateaux differentiability has to be considered and corresponding to it a different set-up, in particular the notion of weak hyperbolicity has to be developed. Afraimovich and Pesin showed how starting with the uncoupled map F and a hyperbolic point $\underline{x}$ of it one can find sufficient conditions to obtain infinite-dimensional stable and finite-dimensional unstable subspaces which both can be obtained as lattice products of stable and unstable subspaces for f. Moreover it is shown how to obtain weakly hyperbolic attractors for Φ_ε from the ones for F.

4. Spatio-Temporal Chaos

Amongst the attractors for spatially extended systems the ones of high dimension are of particular interest, as they govern spatially non-coherent dynamics. The most extreme non-coherent behaviour is known as *spatio-temporal chaos* and is sometimes associated with *fully developed turbulence.* For CML it is characterized by a temporally chaotic motion in every lattice site together with a fast decay of spatial correlations. Mathematically we can give the following definition.

Definition 2. *A coupled map lattice Φ is said to exhibit spatio-temporal chaos, if there exists a unique Gibbs measure invariant under Φ and translations in the lattice, absolutely continuous with respect to Lebesgue measure on finite truncations of the lattice and shows strong-mixing properties for these two mappings.*

Spatio-temporal chaos can be found in CML with weak diffusive coupling and local chaotic dynamics. The first to prove this were Sinai and Bunimovich [4], but they had to use an unnatural coupling for local expanding dynamics f on $Y = [0,1]$ in order to obtain their result. This so-called *SB-coupling* is given by

$$a_{\varepsilon}^{(i)}(\underline{x}) = (1-\alpha(x_i))x_i + \frac{1}{2}\alpha(x_{i-1}+x_{i+1})$$

with a $C^2([0,1])$ function α satisfying

(1) $\alpha(y) = \varepsilon$ for $\delta \le y \le 1-\delta$ with $\delta > 0$ being small;
(2) $\alpha(0) = \alpha(1) = 0$;
(3) $\alpha'(y) \ge 0$ for $0 \le y \le \delta$, $\alpha'(y) \le 0$ for $1-\delta \le y \le 1$.

The motivation for this coupling was as follows. As f provides a *Markov partition* (i.e. there exist open intervals S_i, $1 \le i \le d$, with $\bigcup_i \overline{S_i} = [0,1]$ and$f\overline{S_i} = [0,1]$), one chooses the SB-coupling to use a lattice product of this partition as a respective one for the CML. This works as the BS-coupling in contrast to standard couplings guarantees the necessary condition that the projection of every rectangle onto one site gets mapped to $[0,1]$ under one iteration independently of the other sites.

With a Markov partition it is possible to establish symbolic dynamics on a higher-dimensional shift space (in the case of expanding local dynamics on $\Sigma = \{1,\dots,d\}^{\mathcal{S}}$ with $\mathcal{S} := \mathbb{Z} \times \mathbb{N}$) and to reduce the problem to the construction of a suitable potential for the thermodynamic formalism on this shift space. It remains to investigate the convergence of conditional Lebesgue probabilities $\mu_{\varepsilon,N,M}$ for the dynamics in a single site of a finite lattice of size $N+M+1$ for $N, M \to \infty$. This means a reduction to systems of a finite number of degrees of freedom and to studies of the effects of increasing the number of degrees of freedom. It can be done by choosing a fixed point x^* of f, allowing only the points in a finite number of sites i, $-N \le i \le M$, to take different values than x^* and corresponding to this choosing an interaction operator $A_{\varepsilon,N,M}$. Put $\Phi_{\varepsilon,N,M} := A_{\varepsilon,N,M} \circ F$ and denote the space $\bigotimes_{i=-\infty}^{-N-1}\{x^*\} \times \bigotimes_{i=-N}^{M} Y \times \bigotimes_{i=M+1}^{\infty}\{x^*\}$ by $X_{N,M}$. After showing the Hölder continuity of $\lim_{N,M\to\infty}\mu_{\varepsilon,N,M}$ one can adopt the work of Dobrushin and his coworkers (cf. [7]) as demonstrated by Gundlach and Rand [9]. This result can be described as follows for the case of Σ as above: The shift on Σ is a mapping τ induced by the common translation in $\mathcal{S}$. For $\alpha \in (0,1)$, let $\mathcal{C}^\alpha$ consist of those continuous functions f on Σ which, for some constant $C = C(f) > 0$ and any $n_1, n_2 \in \mathbb{N}$ satisfy $|f(\xi) - f(\eta)| \le C(\alpha^{n_1} + \alpha^{n_2})$ whenever $\xi_{\lambda_1,\lambda_2} = \eta_{\lambda_1,\lambda_2}$ for $0 \le \lambda_1 \le n_1$ and $|\lambda_2| \le n_2$. Define for any $V \subset \mathcal{S}$ a symbol space $\Sigma_V := \{1,\dots,d\}^V$. For $B \in \mathcal{C}^\alpha$ and any finite $V \subset \mathcal{S}$ put

$$f_V(\xi) = \frac{\exp(\sum_{x\in\mathcal{S}} B(\tau^x\xi))}{\sum_{\eta\in\Sigma:\eta_{|\mathcal{S}\setminus V}=\xi_{|\mathcal{S}\setminus V}} \exp(\sum_{x\in\mathcal{S}} B(\tau^x\eta))} \tag{1}$$

for $\xi \in \Sigma$. Then a Borel probability measure μ on Σ is a *Gibbs state* for $B \in \mathcal{C}^0$ if for every finite $V \subset \mathcal{S}$,

$$\mu(h) = \int_{\Sigma_{\mathcal{S}\setminus V}} \int_{\Sigma_V} h(\xi) f_V(\xi) d\xi_{|V} d\mu^{(V)}(\xi_{|\mathcal{S}\setminus V}) \quad \text{for all } h \in \mathcal{C} \tag{2}$$

where $\mu^{(V)}$ is the measure on $\Sigma_{\mathcal{S}\setminus V}$ obtained from μ through projection of Σ onto $\Sigma_{\mathcal{S}\setminus V}$. Equations (2) for finite V are called *DLR-equations* in honour of Dobrushin, Lanford and Ruelle. Gundlach and Rand [9] provided the following result.

Theorem 4. *For $B \in \mathcal{C}^\alpha$ there exists a unique τ-invariant Gibbs state μ (which is also an equilibrium state for B) such that τ is strong-mixing. The correlation function for translations in any direction of the lattice $\mathcal{S}$ decays exponentially fast.*

The uniqueness property of Theorem 4 means the exclusion of phase transitions. The speed of mixing is guaranteed to be exponentially fast only in single directions of the lattice, not in general. On the basis of Theorem 4 one obtains the following.

Theorem 5. *If the local dynamics are given by a $C^{1+\alpha}$ expanding mapping f on $[0,1]$ or S^1 and the coupling is of BS-type, then there exists $\varepsilon_1 > 0$ such that the resulting coupled map lattice Φ_ε exhibits spatio-temporal chaos for all $\varepsilon \in [0, \varepsilon_1)$.*

An analogous way to obtain a similar result without the restrictions for the coupling was given by Gundlach and Rand [10]. The improvement was achieved by some abundant work for establishing symbolic dynamics based on the so-called *bundle approach.* It provides a way to deal with CML in the phase of spatio-temporal as products of skew-products where the dynamics on the fibres are low-dimensional and depend weakly on the high-dimensional factor. We explain this method for the case of coupled circle maps. Consider CML where the local dynamics are defined by an expanding $C^{1+\alpha}$ mapping f on S^1 with lift $\tilde{f}$, e.g. $\tilde{f}(x) = sx + A\sin(2\pi x) + C$ with $A, C > 0$, $s \in \mathbb{Z}$ and $s > 1 + A2\pi$. For this kind of maps there exist Markov partitions $\{R_i : 1 \le i \le d\}$. For the interaction we use weak diffusive coupling given by $A_\varepsilon = (a_\varepsilon^{(i)})_{i\in\mathbb{Z}}$ on X. A popular choice is given with the help of a differentiable 1-periodic mapping g on S^1 which satisfies $|g'| \le 1$, in particular $g(x) = (2\pi)^{-1}\sin(2\pi x)$, together with a function $s_\varepsilon : \mathbb{N} \to \mathbb{R}_0^+$ satisfying $s_\varepsilon(n) \le c\exp(-an)$, $s_\varepsilon(n) \le \varepsilon$ for all $n \in \mathbb{N}$, $s_\varepsilon(0) \in [1-\varepsilon, 1]$ for some constants $a, c > 0$. Then one chooses

$$a_\varepsilon^{(i)}(\underline{x}) = x_i + \sum_{i\neq j} s_\varepsilon(|i-j|)g(x_j - x_i) \pmod 1.$$

The most common example is once more given by the nearest neighbour coupling using standard averaging, $a_\varepsilon^{(i)}(\underline{x}) = x_i + \frac{\varepsilon}{2}(g(x_{i+1}-x_i) + g(x_{i-1}-x_i)) \pmod 1$. We consider products $\mathcal{E}^{(i)} := X \times Y$ for $i \in \mathbb{Z}$ and $\mathcal{E} = X \times X = \bigotimes_{i\in\mathbb{Z}} \mathcal{E}^{(i)}$ and call them bundles of Y over X and of X over X, resp. On the bundles $\mathcal{E}^{(i)}$ we consider skew-products or so-called *fibre-preserving bundle maps* given by $(\Phi_\varepsilon, \Psi_\varepsilon^{(i)}) : \mathcal{E}^{(i)} \to \mathcal{E}^{(i)}$ where $\Psi_\varepsilon^{(i)}$ is given on the *fibre* $\mathcal{E}_{\underline{x}}^{(i)} = \underline{x} \times Y$ for $y \in Y$ by

$$\text{(3)} \quad \Psi_{\varepsilon,\underline{x}}^{(i)}(y) = a_\varepsilon^{(i)}(\ldots, f(x_{i-k}), \ldots, f(x_{i-1}), f(y), f(x_{i+1}), \ldots, f(x_{i+k}), \ldots).$$

In particular we have $\Psi_{0,\underline{x}}^{(i)}(z) = f(z)$ for all $\underline{x} \in X$, $z \in S^1$ and $i \in \mathbb{Z}$. Thus $\Psi_{0,\underline{x}}^{(i)}$ is an expanding map from $\underline{x} \times S^1$ to $F(\underline{x}) \times S^1$ with Markov partition $\mathcal{R} = \{R_j\}_{j=1}^d$ for any $\underline{x} \in X$. Since $\Psi_\varepsilon^{(i)}$ is ε-close to $\Psi_0^{(i)}$ in the C^1-topology, one can find a number ε_2 with $0 < \varepsilon_2 \le \varepsilon_1$ such that Φ_ε is expanding on X and $\Psi_{\varepsilon,\underline{x}}^{(i)}$ is an expanding map from $\underline{x} \times S^1$ to $\Phi_\varepsilon(\underline{x}) \times S^1$ for every $\underline{x} \in \mathcal{Y}$, $\varepsilon < \varepsilon_2$ and the latter has a Markov partition $\mathcal{R}_{\varepsilon,i}(\underline{x})$ of open segments $R_{\varepsilon,i}^j(\underline{x})$, $1 \le j \le d$, being close to R_j. This allows us to give a symbolic representation $\{w_j^i(\underline{x})\}_{j=-\infty}^{\infty}$ for every point in $\mathcal{E}_{\underline{x}}$ with respect to the iteration under $\Psi_\varepsilon^{(i)}$: we can associate to each $\underline{x}\in X$ a sequence $\{w^i\}_{i=-\infty}^{\infty}$ of symbol sequences $\{w_j^i\}_{j=0}^{\infty}$ by choosing the w_k^i such that $\Psi_{\varepsilon,\Phi_\varepsilon^k(\underline{x})}^{(i)} \circ \ldots \circ \Psi_{\varepsilon,\underline{x}}^{(i)}(x_i) \in R_{\varepsilon,i}^{w_i^k}(\Phi_\varepsilon^k(\underline{x}))$. Due to $\Phi_\varepsilon(\underline{x}) = \bigotimes_{i=-\infty}^{\infty} \Psi_{\varepsilon,\underline{x}}^{(i)}(x_i)$ and the expansiveness of Φ_ε for $\varepsilon < \varepsilon_2$ each sequence $\{w^i\}_{i=-\infty}^{\infty}$ determines exactly one point $\underline{x} \in X$. So we have arrived at a higher-dimensional shift space with shift τ. As the different directions in this shift space also have different meanings, we denote $\tau^{(1,0)} =: \sigma$ and $\tau^{(0,1)} =: s$. Consequently one has the following theorem.

Theorem 6. *For $\varepsilon > 0$ small there exists a continuous mapping $\theta : \Sigma \to L_\varepsilon$ which conjugates the temporal shift σ (resp. spatial shift s) on Σ to Φ_ε (resp. spatial translation S) on X and defines symbolic dynamics for Φ_ε on X.*

An analogous result can be obtained for $\Phi_{\varepsilon,N,M}$ and then one can proceed examining the limit of the conditional probabilities $\mu_{\varepsilon,N,M}$ for $N, M \to \infty$ as in the work of Sinai and Bunimovich [4]. Let us remark that the bundle approach also makes this part of the work easier (cf. Gundlach and Rand [10]). Anyway, the following theorem can be deduced by this procedure.

Theorem 7. *If the local dynamics are given by a $C^{1+\alpha}$ expanding mapping f on S^1 and the coupling is diffusive, then there exists $\varepsilon_1 > 0$ such that the resulting CML Φ_ε exhibits spatio-temporal chaos for all $\varepsilon \in [0, \varepsilon_1)$. Moreover, for any $N, M \in \mathbb{N}$ the conditional measure $\nu_{\varepsilon,[-N,M]}$ induced on $X_{N,M}$ by the corresponding unique Gibbs state ν_ε is absolutely continuous with respect to $\tilde{\nu}_{\varepsilon,N,M}$ on $X_{N,M}$, where the last measure is the usual SRB measure for the mapping $\Phi_{\varepsilon,N,M}$ (whose existence is guaranteed by usual theory of Gibbs states for hyperbolic diffeomorphisms).*

In the case of local dynamics given by a $C^{1+\alpha}$ diffeomorphism f on a smooth manifold with a hyperbolic invariant set Λ one can proceed similarly starting with the hyperbolic invariant set $L = \Lambda^{\mathbb{Z}} \subset X$ for F. On the basis of bundle versions of Theorem 3, a stable manifold theory and a shadowing lemma it is possible to construct Markov partitions for suitable fibre-preserving mappings defined by (3), hence symbolic dynamics for Φ_ε analogously to the case of coupled circle maps (cf. Gundlach and Rand [8]) as a first step towards the proof of space-time chaos.

A different method was used by Pesin and Sinai [14] as well as Campbell and Rand [5]. It can be seen as an infinite-dimensional analogon of the standard theory of hyperbolic systems. This transfer is possible, as one can show that there is a weakening of the infinite dimensionality of the problem: due to the diffusive coupling all important quantities show the same exponentially weak dependence on distant sites. This enables the use of a direct product structure via conjugacies and hence an approach as for the finite-dimensional problem. The result is as follows.

Theorem 8. *If the local dynamics are given by a $C^{1+\alpha}$ diffeomorphism f on a smooth manifold with a hyperbolic invariant set Λ and the coupling is diffusive, then there exists $\varepsilon_1 > 0$ such that the corresponding coupled map lattice Φ_ε exhibits spatio-temporal chaos for all $\varepsilon \in [0, \varepsilon_1)$.*

An amendment to the theorems 5, 7, 8 is due to Volevich [15], who showed that the unique invariant Gibbs measure ν_ε characterizing the spatio-temporally chaotic behaviour is naturally attracting. In order to formulate this result let $V(L, X)$

denote the set of measures m on $(X, \mathcal{B}(X))$ whose restrictions to $X_{N,M}$ for any $N, M \in I\!N$ are absolutely continuous with respect to the Lebesgue measure L on $X_{N,M}$ with Radon-Nikodyn derivative $m_{N,M}$ satisfying the following spatial correlation condition: if for any $\underline{x}, \underline{y} \in X$ and $r \in \mathbb{Z}$ one has $|x_n - y_n| \leq hq^{|n-r|}$ for all $n \in \mathbb{Z}$ for some $q \in (0,1)$ and $h > 0$, then it follows that

$$\exp(-ch^{\gamma}) \leq \frac{m_{N_1,N_2}(x_{-N_1}, \dots, x_{N_2})}{m_{N_1,N_2}(y_{-N_1}, \dots, y_{N_2})} \leq \exp(ch^{\gamma})$$

for some $c = c(q) > 0$ and $\gamma = \gamma(q) > 0$ independent of N, M.

Corollary 1. *If $m \in V(L, X)$, then the measures $(\Phi_{\varepsilon}^*)^n(m)$ for $n \in I\!N$ converge weakly to the Gibbs measure ν_{ε}.*

Finally let us mention that there exists the possibility of obtaining ν_{ε} via a bundle version of the Ruelle-Perron-Frobenius Theorem (cf. Gundlach and Rand [11]). This option is given for the case of coupled expanding circle maps Φ as follows.

Theorem 9. *For each $\underline{x} \in X$, $i \in \mathbb{Z}$ there exist measures $\mu_{\underline{x}}^{(i)}$ on S^1 and $\mu_{\underline{x}} = \bigotimes_{i=-\infty}^{\infty} \mu_{\underline{x}}^{(i)}$ on X with the following properties:*

(1) *the $\mu_{\underline{x}}^{(i)}$ are absolutely continuous with respect to Lebesgue measure on S^1 and the restriction of $\mu_{\underline{x}}$ to $X_{N,M}$ for any $N, M \in I\!N$ is absolutely continuous with respect to Lebesgue measure on $X_{N,M}$;*

(2) *the $\mu_{\underline{x}}^{(i)}$ are equivariant, i.e. $\mu_{\underline{x}} \circ \Phi_{\varepsilon} = \mu_{\Phi_{\varepsilon}(\underline{x})}$, and strong mixing in the sense that for any $w_{\underline{x}} \in L^2(\mu_{\underline{x}})$ and $v_{\Phi^n(\underline{x})} \in L^2(\mu_{\Phi^n(\underline{x})})$, $n \in I\!N$,*

$$\lim_{n \to \infty} \Big| \int v_{\Phi^n(\underline{x})} \circ \Psi_{\Phi^{n\;1}(\underline{x})} \circ \dots \circ \Psi_{\underline{x}} \times w_{\underline{x}} \, d\mu_{\underline{x}} \quad \int v_{\Phi^n(\underline{x})} d\mu_{\Phi^n(\underline{x})} \times \int w_{\underline{x}} d\mu_{\underline{x}} \Big| = 0;$$

(3) *the $\mu_{\underline{x}}^{(i)}$ are the conditional probabilities of ν_{ε} in the site i in the sense that $\int \mu_{\underline{x}} \, d\nu_{\varepsilon}(\underline{x}) = \nu_{\varepsilon}$ and $\lim_{n \to \infty} \frac{1}{n} \sum_{i=0}^{n-1} \mu_{\Phi_{\varepsilon}^i(\underline{x})} = \nu_{\varepsilon}$ for ν_{ε}-almost all $\underline{x} \in X$.*

All the results stated so far are only valid for local dynamics given by expanding maps or diffeomorphisms. Popular examples like coupled logistic maps are not contained in the theory developed by Sinai and coworkers as well as by Rand and his coworkers. However, Keller and Künzle [13] found a way to deal with CML where the coupling is reasonable and the local dynamics are given by mappings of particular interest for simulations. Their method is based on a *Perron-Frobenius operator* theory for *functions of bounded variation.* It turned out that the respective method for Φ_{ε} and their finite versions $\Phi_{\varepsilon,N,M}$ are not completely analogous, as in inifinite dimension nice compactness properties inside the space of functions of bounded variations get lost. This prevents the proof of the uniqueness of the invariant measure for Φ_{ε}. Anyway, Keller and Künzle obtained the following result.

Theorem 10. *Assume that the local dynamics are given by a piecewise monotonic mapping f on $[0,1]$ such that each monotone branch of f is of class C^2 and has slope strictly greater than 2, and the coupling is diffusive. Then one can find $\tilde{\varepsilon} > 0$ and for every $N \in I\!N$ an $\varepsilon_N > 0$ with $\varepsilon_N \to 0$ for $N \to \infty$ such that for every $\varepsilon \in [0, \tilde{\varepsilon})$ there exists a Gibbs state ν_{ε} which is invariant under $\Phi_{\varepsilon,N,M}$ and S, absolutely continuous with respect to Lebesgue measure on $X_{N,M}$ and which is mixing, if the respective measure for Φ_0 is mixing. This measure ν_{ε} is unique, if $\varepsilon \in [0, \varepsilon_{N+M})$,* *and for every $\varepsilon \in [0, \tilde{\varepsilon})$ there exists a Gibbs state $\tilde{\nu}_\varepsilon$ which is invariant under Φ_ε and S, absolutely continuous with respect to Lebesgue measure on finite truncations of the lattice and which is mixing, if the respective measure for Φ_0 is mixing.*

5. Attractors from Reaction-Diffusion Equations

Though CML are more or less artificial model systems, it is hoped that they can be related to other spatially extended systems. This holds in particular for reaction-diffusion equations (RDE). The architectures of both systems are similar in their combination of local nonlinear dynamics (reaction) and spatial coupling (diffusion). In general one should not expect to find the same dynamical behaviour for partial differential equations and their (discrete) lattice versions. We would like to show an example where the connection between solutions of an RDE and orbits of a CML is deeper than just similarity.

Kaspar and Schuster [12] have suggested to consider lattice versions of reaction-diffusion equations with nonlinear impacts like

$$\frac{\partial u}{\partial t} = c\frac{\partial^2 u}{\partial x^2} + \sum_n \delta(t-n)\frac{\partial V}{\partial x}, \tag{4}$$

where c is a coupling constant and V a nonlinear potential. Equation (4) has got the nice property that a corresponding time-one mapping can be given explicitly:

$$(\Theta(u))(x) = \frac{1}{4\pi c}\int_{-\infty}^{+\infty} e^{-\frac{(x-\xi)^2}{4c}} f(u(\xi))d\xi \tag{5}$$

where f is a nonlinear function defined by $f(u) := u + V'(u)$. Kaspar and Schuster used equation (4) as a motivation to study CML of the form

$$x_{n+1}(i) = \frac{1}{1+2\varepsilon}f(x_n(i)) + \frac{\varepsilon}{1+2\varepsilon}[f(x_n(i+1)) + f(x_n(i-1))]$$

(which is the discrete analogon of (4)) for some interesting mappings f. As we are interested in comparing the dynamics of (4) and an approriate CML, let us propose a discretized version of (5). Choose for sufficiently small c a scheme for a numerical integration like the Newton-Cotes formula for sufficiently many steps $2n$ described by weights $w_{i,n}$ as well as a number b such that for a given $\alpha > 0$

$$\frac{\|f\|}{4\pi c}\left(\int_{-\infty}^{-b} e^{-\frac{(x-\xi)^2}{4c}} d\xi + \int_{b}^{+\infty} e^{-\frac{(x-\xi)^2}{4c}} d\xi\right) \leq \frac{\alpha}{2}, \tag{6}$$

$$\frac{1}{4\pi c}\left|\int_{-b}^{b} e^{-\frac{(x-\xi)^2}{4c}} f(u(\xi))d\xi - \frac{b}{n}\sum_{j=-n}^{n} w_{j,n}e^{-\frac{j^2b^2}{4cn^2}} f(u(x+j\frac{b}{n}))\right| \leq \frac{\alpha}{2}. \tag{7}$$

Suppose $\varphi := \frac{b}{4\pi cn}w_{0,n}f$ possesses a hyperbolic set Λ. By setting $\varepsilon_i := \exp(-j^2b^2/4cn^2)w_{i,n}/w_{0,n}$ we are led to consider Φ_ε for some $\varepsilon = (\varepsilon_1, \dots, \varepsilon_n)$ of the form

$$\Phi_\varepsilon(\underline{x})_i = f(x_i) + \sum_{j=1}^{n} \varepsilon_j[f(x_{i-j}) + f(x_{i+j})]. \tag{8}$$

For small c and a suitable scheme for the numerical integration, Φ_ε also has a hyperbolic invariant set L_ε by Theorem 3. According to Theorem 1 we obtain the following result.

Theorem 11. *For a given $\gamma > 0$ there exists $\alpha > 0$ such that any time-1-solution u of (4) with $dist((\Theta^n(u)(i\frac{b}{n}))_{i\in\mathbb{Z}}, L_\varepsilon) < \alpha$ is γ-shadowed by an orbit of the CML (8), i.e. there exists a point $\underline{x}$ such that*

$$\|(\Theta^n(u))(\frac{b}{n}i) - (\Phi_\varepsilon^n(\underline{x}))\| \le \alpha \text{ for all } i \in \mathbb{Z},$$

if the numerical scheme satisfies (6) and (7).

Though this is a rather particular example, it is hoped that there might be more "natural" spatially extended systems like this RDE which can be related to CML, and hence increase the importance of CML as relevant systems for studying spatio-temporal phenomena.

References

1. V. Afraimovich and L.A. Bunimovich, Simplest Structures in Coupled Map Lattices and Their Stability. Preprint, Georgia Tech. 1992.
2. V. Afraimovich and Ya. Pesin, Hyperbolicity of Infinite Dimensional Drift Systems. *Nonlinearity 3, 1-19* (1990).
3. V. Afraimovich and Ya. Pesin, Travelling Waves in Lattice Models of Multi-Dimensional und Multi-Component Media. Preprint, Georgia Tech. 1992.
4. L.A. Bunimovich and Ya.G. Sinai, Spacetime Chaos in Coupled Map Lattices. *Nonlinearity 1, 491-516* (1988).
5. K. Campbell and D.A. Rand, A Natural Measure for Coupled Map Lattices. Preprint (1993).
6. S.N. Chow, Lattice Dynamics. Talk at the Chaotic Numerics Workshop, Deakin University, Geelong (1993).
7. R.L. Dobrushin and S.B. Shlosman, Completely Analytical Interactions: Constructive Description. *Journ. Stat. Phys. 46, 983-1014* (1987).
8. V.M. Gundlach and D.A. Rand, Spatio-Temporal Chaos: 1. Hyperbolicity, Structural Stability, Spatio-Temporal Shadowing and Symbolic Dynamics. *Nonlinearity 6, 165-200* (1993).
9. V.M. Gundlach and D.A. Rand, Spatio-Temporal Chaos: 2. Unique Gibbs States for Higher-Dimensional Symbolic Systems. *Nonlinearity 6, 201-213* (1993).
10. V.M. Gundlach and D.A. Rand, Spatio-Temporal Chaos: 3. Natural Spatio-Temporal Measures for Coupled Circle Map Lattices. *Nonlinearity 6, 215-230* (1993).
11. V.M. Gundlach and D.A. Rand, Existence and Uniqueness of Gibbs Bundles and a Bundle Version of the Ruelle-Perron-Frobenius Theorem. *Preprint* (1993).
12. F. Kaspar and H.G. Schuster, Scaling at the Onset of Spatial Disorder in Coupled Piecewise Linear Maps. *Phys. Lett. A 113, no. 9, 451-453* (1989).
13. G. Keller and M. Künzle, Transfer Operators for Coupled Map Lattices. *Ergod. Th. & Dynam. Sys. 12, 297-318* (1992).
14. Ya. Pesin and Ya. Sinai, Space-Time Chaos in Chains of Weakly Hyperbolic Mappings. *Advances in Soviet Math. 3* (1991).
15. V.L. Volevich, Kinetics of Coupled Map Lattices. *Nonlinearity 4, 37-48* (1991).

Institut für Dynamische Systeme, Universität, Postfach, 28334 Bremen, Germany
E-mail address: gundlach@mathematik.uni-bremen.de

Contemporary Mathematics
Volume **172**, 1994

Effective Chaos in the Nonlinear Schrödinger Equation

M.J. ABLOWITZ AND C.M. SCHOBER

ABSTRACT. In this paper we provide a concrete representation of parameter regimes for which the NLS is "effectively" chaotic. Using inverse spectral theory we establish that for these parameter regimes we are exponentially close to homoclinic manifolds. This proximity is responsible for the numerically induced instabilities and chaos which are computationally unavoidable. The ideas presented here also apply to related equations which share a similar phase space structure.

1. Introduction

The focusing Nonlinear Schrödinger equation (NLS)

$$iu_t + u_{xx} + 2|u|^2 u = 0 \tag{1}$$

with periodic boundary conditions, $u(x + L, t) = u(x, t)$, is a prototypical equation for the study of numerically induced chaos in nonlinear wave equations. The NLS equation is an integrable infinite-dimensional Hamiltonian system and, when periodic boundary conditions are imposed, possesses a complicated phase space structure. The flow of the NLS equation may be viewed as evolving on level sets of constants of motion which are typically infinite dimensional tori of maximal dimension and are stable. However the NLS phase space also contains critical level sets or critical tori which are not of maximal dimension and which may be either stable or unstable. The quasi-periodic flow on an unstable torus contains a finite number M of linearly unstable modes. The unstable flow consists

1991 *Mathematics Subject Classification.* Primary 35F20, 34C35; Secondary 34A50, 65CXX.

This work was partially supported by the Air Force Office of Scientific Research under grant AFOSR-90-0039, the NSF grant DMS-9024528, and the office of Naval Research grants N00014-91-J-4037 and N00014-92-J-1274.

We wish to thank E.A. Overman for use of his spectral solver package.

This paper is in final form and no version of it will be submitted for publication elsewhere.

of a tori component (quasi-periodic solution) and a finite M-dimensional manifold of orbits homoclinic to the quasi-periodic solution and have been termed whiskered tori [**13**] or homoclinic manifolds [**5**]. Such degenerate tori have proven to be extremely sensitive to the perturbations induced by numerical schemes [**1, 10, 14, 15, 5**]. Specifically we have shown that when using Hamiltonian numerical schemes, the presence of nearby whiskered tori is responsible for the generation of numerical instabilities and chaos. In fact, for certain parameter regimes, we have demonstrated that errors on the order of round-off error alone are capable of producing spurious chaotic solutions [**5**].

Numerical chaos in the NLS equation can be generated using initial data near low dimensional whiskered tori [**1**]. When one employs a "standard" conservative finite difference scheme (see Appendix), numerically induced chaos is observed for coarse meshes (i.e. triggered by truncation effects). Integrable equations like the NLS are natural candidates for study since we have the analytical theory available to recognize and interpret numerical chaos in nearby systems. Moreover we do not need to differentiate numerical chaos from physical chaos. Using the associated spectral theory of the NLS equation we can numerically monitor time series for crossings of unstable critical level sets of the constants of motion (homoclinic crossings) and correlate the homoclinic crossings with the appearance of instabilities in the system and transitions from quasi-periodic to chaotic motion. In addition, we employ dynamical systems diagnostics such as divergence of nearby trajectories, etc., in order to quantify our observations. For example, in [**14**], it was observed that the time series contains many and continual homoclinic crossings, indicating that the flow associated with the standard scheme is chaotic in response to the break up of the homoclinic orbit under the perturbation induced by the discretization. An "integrable" finite difference scheme (see Appendix) does not indicate any numerical instabilities or chaos in this parameter regime.

From a different point of view, it was shown that symplectic discretizations of planar Hamiltonian systems (ODE's) restore the qualitative features of the system exponentially fast [**11**]. By analogy, it might be thought that if an integrable or symplectic integrator which preseves the underlying geometric structure is used, then the numerical chaos can be avoided. However, there are important differences between low dimensional (planar systems) and infinite dimensinal PDE's. For NLS there are significant parameter regimes where errors on the order of round-off grow rapidly. These errors saturate eventually at values comparable to the main wave. A serious phase instability is created in the solution which cannot be eliminated by refinement (in the presence of round-off). Increasing the accuracy of the simulations (by increasing the tolerance, refining the mesh or switching schemes) moves the transition from a quasi-periodic solution to a chaotic solution characterized by random phases to a somewhat different (perhaps later) time.

As a typical example of round-off "triggered" instability consider the following

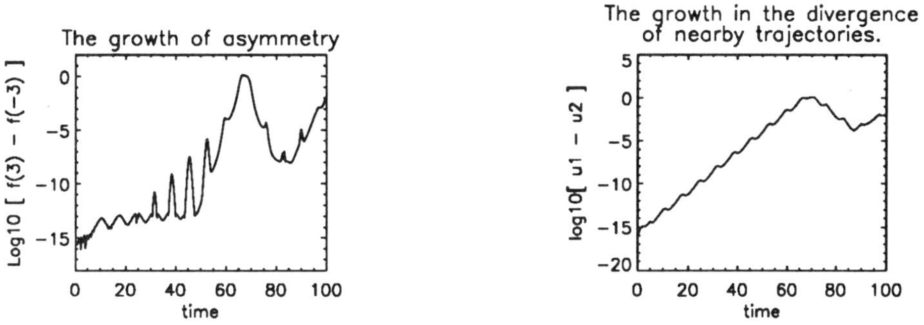

FIGURE 1. The solution of IDNLS with initial data $u(x, 0) = 0.7(1 + 10^{-1} \cos 3\mu x)$, $N = 128$, and $0 \leq t \leq 100$.

"4-phase" solution of the NLS equation corresponding to the initial data:

$$u(x, 0) = a(1 + \epsilon \cos 3\mu x) \tag{2}$$

with $L = 4\sqrt{2}\pi$, $a = 0.7$, $\mu = 2\pi/L$, $\epsilon = 0.1$.

From linear theory (see §2) we know that for (2), $u(x, 0)$ is nearby a plane wave state with 3 unstable modes. We numerically integrated the NLS equation using the integrable finite difference scheme with N=128 which adequately resolves the spatial structure (the scheme is integrable for all N). In time we use a fourth order adaptive Runge-Kutta algorithm (a subroutine in the NAG sofware library) with a low error/tolerance so that temporal truncation errors are on the order of round-off. We find that the growth of asymmetry is exponential and reaches $\mathcal{O}(1)$ before it saturates and that after a moderate time interval ($t \approx 60$) two mathematically equivalent schemes (the differences are only due to round-off effects, this is sometimes referred to as divergence of trajectories) have differences which also reach $\mathcal{O}(1)$ (Figure 1). This phenomena is triggered by errors on the order of round-off (10^{-15}); the only difference in the calculation of the two trajectories is the precision of machine arithmetic. In these calculations, the deviation in the first 3 actions is $\mathcal{O}(10^{-3})$ or smaller. There is no drift away from the fixed point. Associated with each action is a conjugate phase variable (provided by the Dirichlet spectrum). For the 3 actions corresponding to the 3 unstable modes, the associated phase variables may be locked or they may be free to move. This property of the phases is important in the structure of the level sets and is responsible for the extreme sensitivity to small errors. Very small perturbations may or may not activate the degrees of freedom associated with the whisker. When the perturbation is produced by roundoff errors, there is no preference one way or the other and an $\mathcal{O}(1)$ random change in the phase variables results.

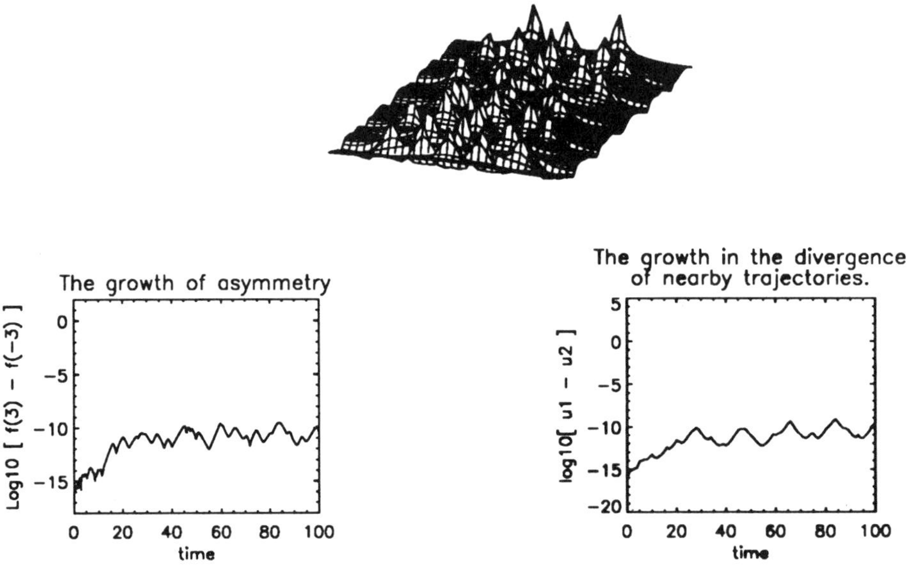

FIGURE 2. The solution of IDNLS with initial data $u(x,0) = 0.7(1+10^{-1}\cos\mu x)$, $N = 128$ and $0 \leq t \leq 100$.

However if we use the following initial data (L, a, μ, ϵ as before),

$$u(x,0) = a(1+\epsilon\cos\mu x) \tag{3}$$

and keep the parameters of the numerical simulation the same as before we obtain substantially different results. There are not striking differences in initial data (2) and (3) (the only difference being a perturbation of $\cos 3\mu x$ in (2) vs. $\cos\mu x$ in (3); ϵ is large for both and the first two Hamiltonians for data (2) and (3) are within 10^{-1}). In Figure 2 the surface, growth of asymmetry and the divergence of nearby trajectories is presented for data (3). There is a slow drift in the divergence of trajectories and in the growth of asymmetry.

This example serves to highlight the necessity for precise apriori knowledge of the proximity to homoclinic manifolds. However, it is not obvious from physical data when one is actually close, or how to determine proximity to homoclinic manifolds from a few global constants. For the family of initial data under consideration, as the energy levels are varied (even slightly), there are alternating windows of tori and whiskered tori. In this paper we show that initial data (3) is representative data for a family of solutions that is exponentially close (in spectral space) to homoclinic manifolds/whiskered tori. We believe that the geometry of the level sets is the significant feature for a numerical simulation to preserve. Obviously this exponentially small splitting cannot be maintained and thus errors on the order of round-off destroy the solution. Further, we obtain estimates of proximity for the neutrally stable tori that have initial data nearby to the plane wave state.

Homoclnic manifolds associated with the NLS equation can be obtained with the Inverse Scattering Transform. There is a known class of quasi-periodic solutions which are given in terms of genus-N Riemann theta functions and which may be stable or unstable. Given a critical torus, the homoclinic manifold can be constructed either with a Backlund transformation of the unstable quasi-periodic solution [**8**] or by transforming known soliton solutions [**10**]. In §2 we recall a few spectral preliminaries and show the correspondence between whiskered tori and complex double points of the periodic spectrum . In §3 we investigate the splitting of the double points under small perturbations in the potential via a perturbation analysis. Armed with this information one then has a means of determining the feasibility of accurate numerical simulations.

2. Spectral Background

The NLS equation is equivalent to consistency of the associated linear systems ("Lax Pairs"):

$$L^{(x)}\mathbf{v} = \begin{pmatrix} \partial/\partial x + i\zeta & -u \\ u^* & \partial/\partial x - i\zeta \end{pmatrix} \mathbf{v} = 0 \tag{4}$$

$$L^{(t)}\mathbf{v} = \begin{pmatrix} \partial/\partial t - i(|u|^2 - 2\zeta^2) & -iu_x - 2\zeta u \\ -iu_x^* + 2\zeta u^* & \partial/\partial t + i(|u|^2 - 2\zeta^2) \end{pmatrix} \mathbf{v} = 0 \tag{5}$$

The solutions of the NLS equation are characterized in terms of the spectrum of $L^{(x)}$ which is defined as

$$\sigma(L^{(x)}) = \{\zeta | L^{(x)}\mathbf{v} = 0, |\mathbf{v}| \text{ bounded } \forall x\} \tag{6}$$

Since the potential $u(x,t)$ is periodic in x, the spectrum is obtained using Floquet theory.

The fundamental matrix, $\mathcal{M}(x;\zeta,u)$, is defined by the IVP

$$L^{(x)}\mathcal{M} = 0 \tag{7}$$

$$\mathcal{M}(0;\zeta,u) = \begin{pmatrix} 1 & 0 \\ 0 & 1 \end{pmatrix}. \tag{8}$$

and the discriminant, $\Delta(\zeta,u)$, of the monodromy matrix, $\mathcal{M}(L;\zeta,u)$, is given by

$$\Delta(\zeta,u) = Tr\,\mathcal{M}(l;\zeta,u) \tag{9}$$

Condition (6) translates into the following condition on the discriminant:

$$\sigma(L^{(x)}(u,u^*)) = \{\zeta \in C | \Delta(\zeta,u) \in R, -2 \le \Delta \le 2\} \tag{10}$$

The spectral quantities to which we will refer are:

(i) The simple periodic spectrum

$$\sigma^s = \{\zeta_j^s | \Delta(\zeta,u) = \pm 2,\ d\Delta/d\zeta \ne 0\} \tag{11}$$

and

(ii) The double points of the spectrum

$$\sigma^d = \{\zeta_j^d | \Delta(\zeta, u) = \pm 2,\ d\Delta/d\zeta = 0,\ d^2\Delta/d\zeta^2 \neq 0\} \tag{12}$$

It is possible to obtain critical points of higher multiplicity, however we will consider just the case of double points in the spectrum. Also, for convinience, in this discussion we restrict our attention to even potentials.

The spectrum of $L^{(x)}$ is invariant under the NLS flow and the periodic/ antiperiodic eigenvalues $\{\zeta_i\}_1^\infty$ provide sufficient invariants to establish the integrability of the NLS equation. $L^{(x)}$ is not self-adjoint; the spectrum consists of curves in the complex ζ-plane. The simple periodic eigenvalues determine the end points of the curves of spectrum and the double points must lie on the bands of spectrum.

The nonlinear mode content of solutions of the NLS is completely determined by the location of the periodic eigenvalues ζ_j. The spectrum of $L^{(x)}$ is symmetric with respect to the real axis, hence we can characterize the solution in terms of spectrum in the upper half ζ-plane. If $u(x,t)$ is an N-phase solution, $u(x,t) = u(\theta_1, \dots, \theta_N)$, the phases evolve according to $\theta_j = \kappa_j x + \omega_j t + \theta_j^{(o)}$ where κ_j and ω_j are completely determined by σ^s. The isospectral class of an N-phase solution (solutions with the same spectrum) is an N-dimensional torus T^N coordinatized by the phases (T^N is the product of N circles and resides in the NLS phase space). Instability of an N-phase solution depends on the existence of complex double points. There are no instabilities associated with simple spectrum or real double points. However, there is a correspondence between the number of complex double points, the number of unstable modes of the linearized NLS equation, and the dimension of the manifold homoclinic to the N-phase torus. For N bands of spectrum (containing M complex double points) we obtain an unstable N-phase solution corresponding to flow on a torus T^N with stable (unstable) manifolds of dimension m, where $m \leq M$. When the complex double points are all pure imaginary, strict equality is obtained i.e. $m = M$. Otherwise, each complex double point must be individually checked for instability. The simplest example to illustrate the correspondence between the double points in the spectrum, the number of linearly unstable modes and the structure of the critical level sets is associated with the plane wave solution $u_0(x,t) = ae^{2i|a|^2 t}$. In the focusing case, the plane wave state can be modulationally unstable. When considering small perturbations of the form $u(x,t) = u_0(x,t)(1 + \epsilon(x,t))$, $|\epsilon| \ll 1$, ϵ satisfies the linearized NLS

$$i\epsilon_t + \epsilon_{xx} + 2|a|^2(\epsilon + \epsilon^*) = 0$$

and consequently $\epsilon(x,t)\alpha e^{i\mu x}e^{\sigma_n t}$, where $\mu_n = 2\pi n/L$ and the growth rate σ_n is given by $\sigma_n = \mu_n\sqrt{4|a|^2 - \mu_n^2}$. Thus the solution is unstable provided $0 < (n\pi/L)^2 < |a|^2$. The number of unstable modes is the largest integer M satisfying $0 < M < |a|L/\pi$. The fastest growth rate corresponds to a value $n(\mu_n)$ for which σ_n is maximal (i.e. closest to $2|a|^2$).

For the plane wave solution, $u_0(x,t); t=0$, the system (4) has constant coefficients and is easily solved. The discriminant $\Delta(\zeta,u_0) = 2\cos L\sqrt{\zeta^2+a^2}$ and the periodic spectrum is given by $\zeta_j^2 = (j\pi/L)^2 - a^2$. Each of these points (except for $j=0$) is a double point. The spectrum of the plane wave consists of the real axis and a single band of spectrum along the imaginary axis terminating at $\pm ia$. There is an important distinction between double points on the real axis, $0 < a^2 < (j\pi/L)^2$, and complex double points, $0 < (j\pi/L)^2 < a^2$. The condition for complex double points is exactly the condition for linear instability. The number of complex double points corresponds exactly to the number of linearly unstable modes. There is no instability associated with the real double points. There is one band of spectrum in the upper half plane for the plane wave solution and the tori, T^1, in the level sets are actually circles. For M complex double points embedded in this band of spectrum, these circles have stable (unstable) manifolds of dimension M. If the double points are all real then the tori, T^1, are stable.

How do the nearby states to the plane wave solution behave and what are their associated spectral configurations? Here is an example: consider the initial condition $u(x,0) = a(1+\epsilon_0\cos(\mu x))$, with $\mu = 2\pi/L$, $a = 1/2$, $L = 2\sqrt{2}\pi$ (u(x,0) contains one unstable mode) and $\epsilon_0 = |\epsilon_0|e^{i\phi_0}$, $|\epsilon| \ll 1$. There are three possibilities for this type of even initial data which are illustrated by the following: **(a)** $\phi_0 = \pi/4$ (see Figure 1) which corresponds to the "purely unstable" case (where the coefficient of $e^{-\sigma_n t}$ vanishes (recall $\sigma_n = \mu_n\sqrt{4a^2-\mu_n^2}$, there are two signs for the growth rate)); **(b)** $\phi_0 = 0$ (see Figure 2); and **(c)** $\phi_0 = \pi/2$ (see Figure 3). Case (a) is an example of a solution homoclinic to the plane wave state. The homoclinic solution is characterized by a single mode and its associated spectral configuration is given in Figure 3. For this special form of the perturbation, the double point is not split. The isospectral set of the plane wave solution includes this homoclinic solution which is an explicit representation of the unstable manifold of the plane wave and which approaches the unstable tori T^1 as $t \to \pm\infty$. We call cases (b) and (c) "inside" and "outside" the homoclinic orbit: $\phi_0 = \pi/4$. Cases (b) and (c) are examples of 2-phase solutions and their spectral configurations are given in Figures 4 and 5, respectively. In the surfaces (Figures 4,5), only one phase is observable (for a general N-phase solution only N-1 phases are observable) since $|u(x,t)|$ is represented and not $u(x,t)$. The double point has split into two simple points and has opened either into a "gap" in the spectrum (gap state -Figure 4) or has formed a "cross state" in the spectrum (cross state -Figure 5). The corresponding level sets are two dimensional stable tori, T^2. The case $\phi_0 = \pi/4$ acts as a "separatrix" between two behaviors. Note that the frequency of maxima (at $x=0$) in case (c) is roughly half that of case (b), much the same way as it is for the well known cubic nonlinear oscillator (i.e. Duffing's equation). Nearby to the plane wave state, there is a rich class of nontrivial solutions to the NLS which are N-phase solutions and in the degenerate case reduce to homoclinic solutions.

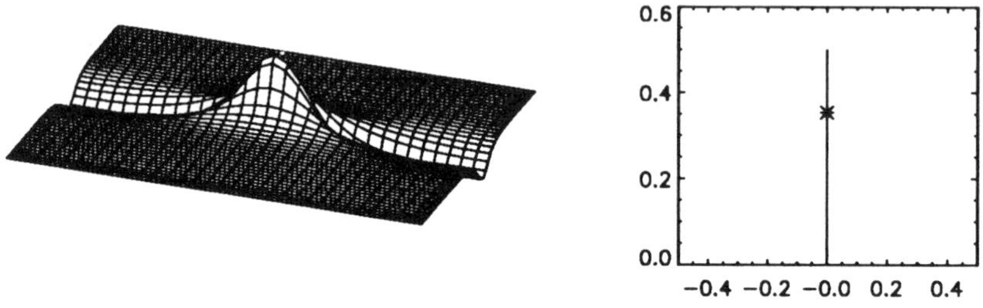

FIGURE 3. The surface $|u(x,t)|$ and the nonlinear spectrum with one double point for the homoclinic solution of the NLS equation with initial condition $u(x,0) = 0.5(1 + 10^{-5}(1 + i)\cos\mu x)$, $0 \leq t \leq 50$.

3. Proximity to Homoclinic Manifolds via Perturbation Analysis

Since the initial data under consideration is within an ϵ neighborhood of the plane wave, we can calculate the periodic/antiperiodic spectrum of (2-3) via perturbation analysis [**5, 2**]. Recall that the plane wave spectrum contains an infinite number of double points. We will be concerned with the splitting of each of the double points under the perturbation. Let ζ_j, $(j = 1, N)$, denote the associated spectrum of the homoclinic manifold $u_0(x,t)$. For the perturbed potential, $u^\epsilon(x,t)$, the spectral elements deform to $\zeta_j^{(\epsilon)}$. The proximity to the homoclinic manifold can be measured by

$$\min_{1\leq k\leq N} |\zeta_k^{(\epsilon)} - \zeta_k| \tag{13}$$

This is a simplification of the generic case of an arbitrary potential, where to determine proximity, it is neccessary in addition, to minimize over all the nearby spectral configurations that contain complex double points. The spectrum is invariant under the NLS flow, hence we consider $u(x,0)$, and supress the time dependence of the eigenfunctions etc., in the following analysis. The initial data under consideration is of the form

$$u = a + \epsilon u^{(1)} \tag{14}$$

$$= a + \epsilon \exp(i\phi)\cos(\mu_n x), \tag{15}$$

where a is a real constant and $\mu_n = 2\pi n/L$ (for integer n). At the double points of $\Delta(\zeta, u)$, the discriminant Δ and the eigenfunctions are analytic functions of u, so it is natural to assume the following perturbation expansions,

$$\zeta_j = \zeta_j^{(0)} + \epsilon\zeta_j^{(1)} + \epsilon^2\zeta_j^{(2)} + \cdots \tag{16}$$

$$v_j = v_j^{(0)} + \epsilon v_j^{(1)} + \epsilon^2 v_j^{(2)} + \cdots \tag{17}$$

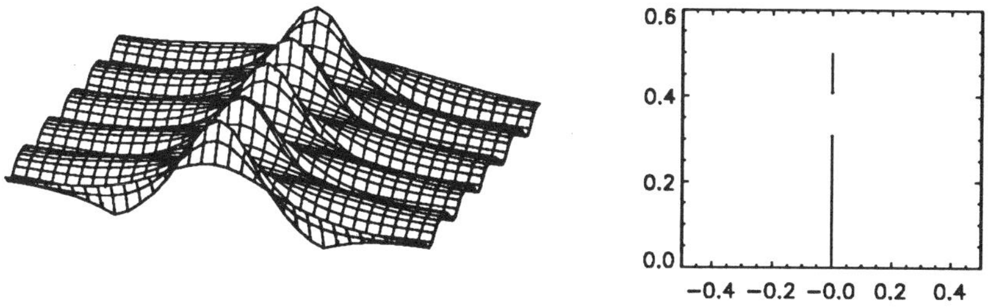

FIGURE 4. The surface $|u(x,t)|$ and the nonlinear spectrum with one imaginary gap for a 2-phase solution of the NLS equation with initial condition $u(x,0) = 0.5(1 + 10^{-1}\cos\mu x)$, $0 \leq t \leq 50$.

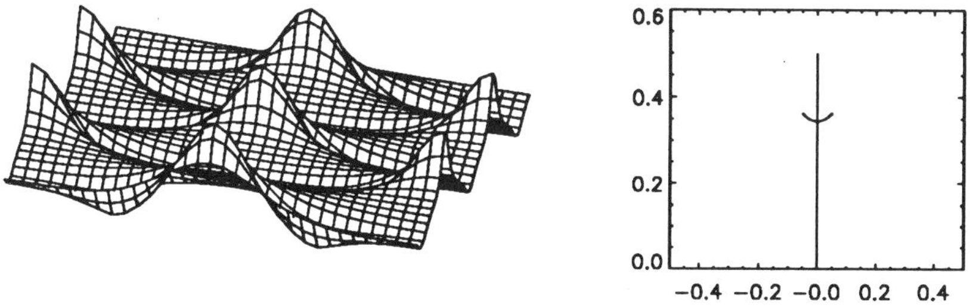

FIGURE 5. The surface $|u(x,t)|$ and the nonlinear spectrum with one cross for a 2-phase solution of the NLS equation with initial condition $u(x,0) = 0.5(1 + 10^{-1} i \cos\mu x)$, $0 \leq t \leq 50$.

where

$$v_j = \begin{pmatrix} v_{j1} \\ v_{j2} \end{pmatrix}.$$

This type of perturbation analysis was used in the computation of splitting of eigenvalues such as those described at the end of the previous section on the 2-phase solution [**2**].

Substituting these expansions into the operator, (4), and equating the various orders of ϵ, we obtain the following :

$$\mathcal{O}(\epsilon^0) \to \mathcal{L}v_j^{(0)} = 0, \tag{18}$$

$$\mathcal{O}(\epsilon) \to \mathcal{L}v_j^{(1)} = \begin{cases} -i\zeta_j^{(1)} v_{j1}^{(0)} + u^{(1)} v_{j2}^{(0)} \\ -i\zeta_j^{(1)} v_{j2}^{(0)} + u^{(1)*} v_{j1}^{(0)} \end{cases} \tag{19}$$

$$\mathcal{O}(\epsilon^2) \to \mathcal{L}v_j^{(2)} = \begin{cases} -i\zeta_j^{(1)} v_{j1}^{(1)} - i\zeta_j^{(2)} v_{j1}^{(0)} + u^{(1)} v_{j2}^{(1)} \\ -i\zeta_j^{(1)} v_{j2}^{(1)} - i\zeta_j^{(2)} v_{j2}^{(0)} + u^{(1)*} v_{j1}^{(1)} \end{cases} \tag{20}$$

$$\vdots$$

$$\mathcal{O}(\epsilon^m) \to \mathcal{L}v_j^{(m)} = \begin{cases} -i\sum_{k=0}^{m-1} \zeta_j^{(m-k)} v_{j1}^{(k)} + u^{(1)} v_{j2}^{(m-1)} \\ -i\sum_{k=0}^{m-1} \zeta_j^{(m-k)} v_{j2}^{(k)} + u^{(1)*} v_{j1}^{(m-1)} \end{cases} \tag{21}$$

where

$$\mathcal{L} = \begin{pmatrix} 1 & 0 \\ 0 & -1 \end{pmatrix} \frac{\partial}{\partial x} - \begin{pmatrix} 0 & a \\ a & 0 \end{pmatrix} + i\zeta_j \begin{pmatrix} 1 & 0 \\ 0 & 1 \end{pmatrix}. \tag{22}$$

The leading order problem yields the spectrum and eigenfunctions for the plane wave state. At the double points, ζ_j (for simplicity, hereafter we use ζ_j instead of $\zeta_j^{(0)}$), the eigenspace is two dimensional and is spanned by the eigenfunctions

$$\phi_j^{\pm} = \exp(\pm ik_j x) \begin{pmatrix} 1 \\ (i/a)(\pm k_j + \zeta_j) \end{pmatrix}$$

and the general solution is given by

$$v_j^{(0)} = A^+ \phi_j^+ + A^- \phi_j^-, \tag{23}$$

where $\zeta_j^2 = k_j^2 - a^2$, $k_j = j\pi/L$

Assuming periodic/antiperiodic eigenfunctions of period L, the solvability condition for the system

$$\mathcal{L}v = F, \tag{24}$$

with

$$F = \begin{pmatrix} F_1 \\ F_2 \end{pmatrix}$$

is given by the orthogonality condition

$$\int_0^L (F_1 w_1^* + F_2 w_2^*)\, dx = 0$$

for all w in the nullspace of the Hermitian operator, $\mathcal{L}^H$, where,

$$\mathcal{L}^H = -\begin{pmatrix} 1 & 0 \\ 0 & -1 \end{pmatrix} \frac{\partial}{\partial x} - \begin{pmatrix} 0 & a \\ a & 0 \end{pmatrix} - i\zeta_j^* \begin{pmatrix} 1 & 0 \\ 0 & 1 \end{pmatrix}. \tag{25}$$

Noting that the nullspace of $\mathcal{L}^H$, at the double points, is spanned by

$$(26)\qquad \begin{pmatrix} \left(\phi_{j2}^{\pm}\right)^* \\ \left(\phi_{j1}^{\pm}\right)^* \end{pmatrix},$$

the general solvability condition assumes the following form:

$$(27)\qquad \int_0^L \left(F_1\phi_{j2}^+ + F_2\phi_{j1}^+\right) dx = 0,$$

$$(28)\qquad \int_0^L \left(F_1\phi_{j2}^- + F_2\phi_{j1}^-\right) dx = 0.$$

The solvability conditions (27,28) applied to (19) yields the system of equations,

$$(29)\qquad \begin{pmatrix} T_+ & T \\ T & T_- \end{pmatrix} \begin{pmatrix} A^+ \\ A^- \end{pmatrix} = 0,$$

where

$$(30)\qquad T = 2\zeta_j\zeta_j^{(1)}/a$$

$$(31)\qquad T_\pm = \begin{cases} -\frac{1}{2a^2}(\pm k_j + \zeta_j)^2 \exp(i\phi) + \frac{1}{2}\exp(-i\phi) & j = n \\ 0 & j \neq n \end{cases}.$$

Consequently the system can be solved nontrivially for $A^\pm$ if

(32)

$$\left(\zeta_j^{(1)}\right)^2 = \begin{cases} \frac{a^2}{16\zeta_j^2}\left(e^{-i\phi} - \frac{1}{a^2}(k_j+\zeta_j)^2 e^{i\phi}\right)\left(e^{-i\phi} - \frac{1}{a^2}(-k_j+\zeta_j)^2 e^{i\phi}\right) & j = n \\ 0 & j \neq n \end{cases}.$$

At $\mathcal{O}(\epsilon)$, there is a correction only to the double point, ζ_j, $(j = n)$. A specific double point is selected in resonance with the perturbation in the eigenfunction. The other double points do not experience an $\mathcal{O}(\epsilon)$ correction.

The behavior of the correction $\zeta_j^{(1)}$, $(j = n)$, depends on whether the double point ζ_j is real or imaginary:

$$(33)\qquad \left(\zeta_j^{(1)}\right)^2 = \begin{cases} -\frac{a^2}{4\zeta_j^2}\sin(\phi+\theta)\sin(\phi-\theta) & \text{for } \zeta_j \text{ imaginary} \\ -\frac{a^2}{8\zeta_j^2}\left[\cos 2\phi + 1 - 2k_j^2/a^2\right] & \text{for } \zeta_j \text{ real} \end{cases}$$

where $\tan\theta = Im(\zeta_j)/k_j$.

For imaginary ζ_j, the correction $\zeta_j^{(1)}$ can be real, zero or pure imaginary [**2**], depending on the choice of ϕ in the perturbed potential. Complex double points can split into either crosses or gaps in the spectrum (see figs. 4-5 for a representation of this splitting). This is a realization of the saddle structure of the real part $\Delta(\zeta_j, u)$ when ζ_j is imaginary.

For real ζ_j, the perturbation can only be imaginary ($a^2 < k^2$ for real double points). Gaps cannot appear on the real axis in the spectrum of $\mathrm{L}^{(x)}$. Hence the situation with real double points is very different from from that of imaginary double points. Splitting of the real double points introduces additional degrees of freedom into the spatial structure but does not introduce any instability as homoclinic manifolds are not associated with them.

Let's examine the second order correction to the double points for which $j \neq n$. At these double points, $\zeta_j^{(1)} = 0$, so

$$\mathcal{L}v_j^{(2)} = \left\{ \begin{array}{l} -i\zeta_j^{(2)} v_{j1}^{(0)} + u^{(1)} v_{j2}^{(1)} \\ -i\zeta_j^{(2)} v_{j2}^{(0)} + u^{(1)*} v_{j1}^{(1)} \end{array} \right. \tag{34}$$

To apply the solvability condition to $\mathcal{L}v_j^{(2)} = F$ we need the $\mathcal{O}(\epsilon)$ correction to the eigenfunctions, $v_j^{(1)}$. We find these to be given by

$$v_j^{(1)} = Ae^{i(k_j+\mu_n)x} + Be^{i(k_j-\mu_n)x} + Ce^{i(-k_j+\mu_n)x} + De^{-i(k_j+\mu_n)x} \tag{35}$$

where

$$A = \frac{A^+}{\mu_n^2 + 2k_j\mu_n} \begin{pmatrix} (2a^2 + k_j\mu_n + \zeta_j\mu_n)/a \\ i(2k_j + 2\zeta_j + \mu_n) \end{pmatrix} \tag{36}$$

$$B = \frac{A^+}{\mu_n^2 - 2k_j\mu_n} \begin{pmatrix} (2a^2 - k_j\mu_n - \zeta_j\mu_n)/a \\ i(2k_j + 2\zeta_j - \mu_n) \end{pmatrix} \tag{37}$$

$$C = \frac{A^-}{\mu_n^2 - 2k_j\mu_n} \begin{pmatrix} (2a^2 - k_j\mu_n + \zeta_j\mu_n)/a \\ i(-2k_j + 2\zeta_j + \mu_n) \end{pmatrix} \tag{38}$$

$$D = \frac{A^-}{\mu_n^2 + 2k_j\mu_n} \begin{pmatrix} (2a^2 + k_j\mu_n - \zeta_j\mu_n)/a \\ i(-2k_j + 2\zeta_j - \mu_n) \end{pmatrix} \tag{39}$$

Returning to the solvability condition at $\mathcal{O}(\epsilon^2)$ we obtain the following system

$$\begin{pmatrix} \alpha & \zeta_j^{(2)} - \frac{\mu_n^2\zeta_j}{\mu_n^4 - 4\mu_n^2 k_j^2} \\ \zeta_j^{(2)} - \frac{\mu_n^2\zeta_j}{\mu_n^4 - 4\mu_n^2 k_j^2} & \alpha \end{pmatrix} \begin{pmatrix} A^+ \\ A^- \end{pmatrix} = 0. \tag{40}$$

where

$$\alpha = \left\{ \begin{array}{ll} \frac{k_j - \zeta_j}{\mu_n^2 - 2\mu_n k_j} & j = 2n \\ 0 & j \neq 2n \end{array} \right. . \tag{41}$$

Nontrivial solutions implies

$$\zeta_j^{(2)} = \frac{\mu_n^2\zeta_j}{\mu_n^4 - 4\mu_n^2 k_j^2} \pm \alpha. \tag{42}$$

Consequently only the double point ζ_j, $(j = 2n)$, experiences an ϵ^2-splitting of 2α. The other double points, $j \neq n, 2n$, experience just a translation of $\zeta_j^{(2)}$ at $\mathcal{O}(\epsilon^2)$.

At $\mathcal{O}(\epsilon^m)$, to examine the splitting of the double points ζ_j, $j \neq n, 2n, \ldots, (m-1)n$, note that the general form of the correction to the eigenfunction, $v_j^{(m-1)}$, is

$$\begin{aligned} v_j^{(m-1)} &= A^{(m-1)} e^{i(k_j+(m-1)\mu_n)x} + B^{(m-1)} e^{i(k_j-(m-1)\mu_n)x} \\ &\quad + C^{(m-1)} e^{i(-k_j+(m-1)\mu_n)x} + D^{(m-1)} e^{-i(k_j+(m-1)\mu_n)x} \\ &\quad + v_j^{(m-2)} \end{aligned} \tag{43}$$

where $A^{(m-1)}, B^{(m-1)}, C^{(m-1)}, D^{(m-1)}$ are column vectors. The solvabilty condition at $\mathcal{O}(\epsilon^m)$ yields the following system

$$\begin{pmatrix} \alpha^+ & 2\zeta_j\zeta_j^{(m)}/a \\ 2\zeta_j\zeta_j^{(m)}/a & \alpha^- \end{pmatrix} \begin{pmatrix} A^+ \\ A^- \end{pmatrix} = 0. \tag{44}$$

where

$$\alpha^\pm = \begin{cases} \alpha_m^\pm & j = mn \\ 0 & j \neq mn \end{cases} \tag{45}$$

with

$$\alpha_m^\pm = \frac{1}{2}\left(\frac{\pm k_{mn} + \zeta_{mn}}{a} A_2^{(m-1)} - i A_1^{(m-1)} \right).$$

Nontrivial solutions implies

$$\zeta_j^{(m)} = \pm\sqrt{\alpha_m^+ \alpha_m^-}. \tag{46}$$

Only the double points, ζ_j, which are m-fold multiples of the fundamental mode, $(j = mn)$ will experience an ϵ^m-splitting of $2\alpha_m$. The splitting distance of the remaining double points $(j \neq mn)$ is beyond all orders in ϵ! When there is more than one complex double point in the initial spectral configuration, and other than the fundamental mode is perturbed, the result is "exponential closeness" to the homoclinic manifold (in numerical calculations we have established that in general this type of perturbation results in an exponentially small splitting of the eigenvalues).

To confirm this perturbation analysis,the spectrum of the following initial data

$$q = 1.5(1 + 0.1\cos\mu_8 x) \tag{47}$$

was computed numerically and compared with the theoretical estimates. Our analysis predicts that at $\mathcal{O}(\epsilon)$, the 8th double point will split and the splitting is 0.15. At $\mathcal{O}(\epsilon^2)$ the other 7 double points should translate according to the formula

$$\zeta_j^{(2)} = \frac{\mu_n^2\zeta_j}{\mu_n^4 - 4\mu_n^2 k_j^2}. \tag{48}$$

j	Predicted	Computed
1	0.00425	0.00425
2	0.00437	0.00437
3	0.00459	0.00452
4	0.00496	0.00497
5	0.00559	0.00560
6	0.00681	0.00683
7	0.01017	0.01017
8	0.15000	0.15001

TABLE 1.

Table 1 shows a comparison of the predicted versus obtained splitting distance and translation. The corrections are $\mathcal{O}(10^{-3})$, so with four decimal agreement we correctly capture them.

As another example we discuss these results with regard to the initial data (2): $u(x,0) = a(1 + \epsilon \cos 3\mu x)$ with ϵ small. We find that at leading order there are 3 complex double points. Each one of these double points corresponds to an unstable mode. Under perturbation the double point corresponding to the perturbation 3μ (j=3)is split by $\mathcal{O}(\epsilon)$. The double points which correspond to multiples of the perturbation (i.e. j=3m) are split $\mathcal{O}(\epsilon^m)$. The first two complex double points, ζ_j, ($j = 1,2$) while translating by $\mathcal{O}(\epsilon^2)$ nevertheless have a "splitting distance" which is smaller than any power of ϵ. Consequently we are "exponentially close" to the homoclinic manifolds, and small deviations on the order of round-off can lead to homoclinic crossings and the observed chaos. The extreme proximity of these data values to the unperturbed homoclinic manifolds, prevents us from numerically calculating the "true" solution after moderate time. The calculations are influenced by minuscule errors; computationally the NLS equation is "effectively chaotic" in this range of parameter space despite it being "integrable". Extreme care must be exercised in the physical interpretation of numerical solutions. On the other hand, since NLS is so special, we expect an underlying structure to the temporal disorder and we believe that a statistical study would reveal this structure.

4. Conclusions

The parameter regimes which yield extremely small splitting of the eigenvalues and therefore a chaotic flow should be possible to simulate in laboratory experiments (e.g. nonlinear optics, water waves). Such laboratory experiments might be extremely sensitive and could result in a chaotic response (the response would be heavily influenced by the next order terms which are not included in the original NLS derivation).

We note that given general initial data it is unclear how to recognize stable/unstable data by only examining global quantities like the constants of motion. By looking carefully at spectral configurations we have shown that the data (2) is typical of more general data which will inevitably put one exponentially close to homoclinic manifolds. This concrete representation of data which are $\mathcal{O}(\epsilon^m)$ or exponentially close to unstable tori is accessible: it is certainly easy to reproduce numerically; and we believe experimentally as well.

It is apparent that a central issue to consider in order to obtain reliable results is the proximity of the initial data to complex double points. We also note that it is not just homoclinic manifolds that cannot be reliably simulated, but a non-vanishing measure of the quasiperiodic solutions. A priori estimates regarding proximity to homoclinic manifolds will be developed for other general classes of initial data (e.g. perturbations of general N-phase solutions). This will enable one to determine the feasibility of moderate to long time simulations. Such estimates will help us develop criteria that will ensure that level sets will not change their geometry.

5. Appendix

In order to preserve important properties and constants of the flow, conservative (in space) difference schemes have been used in connection with the NLS equation and similar equations.

The two finite difference schemes discussed are:

$$i\dot{u}_n + (u_{n+1} + u_{n-1} - 2u_n)/h^2 + 2|u_n|^2 u_n = 0 \tag{49}$$

$$i\dot{u}_n + (u_{n+1} + u_{n-1} - 2u_n)/h^2 + |u_n|^2 (u_{n+1} + u_{n-1}) = 0 \tag{50}$$

where $h = L/N$ and $u_{j+N} = u_j$. These schemes are referred to as the diagonal (DDNLS) and the integrable (IDNLS) discretizations of the NLS equation, respectively. Both schemes are Hamiltonian and second order accurate. There are two constants of the motion associated with DDNLS, (the norm and Hamiltonian), whereas IDNLS is a completely integrable system [**3**].

References

1. M.J. Ablowitz, and B.M. Herbst, *On homoclinic structure and numerically induced chaos for the nonlinear Schrödinger equation*, SIAM J. Appl. Math. **50** (1990), 339-351.
2. ———, *On homoclinic boundaries in the nonlinear Schrödinger equation, in "Hamiltonian Systyems, Transformation Groups and Spectral Transform"*, eds. J. Harnad and J. Marsden, 121-131, Les Publications de Centre Recherches Mathematiques, Montreal.
3. M.J. Ablowitz, and J.F. Ladik, *Nonlinear differentia-difference equations and Fourier analysis*, J. Math. Phys. **17** (1976), 1011-1018.
4. M.J. Ablowitz, and C.M. Schober (in preparation).
5. M.J. Ablowitz, C.M. Schober, and B.M. Herbst, *Numerical chaos, roundoff errors, and homoclinic manifolds*, Phys. Rev. Lett. **71** (1993), 2683-2686.
6. A.R. Bishop, M.G. Forest, D.M. McLaughlin, and E.A. Overman II, *A quasiperiodic route to chaos in a near-integrable PDE*, Physica D **23** (1986), 293-328.
7. J. De Frutos, and J.M. Sanz-Serna, *Split-step spectral schemes for nonlinear Dirac Systems*, J. Comput. Phys. **83** (1989), 407-423.
8. N. Ercolani, M.G. Forest, and D.W. McLaughlin, *Geometry of modulational instability*, Physica D **43** (1990), 349-384.

9. G. Forest, C. Goedde, and A. Sinha, *Instability-driven energy transport in near-integrable, many degree-of-freedom, Hamiltonian systems*, Phys. Rev. Lett. **68** (1992), 2722-2725.
10. B.M. Herbst, and M.J. Ablowitz, *Numerical induced chaos in the nonlinear Schrödinger equation*, Phys. Rev. Lett. **62** (1989), 2065-2068.
11. ———, *Numerical chaos, symplectic integrators, and exponentially small splitting distances*, J. Comput. Phys. **105** (1993), 122-132.
12. A.R. Its, and V.P. Kotlarov, *Explicit formulas for solutions of a nonlinear Schrödinger equation*, Dokl. Akad. Nauk Ukain. SSR Ser. A **11** (1976), 965-968.
E.R Tracy, and H.H. Chen, *Nonlinear self-modulational: an exactly solvable model*, Phys. Rev. A **37** (1988), 815-839.
13. D.W. McLaughlin, *Whiskered Tori for nonlinear Schrödinger equations,* Princeton University, preprint, (1991).
14. D.W. McLaughlin, and C.M. Schober, *Chaotic and homoclinic behavior for numerical discretizations of the nonlinear Schrödinger equation*, Physica D **57** (1992), 447-465.
15. C.M. Schober, *Numerical and analytical studies of the discrete nonlinear Schrödinger equation*, Ph.D. Thesis, University of Arizona, (1991).
A. Calini, N. Ercolani, D.W. McLaughlin, and C.M. Schober, *Melnikov analysis of numerically induced chaos in the nonlinear Schrödinger equation* (in preparation).

Program in Applied Mathematics, University of Colorado at Boulder, Boulder, CO 80309

E-mail address: schober@boulder.colorado.edu

Contemporary Mathematics
Volume **172**, 1994

Discretisation Effect on a Dynamical System with Discontinuity

XINGHUO YU

ABSTRACT. Two case studies are presented to show discretisation effects on a dynamical system with discontinuity – the Variable Structure Control system. It is shown that discretisation actually introduces an extra dimension of freedom, namely the sampling period, whose value is crucial to the occurrence of irregular and unpredictable behaviour and instability. Computer simulations are presented to demonstrate the various discretisation effects.

1. Introduction

The dynamical system we consider here is a control system which is defined as

$$\dot{x} = f(x, u) \tag{1}$$

where $x \in R^n$ is the system state, and $u \in R^1$ is a scalar control input to be determined. The control u is a closed loop feedback control which has the form $u = u(x)$. Once such a control is specified the control system (1) becomes an autonomous continuous-time dynamical system $\dot{x} = f(x, u(x))$.

One prominent control strategy is by means of Variable Structure Control (VSC), characterised by the control structure

$$u(x) = \begin{cases} u^+(x) & s(x) > 0 \\ u^-(x) & s(x) < 0 \end{cases} \tag{2}$$

where $u^+(x) \neq u^-(x)$ and $s(x) = 0$ is the switching manifold on which the discontinuity takes place. The system (1) under the control (2) is actually a differential equation with right hand side discontinuity. The task is to design a control law satisfying (2), such that the sytem trajectories reach and reside on the prescribed switching manifold $s(x) = 0$ for $t > t_0$ where t_0 is a particular moment. The prescribed switching manifold characterises the desired dynamics to be achieved.

To ensure the sliding on $s(x) = 0$, the condition

$$s\dot{s} < 0 \tag{3}$$

1991 *Mathematics Subject Classification.* Primary 93B12; Secondary 34A60.
The auther was supported by a grant from Australian Research Council.
This paper is in final form and no version of it will be submitted for publication elsewhere.

is needed in the neighbouthood of $s(x) = 0$ for the design of u such that the state x crosses the switching manifold $s = 0$, for example, with $u^+(x)$ from $s > 0$ to $s < 0$, and recrosses back from $s < 0$ to $s > 0$ as soon as $u^-(x)$ takes place, and so on and so forth, resulting in the value of u being altered between u^+ and u^-. The resulting motion is called *sliding mode* because of the resulting chattering along $s = 0$.

Due to the discontinuity of the control on the switching manifold $s = 0$, the solution of the differential equation with right-hand side discontinuity should be redefined. Assuming $s = 0$ is a regular and smooth manifold, a function $x(t)$ is a *solution* to (1) with closed loop feedback control $u = u(x)$ in the sense of Filippov ([**5**], [**12**]) if

$$\dot{x}(t) = \alpha f_0^+(x) + (1 - \alpha) f_0^-(x) \tag{4}$$

where $f_0^+(x) = \langle \nabla s, f^+(x, u^+(x)) \rangle$, $f_0^-(x) = \langle \nabla s, f^-(x, u^-(x)) \rangle$, and a proper α $(0 \leq \alpha \leq 1)$ such that the $\dot{x}$ is orthogonal to the tangent of $s = 0$, so the solution remains on the manifold. Here $f^+(x, u^+(x))$ and $f^-(x, u^-(x))$ are limit values at $s = 0$ approaching from sides $s > 0$ and $s < 0$ respectively.

When in sliding mode, the system satisfies equations

$$s(x) = 0 \quad \text{and} \quad \dot{s}(x) = 0 \tag{5}$$

and exhibits invariance properties [**12**] yielding motion which is independent of certain system parameters and disturbances.

The invariance properties are a very important aspect of control systems as the systems must perform well regardless of certain disturbances in their work environments. For further details on the theory of VSC, readers are referred to the book and survey papers ([**3**], [**1**], [**12**]).

The theory of continuous-time VSC presents elegant results which are very attractive to control engineers. However when coming to implementation, especially digital implementation, due to the physical limitation of switching equipments, the frequency of switching is fixed and not very high. The discretisation schemes may hardly approximate the continuous-time VSC at all because of the comparatively low switching frequency. This in turn may cause problems, such as severe chattering (or zigzagging), chaos and instability etc.

There has been little research done on discretisation chaos of control systems except the work of Ushio and Harai [**11**] who investigated chaos in a class of sampled-data control system with nonlinear controllers. Rubio *et al* [**10**] studied chaotic behaviour in an adaptive control system with the use of power spectrum. Grantham *et al* [**2**] investigated the discretisation chaos of prey-predator feedback control with discontinuous control variable derivatives using Lyapunov exponents approach and fractal dimensions.

However there has been virtually no report on discretisation effects on dynamical systems with discontinuous control except the work [**7**] in which the linear systems with discontinuous control, discretised using the exponential discretisation scheme, were studied to show the existence of discretisation chaos in the systems.

This article further explores the aspects of discretisation effects on the dynamics with discontinuity by studying two example cases. The discretisation schemes we use here are, for the linear system, the zero-order-hold sampler (ZOH) (i.e. freezing

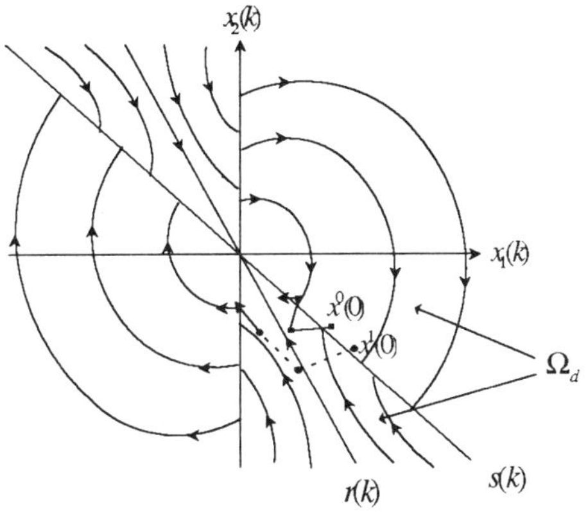

FIGURE 1. Phase plane portrait

the control value during each sampling period) [**4**], for the nonlinear system, the fourth order Runge-Kutta algorithm. The case studies are done experimentally, intending to present phenomena and analysis to show the properties peculiar to the dynamics. The purpose of this article is to raise research interest in the study of discretisation effects on the dynamics with discontinuity. Many questions remain open and need to be addressed.

2. Case 1 : Discretisation effect on a linear control system with discontinuity

Consider the two-dimensional continuous-time VSC system

$$
\begin{aligned}
\dot{x}_1 &= x_2 \\
\dot{x}_2 &= -fx_2 + u
\end{aligned}
\tag{6}
$$

where $u = -\alpha_1|x_1|\text{sgn}(s)$, $\alpha_1 > 0$ is a typical VSC, in which the switching line is defined by

$$s = cx_1 + x_2 = 0, \quad c > 0 \tag{7}$$

which is asymptotically stable. It is well-known that the necessary and sufficient condition [**3**] for $s = 0$ to be a sliding mode in its neighbourhood, characterised by $\dot{x}_1 = -cx_1$, is

$$c^2 - \alpha_1 < cf < c^2 + \alpha_1 \tag{8}$$

It needs to point out that the condition (8) may allow a switching between two unstable systems.

Using the ZOH with a sampling period h (i.e. $u(t) = u(k)$, for $kh \leq t < (k+1)h$), the system is discretised as [**6**]

$$x(k+1) = \Phi x(k) + \Gamma u(k) \tag{9}$$

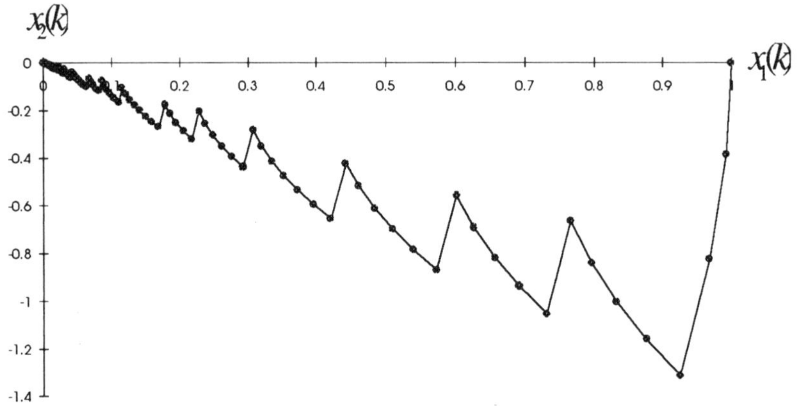

FIGURE 2. System trajectory with $h = 0.04$, $x_1(0) = 1.0$, $x_2(0) = 0.1$

where $x^T = (x_1\ x_2)$,

$$\Phi = \begin{bmatrix} 1 & (1 - \exp(-fh))/f \\ 0 & \exp(-fh) \end{bmatrix} \tag{10}$$

$$\Gamma = [(h/f) + (\exp(-fh) - 1)/f^2 \quad (1 - \exp(-fh))/f]^T \tag{11}$$

and $u(k) = -\alpha_1|x_1(k)|\text{sgn}(s(k))$ with $s(k) = cx_1(k) + x_2(k)$.

The domain of attraction towards the sliding mode, denoted as Ω_d and depicted in Figure 1, is constructed as

$$\Omega_d = \{(x_1, x_2) \mid x_1 r(x) < 0\} \tag{12}$$

where $r(x) = 0$ is the stable asymptote of the control system with $u = \alpha_1 x_1$ [**6**]. The sampling period should then be chosen such that with one sampling period the system state does not overshoot the domain of attraction Ω_d (see the trajectory starting from $x^0(0)$ in Figure 1). An algorithm determining the upper bound of the sampling period, which guarantees the existence of pseudo-sliding mode, has been developed [**6**]. A typical pseudo-sliding mode is depicted in Figure 2, with $\alpha_1 = 9$, $f = -3.49$, $c = 1$, and $h = 0.04$ which is less than the upper bound $H = 0.466$. The system trajectory appears zigzagging along the sliding line.

The problem of interest is the case when h is larger than the upper bound of sampling period. In Figure 1, assume the system state at instant kh, $x(k)$, is just on $x_1(k)s(k) = 0^+$, with one sampling period, the system state $x(k+1)$ may land just below the line determined by $x_1 r(x) = 0^-$. Since the system trajectories diverge from the line $r(x) = 0$, the state $x(k+2)$ may then move away from the line $r(x) = 0$, which means it goes away from $s = 0$ as well (see the trajectory starting from $x^1(0)$ in Figure 1). This behaviour may repeat many times, resulting in appearance of divergence from $s = 0$ such that the pseudo-sliding mode does not exist. Such divergence of trajectories from the sliding line is illustrated in Figure 3, the data of which is the same as for Figure 2 except $h = 0.055$.

The Lyapunov exponents method is used here to further study the change of system behaviour with respect to the sampling period.

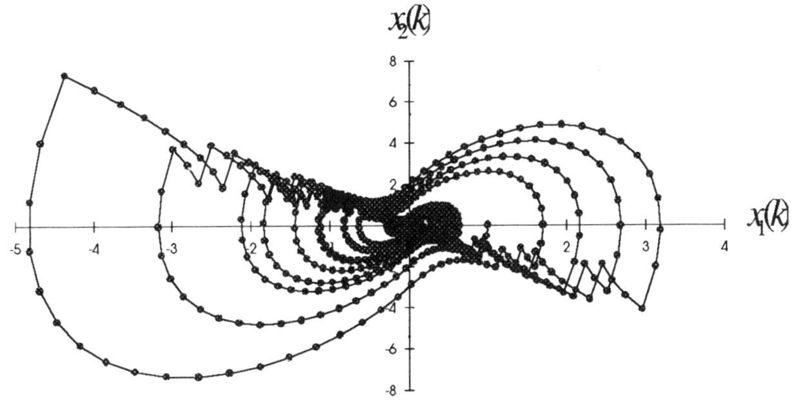

FIGURE 3. System trajectory with $h = 0.055$, $x_1(0) = 1.0$, $x_2(0) = 0.1$

The Lyapunov exponents method is often used to measure the growth rates of the distance between neighbouring trajectories of nonlinear chaotic dynamics. Let $\delta(t)$ denote the distance between two trajectories for a continuous-time system. If $\delta(0)$ is small and $\delta(t) \to \delta(0)\exp(\varrho t)$ as $t \to \infty$, then ϱ is called a Lyapunov exponent. The distance between trajectories grows, shrinks, or remains constant depending on whether ϱ is positive, negative, or zero respectively. The definition of a Lyapunov exponent for discrete-time systems is the same as for continuous-time systems except that t is replaced by kh. However in this study, we consider $\delta(kh)$ as a distance between a trajectory and the origin of the phase plane. Therefore we actually study the growth rates of the distance between the system trajectory and the origin.

For the continuous-time system (1), because the system eventually exhibits an asymptotically stable sliding mode governed by $\dot{x}_1 = -c_1 x_1$, the Lyapunov exponent for the continuous-time system is -1, indicating the trajectory shrinks with rate -1. However the Lyapunov exponent for the discrete-time system (9) is not obvious. Using the Gram-Schmidt algorithm [**2**], the Lyapunov exponent versus the sampling period h is calculated and illustrated in Figure 4. While h increases from 0 to about 0.046, the Lyapunov exponent slowly decreases from -1, the slope of the sliding line (7), indicating that the chattering becomes increasingly worse with the progressive increase of h. The chaotic (or, irregular and unpredictable) behaviour starts when h is greater than 0.046. The Lyapunov exponent jumps up sharply and irregularly with little oscillation with respect to the increase of h. Apparently the pseudo-sliding mode does not exist at all. However the system may be stable (but not asymptotically stable), but the trajectories appear irregular and unpredictable. The Lyapunov exponent becomes positive when $h > 0.06$, indicating the trajectory exponentially grows, i.e. the system is unstable.

The analysis above shows the change of the system behaviour from chattering along the sliding line to divergence from the sliding line, and further to instability.

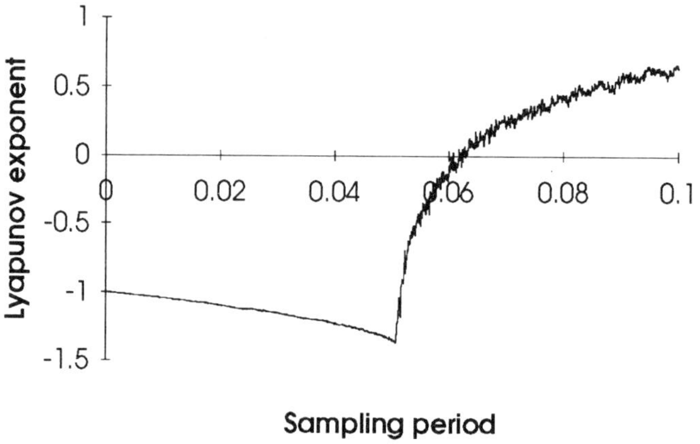

FIGURE 4. Lyapunov exponent versus sampling period

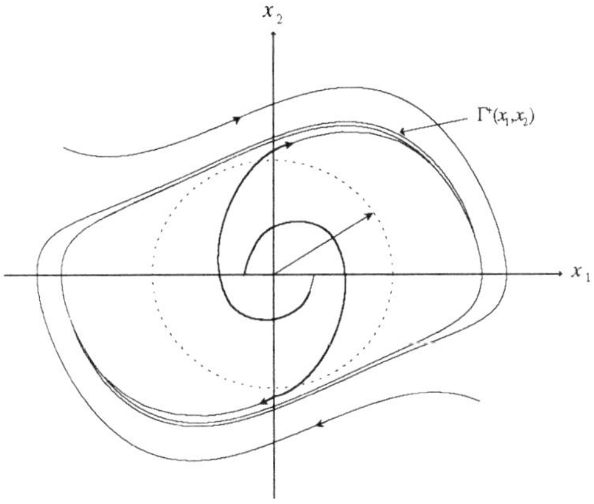

FIGURE 5. Van der Pol's stable limit cycle

3. Case 2: Discretisation effect on a nonlinear control system with discontinuity

The van der Pol equation

$$\dot{x}_1 = x_2 \tag{13}$$

$$\dot{x}_2 = 2\zeta\omega_n(1-\mu x_1^2)x_2 u - \omega_n^2 x_1 \tag{14}$$

with $u = 1$ representing a simple variable damped system in which the negative dampling occurs in the strip $|x_1| < 1/\sqrt{\mu}$ and positive damping in the half-planes $|x_1| > 1/\sqrt{\mu}$. This type of damping leads to a limit cycle, which is denoted by $\Gamma^+(x_1, x_2) = 0$ and depicted in Figure 5. When $u = -1$, the system (14) has a "reverse time and rotated" limit circle, which is denoted by $\Omega^-(x_1, x_2) = 0$ and depicted in Figure 6. Shaping the response of the oscillator is of interest to control engineers to obtain a quasi-sinusoidal response with robustness.

A simple switching between the van del Pol equations with $u = 1$ and $u = -1$ makes it possible to obtain such a robust oscillator [**13**]. Choosing the switching

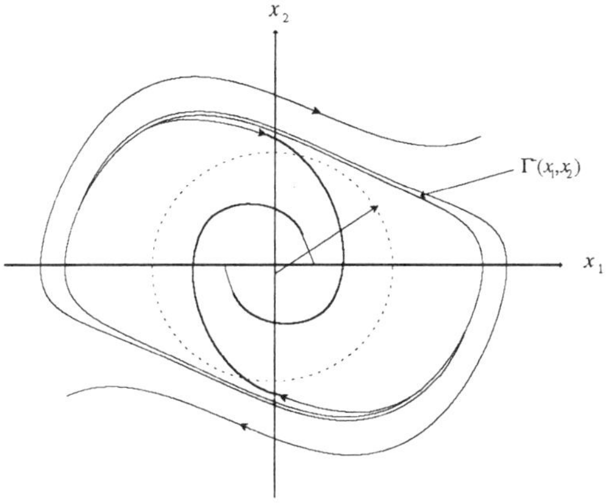

FIGURE 6. Van der Pol's unstable limit cycle

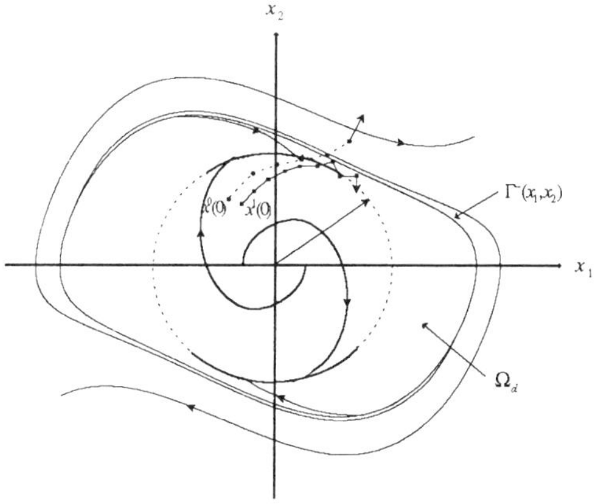

FIGURE 7. VSC controlled van der Pol oscillator

manifold

$$S = \{(x_1, x_2) \in R^2 : s = x_1^2 + \frac{x_2^2}{\omega_n^2} - r^2 = 0 : r < \frac{1}{\sqrt{\mu}}\} \tag{15}$$

and the control law

$$u(x) = \begin{cases} -1 & s(x) > 0 \\ 1 & s(x) < 0 \end{cases} \tag{16}$$

the oscillator approaches the sinusoidal response with the radius r. The sliding harmonic motions are depicted in Figure 7. The existence of the sinusoidal response is guaranteed by the domain of attraction towards the switching manifold, denoted as Ω_d, because of the switching between the van del Pol equations with $u = -1$ and $u = 1$. The domain of attraction Ω_d is actually the intersection of the interior of the "reverse time and rotated " limit cycle $\Gamma^-(x_1, x_2) = 0$, and the band $x_1^2 < 1/\mu$ (see Figure 8). However when discretising the control scheme (16), the control u is fixed, which is either -1 or 1, during the sampling period. We assume here that the trajectories of the discretised system of (14) are exactly on the trajectories for the corresponding continuous-time system (6) except for the points on the switching

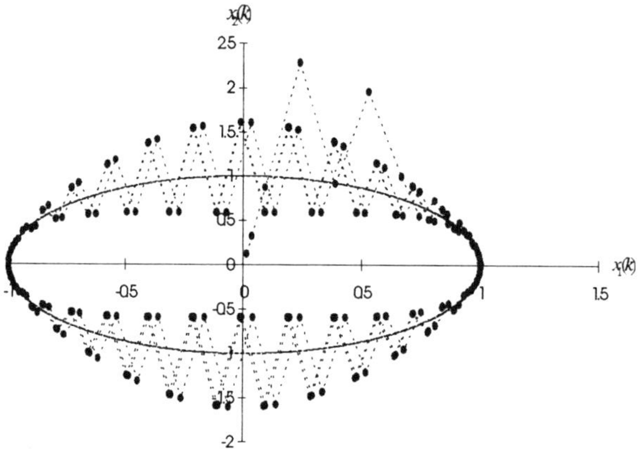

FIGURE 8. Stable discrete VSC controlled van der Pol oscillator

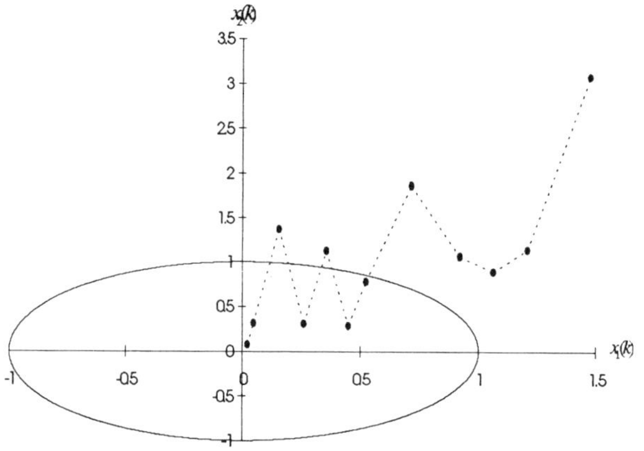

FIGURE 9. Unstable discrete VSC controlled van de Pol oscillator

circle (15). Using the Runge-Kutta algorithm, to simplify our discussion, we assume we can always express the discretised system numerically as

$$x_1(k+1) = f_1^k(x_1(k), x_2(k), h) \tag{17}$$

$$x_2(k+1) = f_2^k(x_1(k), x_2(k), h) \tag{18}$$

i.e. we assume that we know that the discretised system states are almost exactly on the trajectory of the corresponding continuous-time system.

The overshooting of the domain of attraction Ω_d is unavoidable, provided the sampling period is not selected carefully. An overshooting may occur when, as illustrated in Figure 7, the state $x(k)$ with a bigger sampling period, starting from $x^0(0)$, is below the circle defined by (15). The state $x(k+1)$ may overshoot the domain of attraction Ω_d and land outside the limit cycle $\Gamma^-(x_1, x_2) = 0$. $x(k+2)$ will then move away from Ω_d and go to infinite. Divergence from the switching circle then occurs. A smaller sampling period may prevent the system from divergence (see the system state starting from $x^1(0)$ in Figure 7).

To avoid such divergence, we need to limit the selection of the sampling period. Assume in limiting case, a state $x(k)$ is just below (15). The sampling period should be chosen so that with one sampling period $x(k+1)$ just lands on the limiting circle

$\Gamma^-(x_1, x_2) = 0^-$. The upper bound of sampling period H is then determined by

$$H = \text{Sup}\{h : \Gamma^-(f_1^k(x_1(k), x_2(k), h), f_2^k(x_1(k), x_2(k), h)) \leq 0, \\ \text{and } x_1^2(k) + x_2^2(k)/\omega_n^2 \leq r^2, \quad \text{for all } k\} \tag{19}$$

Several simulations have been done. We choose the system parameters as $\mu = 1$, $w_n = 1$, $\zeta = 5$. Calculation of (19) gives $H = 0.132$. In Figure 8, with $h = 0.1$ the system trajectory zigzags along the switching circle, indicating the existence of pseudo-sliding mode. Figure 9 illustrates the system trajectory with $h = 0.15$ moving away from the switching circle towards infinity, indicating instability. This is in conformation with the analysis above.

4. Conclusion and discussion

From the two case studies we can see that discretisation of variable structure control scheme for continuous-time system actually introduces an extra dimension of freedom, namely the sampling period h, whose value is crucial to the occurrence of irregular and unpredictable behaviour and instability. "Low" sampling frequencies may destroy the existence of pseudo-sliding mode. Boundary conditions for the existence of pseudo-sliding mode should then be studied. In this article, only simple cases have been studied. It is not clear what determines the existence of pseudo-sliding mode in the nonlinear systems of high dimension. It is intended to extend the results obtained here to more complex nonlinear systems.

References

1. R. A. DeCarlo, S. H. Zak, and G. P. Matthews, *Variable structure control of nonlinear multivariable systems: a tutorial*, Proceedings of IEEE **76** (1988), 212–232.
2. W. J. Grantham and A. M. Athalye, *Discretization chaos: feedback control and transition to chaos*, Control and Dynamic Systems **34** (1990), 205–277.
3. V. I. Utkin, *Variable structure systems with sliding modes*, IEEE Tranisactions on Automatic Control **AC-22** (1977), 212–222.
4. K. J. Astrom and B. Wittenmark, *Computer Controlled Systems: Theory and Design*, Prentice-Hall, Englewood Cliffs, N.J., 1984.
5. A. G. Fillipov, *Application of the theory of differential right-hand sides to nonlinear problems in automatic control*, Proceedings of 1st IFAC World Congress, 1960, Moscow, 923–927.
6. X. Yu and R. B. Potts, *Computer-controlled variable structure system*, Journal of Australian Mathematical Society Ser. B **34** (1992), 1–17.
7. X. Yu, *Chaos in discrete variable structure systems*, Proceedings of 31st IEEE International Conference on Decision and Control, Tucson, USA, **2** (1992), 1862–1863.
8. ———, *Discrete variable structure control systems*, International Journal of Systems Science **24** (1993), 373–386.
9. ——— *Digital variable structure control with pseudo-sliding mode*, Variable Structure and Lyapunov Control, Ed. A.S.I. Zinober, Spring-Verlag, 1993, 129–154.
10. F. R. Rubio, J. Aracil, and E. F. Camacho, *Chaotic motion in an adaptive control system*, International Journal of control **42** (1985), 353–360.
11. T. Ushio, K. Hirai, *Chaos in nonlinear sampled-data control systems*, International Journal of Control **38** (1983), 1023–1033.
12. A. S. I. Zinober, *Detereministic Control of Uncertain Systems*, Peter Peregrinus Ltd, 1990.
13. H. Sira-Ramirez, *Harmonic response of variable structure controlled Van der Pol oscillators*, IEEE Transactions on Circuits and Systems **34** (1987), 103–106.

Department of Mathematics and Computing, University of Central Queensland, Rockhampton, QLD 4702, Australia
E-mail address: X.Yu@ucq.edu.au

Recent Titles in This Series

(Continued from the front of this publication)

143 **Marvin Knopp and Mark Sheingorn, Editors,** A tribute to Emil Grosswald: Number theory and related analysis, 1993

142 **Chung-Chun Yang and Sheng Gong, Editors,** Several complex variables in China, 1993

141 **A. Y. Cheer and C. P. van Dam, Editors,** Fluid dynamics in biology, 1993

140 **Eric L. Grinberg, Editor,** Geometric analysis, 1992

139 **Vinay Deodhar, Editor,** Kazhdan-Lusztig theory and related topics, 1992

138 **Donald St. P. Richards, Editor,** Hypergeometric functions on domains of positivity, Jack polynomials, and applications, 1992

137 **Alexander Nagel and Edgar Lee Stout, Editors,** The Madison symposium on complex analysis, 1992

136 **Ron Donagi, Editor,** Curves, Jacobians, and Abelian varieties, 1992

135 **Peter Walters, Editor,** Symbolic dynamics and its applications, 1992

134 **Murray Gerstenhaber and Jim Stasheff, Editors,** Deformation theory and quantum groups with applications to mathematical physics, 1992

133 **Alan Adolphson, Steven Sperber, and Marvin Tretkoff, Editors,** p-adic methods in number theory and algebraic geometry, 1992

132 **Mark Gotay, Jerrold Marsden, and Vincent Moncrief, Editors,** Mathematical aspects of classical field theory, 1992

131 **L. A. Bokut′, Yu. L. Ershov, and A. I. Kostrikin, Editors,** Proceedings of the International Conference on Algebra Dedicated to the Memory of A. I. Mal′cev, Parts 1, 2, and 3, 1992

130 **L. Fuchs, K. R. Goodearl, J. T. Stafford, and C. Vinsonhaler, Editors,** Abelian groups and noncommutative rings, 1992

129 **John R. Graef and Jack K. Hale, Editors,** Oscillation and dynamics in delay equations, 1992

128 **Ridgley Lange and Shengwang Wang,** New approaches in spectral decomposition, 1992

127 **Vladimir Oliker and Andrejs Treibergs, Editors,** Geometry and nonlinear partial differential equations, 1992

126 **R. Keith Dennis, Claudio Pedrini, and Michael R. Stein, Editors,** Algebraic K-theory, commutative algebra, and algebraic geometry, 1992

125 **F. Thomas Bruss, Thomas S. Ferguson, and Stephen M. Samuels, Editors,** Strategies for sequential search and selection in real time, 1992

124 **Darrell Haile and James Osterburg, Editors,** Azumaya algebras, actions, and modules, 1992

123 **Steven L. Kleiman and Anders Thorup, Editors,** Enumerative algebraic geometry, 1991

122 **D. H. Sattinger, C. A. Tracy, and S. Venakides, Editors,** Inverse scattering and applications, 1991

121 **Alex J. Feingold, Igor B. Frenkel, and John F. X. Ries,** Spinor construction of vertex operator algebras, triality, and $E_8^{(1)}$, 1991

120 **Robert S. Doran, Editor,** Selfadjoint and nonselfadjoint operator algebras and operator theory, 1991

119 **Robert A. Melter, Azriel Rosenfeld, and Prabir Bhattacharya, Editors,** Vision geometry, 1991

118 **Yan Shi-Jian, Wang Jiagang, and Yang Chung-chun, Editors,** Probability theory and its applications in China, 1991

117 **Morton Brown, Editor,** Continuum theory and dynamical systems, 1991

116 **Brian Harbourne and Robert Speiser, Editors,** Algebraic geometry: Sundance 1988, 1991

115 **Nancy Flournoy and Robert K. Tsutakawa, Editors,** Statistical multiple integration, 1991

(See the AMS catalog for earlier titles)